FOUNDATION MATHEMATICS

L R Mustoe
Loughborough University, Loughborough

M D J Barry
University of Bristol, Bristol

JOHN WILEY & SONS
Chichester · New York · Weinheim · Brisbane · Singapore · Toronto

Copyright © 1998 by John Wiley & Sons Ltd,
Baffins Lane, Chichester,
West Sussex, PO19 1UD, England
National   01243 779777
International (+44) 1243 779777

e-mail (for orders and customer service enquiries): cs-books@wiley.co.uk

Visit our Home Page on http://www.wiley.co.uk or http://www.wiley.com

Other Wiley Editorial Offices

John Wiley & Sons, Inc., 605 Third Avenue,
New York, NY 10158-0012, USA

Wiley-VCH GmbH, Pappelallee 3,
D-69469 Weinheim, Germany

Jacaranda Wiley Ltd, 33 Park Road, Milton,
Queensland 4064, Australia

John Wiley & Sons (Asia) Pte Ltd, 2 Clementi Loop #02-01,
Jin Xing Distripark, Singapore 129809

John Wiley & Sons (Canada) Ltd, 22 Worcester Road,
Rexdale, Ontario M9W 1L1, Canada

British Library Cataloguing in Publication Data

A catalogue record for this book is available from the British Library

ISBN 0 471 970421 (pbk); 0 471 969486 (cloth)

Typeset by Techset Composition Ltd, Salisbury
Printed and bound in Great Britain by Bookcraft (Bath) Ltd
This book is printed on acid-free paper responsibly manufactured from sustainable forestry,
in which at least two trees are planted for each one used for paper production.

To my students, past, present and future

(L. R. Mustoe)

To Felicity

(M. D. J. Barry)

CONTENTS

Preface . . . . **xi**

1 Arithmetic **1**
   1.1 Integers 3
   1.2 Fractions and decimals 8
   1.3 Fractional powers and logarithms 15
   1.4 Errors and the modulus sign 22
        Summary 27
        Answers 28

2 Basic algebra **33**
   2.1 Manipulation of symbols 35
   2.2 Formulae, identities, equations 45
   2.3 Proportionality 59
   2.4 Factorisation 65
   2.5 Simple inequalities 72
        Summary 78
        Answers 80

3 Straight lines **87**
   3.1 Graphs and plotting 89
   3.2 The straight line 94
   3.3 Intersection of two lines 105
   3.4 Linear inequalities 109
   3.5 Reduction to linear form 114
        Summary 122
        Answers 123

4 Quadratics and cubics **133**
   4.1 The quadratic curve 136

4.2 Quadratic equations and roots. 141
4.3 Common problems involving quadratics 153
4.4 Features of cubics 163
   Summary. 170
   Answers 172

5 Geometry **181**
5.1 Shapes in two dimensions 183
5.2 Congruence and similarity 200
5.3 Circles. 211
5.4 Shapes in three dimensions. 222
   Summary. 232
   Answers 234

6 Proof. **239**
6.1 Deductive reasoning and Pythagoras' theorem. 241
6.2 Theorems in classical plane geometry. 247
6.3 Arguments in arithmetic and algebra. 261
6.4 The methodology of proof 265
   Summary. 274
   Answers 275

7 Trigonometry **277**
7.1 Tangent, sine and cosine 279
7.2 Solution of triangles. 290
7.3 Distances and angles in solids 307
7.4 Ratios of any angle and periodic modelling. 318
   Summary. 331
   Answers 333

8 Further algebra **341**
8.1 Arithmetic and geometric progressions 343
8.2 Polynomials. 357
8.3 Factor and remainder theorems. 367
8.4 Elementary rational functions 377
   Summary. 386
   Answers 388

9 Coordinate geometry **403**
9.1 Distances and areas in two dimensions 405

9.2 Loci and simple curves 419

9.3 Circles. 427

9.4 Polar coordinates 436

9.5 Three-dimensional geometry. 447

    Summary. 460

    Answers 462

10 Functions **473**

10.1 Ideas, definitions and graphs. 475

10.2 Translation, scaling, function of a function 484

10.3 Inverse functions 494

10.4 Power laws, exponential and logarithmic functions 505

    Summary. 520

    Answers 521

11 Differentiation **539**

11.1 Rates of change 541

11.2 Simple differentiation. 549

11.3 Tangents and stationary points 562

11.4 Second derivatives and stationary points 576

    Summary. 585

    Answers 587

12 Integration **597**

12.1 Reversing differentiation. 599

12.2 Definite integration and area. 605

12.3 Applications to area, volume and mean values 618

12.4 Centres of gravity 629

    Summary. 641

    Answers 643

Index 651

PREFACE

In today's society, technology plays an increasingly important rôle. Consequently, mathematics, which lies at the heart of technology, is finding ever wider areas of application as we seek to understand more about the way in which the natural world and the man-made environment operate and interact. Traditionally, engineering and science have relied on mathematical models as design tools and for the prediction of the behaviour of many phenomena. Nowadays, mathematics is used also to model situations in many other disciplines including finance, management, politics and geography.

The need for a higher proportion of the population to have a thorough understanding of the foundations of mathematics is greater than it ever has been. Although the widespread availability of computers and pocket calculators has reduced the need for long, tedious calculations to be carried out 'by hand', it is still important to be able to perform simple calculations and to understand the mathematical principles in order to have a sense of control over the processes involved. You cannot learn to drive a car by being a passenger, you cannot be good at playing the piano without a lot of practice and you cannot succeed in sport without consistent attention to a sound basic technique. The same principle is true of mathematics; you need to put in time and effort in order to become proficient.

The book begins with a concise summary of arithmetic and basic algebra, and a discussion of quadratics and cubics strongly emphasising geometric ideas. Then follow the principles of Euclidean and Cartesian geometry and the concept of proof. Next are trigonometry, further algebra and functions and their inverses. Finally, the concepts of differential and integral calculus are introduced

Each chapter starts with a list of learning objectives and ends with a summary of key points and results. A generous supply of worked examples incorporating motivating applications is designed to build knowledge and skill. Practice is essential and the exercises are graded in difficulty for first reading and then revision; the answers at the end of each chapter include helpful hints. Use of a pocket calculator is encouraged where appropriate (indicated by calculator icon). Many of the exercises can be validated by computer algebra and its use is strongly recommended where higher algebraic accuracy can be achieved and drudgery removed. Asterisks indicate difficult exercises. Lecturers'

Resource Guides/Teachers' Manuals will be available in due course; please contact the publisher for details.

This book has been written to give students the means of obtaining a basic grounding in mathematics. *Foundation Mathematics* together with its sequel *Mathematics in Engineering and Science* take the reader forward, in both content and style, from a level close to UK GCSE mathematics, and its international equivalents, to first year university-level mathematics. The concise and focused approach will also help the returning student and will build the necessary confidence to tackle the more advanced ideas of *Mathematics in Engineering and Science*.

The authors have between them a wide experience of teaching mathematics to a broad spectrum of students: undergraduate and secondary students of mathematics and other sciences, engineering students and apprentices, mature students, those with an arts background and those who need to advance from GCSE to first year university level. Their commitment to developing the highest standards in mathematics education includes involvement with the Mathematics Working Group of SEFI (*Société Européenne pour la Formation des Ingénieurs*), whose work has included the formulation of a recommended core curriculum for the mathematical education of engineering undergraduates.

The growing gap between secondary and tertiary mathematics is of increasing concern and the authors believe that there is a need for a no-nonsense textbook to help students cross over with confidence. They hope that this book fulfils that need.

Our thanks are due to Nicci for typing the bulk of the manuscript, to our students whose comments have helped to improve the text and to the staff of John Wiley and Sons Ltd for their help in the preparation of this book.

L R Mustoe
M D J Barry

1 ARITHMETIC

Introduction

Arithmetic is the branch of mathematics concerned with numerical calculations and is therefore at the heart of the subject. Even in these days of computers and pocket calculators, a firm grasp of the principles and processes of arithmetic is fundamental to the rest of mathematics.

Objectives

After working through this chapter you should be able to

- express an integer as the product of its prime factors
- calculate the highest common factor and lowest common multiple of a set of integers
- understand the rules governing the existence of powers of a number
- obtain the modulus of a number
- combine powers of a number
- express a fraction in its lowest form
- carry out arithmetic operations on fractions
- express a fraction in decimal form and vice versa
- carry out arithmetic operations on numbers in decimal form
- round numerical values to a specified number of decimal places or significant figures
- understand the scientific notation form of a number
- carry out arithmetic operations on surds
- manipulate logarithms
- understand errors in measurements and how to combine them

1.1 INTEGERS

Factors

When two or more **integers**, i.e. whole numbers, are multiplied together to form a **product**, each of them is called a **factor** of that product.

Examples

1. $2 \times 7 = 14$, so 2 and 7 are factors of 14.

2. $2 \times 3 \times 5 = 30$, so 2, 3 and 5 are factors of 30. ■

Notice that $30 = 15 \times 2 = 10 \times 3 = 6 \times 5$ and therefore 15, 10 and 6 are also factors of 30.

Furthermore, $14 = 1 \times 14$ and $30 = 30 \times 1$, which means that 1 and 14 are factors of 14, and 1 and 30 are factors of 30. It is true in general that any integer has at least two factors, namely 1 and the integer itself. Any integer which has these factors only is called a **prime number**.

The first seven prime numbers are 2, 3, 5, 7, 11, 13 and 17. *It is a convention not to regard* 1 *as a prime number.* We have seen that $14 = 2 \times 7$ and cannot therefore be regarded as prime; it is a **composite number**.

The most complete **factorisation** of an integer is to write it as the product of prime factors. It is accepted practice to write the factors in ascending order. The factorisation is then unique; for example, 2×7 is the only possible prime factorisation of 14.

Example

We write $12 = 2 \times 2 \times 3$ and $231 = 3 \times 7 \times 11$. ■

Sometimes a factor is repeated; for example, $300 = 2 \times 2 \times 3 \times 5 \times 5$; we say that the prime factors of 300 are 2, 3 and 5.

Common factors and common multiples

If two integers have a factor in common then it is a **common factor** of the two numbers.

Example

24 has the factors 1, 2, 3, 4, 6, 8, 12 and 24 whereas 80 has the factors 1, 2, 4, 5, 8, 10, 16, 20, 40 and 80. The common factors are 1, 2, 4 and 8. ■

The **highest common factor** (HCF) is the largest of these common factors and in the example above it is 8.

> The highest common factor of two or more integers is the largest integer which is a factor of each of the given integers.

The HCF may be a prime number; since $98 = 2 \times 7 \times 7$ and $35 = 5 \times 7$, the HCF of 98 and 35 is 7. Earlier we could have written $24 = 2 \times 2 \times 2 \times 3$ and $80 = 2 \times 2 \times 2 \times 2 \times 5$, identifying $2 \times 2 \times 2$ as the HCF. If the HCF of two integers is 1, the integers are **relatively prime**; examples are 35 and 16.

Note that the HCF of the integers $12 = 2 \times 2 \times 3$ and $72 = 2 \times 2 \times 2 \times 3 \times 3$ is $2 \times 2 \times 3 = 12$ itself. We say that 72 is a **multiple** of 12. Other examples are 14 as a multiple of both 7 and 2, and 30 as a multiple of 10, 6, 5, 3 and 2. *It is customary not to speak of an integer being a multiple of itself or of* 1.

If an integer is a multiple of each of two given integers then it is a **common multiple** of them. For example, some common multiples of 10 and 15 are 30, 60, 90, 120 and 150. The smallest of these, namely 30, is the **lowest common multiple** (LCM).

> The lowest common multiple of two or more integers is the smallest integer which is a multiple of each of the given integers.

A useful way of finding the LCM is to use prime factors.

Example $24 = 2 \times 2 \times 2 \times 3$ and $36 = 2 \times 2 \times 3 \times 3$. The most often that 2 appears is three times and 3 appears at most twice; the LCM is therefore $2 \times 2 \times 2 \times 3 \times 3 = 72$. Note that $72 = 3 \times 24 = 2 \times 36$. ∎

When two integers are relatively prime their LCM is simply the product of these integers.

Example   The integers $40 = 2 \times 2 \times 2 \times 5$ and $21 = 3 \times 7$ are relatively prime and their LCM is $40 \times 21 = 840$. ∎

It is also important to note that the product of two integers is equal to the product of their LCM and HCF.

Example $35 = 5 \times 7$ and $98 = 2 \times 7 \times 7$. The $\text{HCF} = 7$ and the $\text{LCM} = 2 \times 5 \times 7 \times 7 = 490$. $35 \times 98 = 3430 = 7 \times 490 = \text{HCF} \times \text{LCM}$. ∎

So far we have dealt only with positive integers. Knowing that $(-2) \times (-3) = 6$ we could say that the factors of 6 are 2, 3, -2, -3. Similarly we could say that the factors of -6 are 2, 3, -2, -3; we merely note for the moment that -6 can be written as $(-2) \times 3$ or $2 \times (-3)$.

It is useful to remember that the product or the quotient of two negative numbers is positive, e.g. $(-3) \times (-2) = 6$ and $(-15) \div (-5) = 3$, and that the product or the quotient of one positive and one negative number is negative, e.g. $(-3) \times 2 = 3 \times (-2) = -6$ and $(-15) \div 5 = 15 \div (-5) = -3$.

Exercise 1.1

1   Factorise each of the following integers into a product of prime factors:

   (a)   2280   (b)   1210   (c)   4141

   (d)   3087   (e)   14703   (f)   17 963

   (g)   64 900   (h)   3 258 475.

2   Note the following principles of divisibility:

   (a)   If the sum of the digits of an integer is divisible by 3 or 9 then the integer is divisible by 3 or 9, respectively.

   (b)   If the difference of the sums of alternate integer digits is zero or divisible by 11, positive or negative, then so is the integer, (e.g. 616: $6 + 6 - 1 = 11$, and $616 = 2 \times 2 \times 2 \times 7 \times 11$).

   Verify these principles on (a) to (g) of Question 1.

3   What are the LCM and HCF of each of the following sets of integers?

   (a)   210, 720   (b)   1210, 2280   (c)   72, 729

   (d)   96, 168, 240   (e)   48, 84, 144, 720.

4   Verify that 'LCM × HCF = product' for the pairs in parts (a) to (c) of Question 3.

5   It can be shown that the smallest prime factor p of a composite (i.e. non-prime) integer N must be no larger than $\sqrt{N}$. Using this principle one need go no further than 43 in trying to find factors of integers less than 2000. How many of the integers 1990 to 1999 are prime?

6*   Justify the statement in Question 5 that p is less than or equal to $\sqrt{N}$.

Powers and indices

Repeated multiplication of a number by itself leads to the idea of raising a number to some power.

The product $3 \times 3 \times 3 \times 3$ can be written as 3^4; we say that the number 3 has been **raised to the power** 4. We also refer to 4 as the **index** (plural **indices**) and 3 as the **base**.

Raising a number to the power 2 is called **squaring** and raising a number to the power 3 is called **cubing**: hence $16 = 4 \times 4 = 4^2$ is '4 squared' and $27 = 3 \times 3 \times 3 = 3^3$ is '3 cubed'. The laws of handling powers are illustrated by examples.

Examples

1. Any integer raised to the power 1 equals itself (1 is sometimes called unity).

$$5^1 = 5$$

2. When multiplying two powers of the *same* base we add the indices.

$$5^4 \times 5^3 = 5 \times 5 \times 5 \times 5 \times 5 \times 5 \times 5 = 5^7 = 5^{4+3}$$

3. When dividing two powers of the *same* base we subtract the indices.

$$4^5 \div 4^3 = \frac{4 \times 4 \times 4 \times 4 \times 4}{4 \times 4 \times 4} = 4^2 = 4^{5-3}$$

4. Raising a non-zero integer to the power zero gives the result 1.

$$3^2 \div 3^2 = 9 \div 9 = 1, \text{ but } 3^2 \div 3^2 = 3^{2-2} = 3^0, \text{ so } 3^0 = 1$$

5. The power of a power is achieved by multiplying the indices.

$$(3^4)^2 = 3^4 \times 3^4 = 3^{4+4} = 3^8 = 3^{4 \times 2} \ (= (3^2)^4)$$

6. Power of a combination

$$2^4 \times 3^4 = (2 \times 3)^4 = 6^4$$

7. Negative powers

$$3^4 \div 3^6 = \frac{3 \times 3 \times 3 \times 3}{3 \times 3 \times 3 \times 3 \times 3 \times 3} = \frac{1}{3 \times 3} = \frac{1}{3^2}$$

but $3^4 \div 3^6 = 3^{4-6} = 3^{-2}$.

Hence $3^{-2} = \dfrac{1}{3^2}$ and we say that 3^{-2} is the **reciprocal** of 3^2 and vice versa.

Note that $2^{-3} = \dfrac{1}{2^3} = \dfrac{1}{8}$ and $\dfrac{1}{2^{-3}} = 2^3 = 8$. ∎

These seven rules will also apply to powers of negative integers. Note that a negative number raised to an odd power gives a negative result, whereas if is is raised to an even power it gives a positive result. Hence

$$(-3)^5 = -243 \text{ but } (-3)^6 = +729.$$

When several arithmetic operations are involved there is an agreed **order of precedence** with the acronym BODMAS.

In older books the instruction '**of**' was used when multiplying by a fraction, for example 'two-thirds of 12'.

(i) **B**rackets: quantities enclosed in brackets are evaluated separately from other quantities.

(ii) **O**fs, **D**ivisions and **M**ultiplications are carried out next.

(iii) **A**dditions and **S**ubtractions are performed last.

When two or more operations at the same level are to be carried out, the order 'left to right' is followed.

Examples
1. $3 + 4 \div 7$ means 'divide 4 by 7 first, then add 3'.

2. $(3 + 4 \div 7) \times 5 + 2$ means 'evaluate the bracket first, multiply by 5, then add 2'. ■

Brackets can be used to modify the order of precedence.

Example
$(3 + 4) \div 7$ means 'add 3 to 4 then divide the result by 7'. ■

The operation of raising a number to a power is called **exponentiation**. With this operation included, the order of precedence may be renamed BEDMAS where the E stands for exponentiations and 'of' is absorbed into multiplication.

Example
$$3 + 4^2 \times 5 = 3 + 16 \times 5 = 3 + 80 = 83$$

but $(3 + 4)^2 \times 5 = (7)^2 \times 5 = 49 \times 5 - 245.$ ■

Exercise 1.2

1 Expand the following:

(a) $2^3 \times 3^2 \times 5 \times 7^2$

(b) $3^3 \times 5^2 \times 7^3$

(c) $2 \times 3^2 \times 5^3 \times 23 \times 29$

(d) $2^2 \times 3 \times 5 \times 17 \times 19$

(e) $2^{-3} \times 3 \times 5^{-1} \times 41$, as a ratio of integers

(f) $(2^2 \times 3)^{-2} \times (5 \times 2^{-5})^2$, as a ratio of integers

(g) $(2^5 \times 3^2 \times 17^{-10} \times 41^{100})^0$.

2 (a) What is N^0, for $N = 1$, 10^3, 10^6?

   (b) What is N^0 for $N = 10^{-3}, 10^{-6}, 10^{-9}$?

3 Evaluate the following:

(a) $(2^2 + 3) \times 7$ (b) $2^2 + 3 \times 7$

(c) $(3^3 - 2^2) \times 7$ (d) $(2^2 + 3^3) \times (5^2 - 7)$.

1.2 FRACTIONS AND DECIMALS

Fractions

A **fraction** is the ratio of one integer divided by another; for example, $\dfrac{5}{7}$, $\dfrac{-2}{3}$, $\dfrac{18}{7}$. Note that $\dfrac{-2}{3}$ or $\dfrac{2}{-3}$ are both written as $-\dfrac{2}{3}$.

The upper integer is the **numerator** and the lower integer is the **denominator**. Sometimes fractions are called **rational numbers**.

A **proper fraction** has a numerator less than its denominator, e.g. $\dfrac{5}{7}$. A fraction such as $\dfrac{18}{7}$, where the numerator is greater than the denominator, is termed **improper** or **vulgar**.

When $\dfrac{18}{7}$ is written as $2\dfrac{4}{7}$ it is called a **mixed fraction**.

It should be clear that the fractions $\dfrac{3}{5}, \dfrac{6}{10}, \dfrac{12}{20}, \dfrac{30}{50}$ have the same value. They are called **equivalent fractions**. This concept is used to perform a simplification known as **reduction to lowest terms**. As an example

$$\frac{42}{105} = \frac{2 \times 3 \times 7}{3 \times 5 \times 7} = \frac{2}{5}$$

where we have divided the original fraction top and bottom by the common factor $3 \times 7 = 21$. We can proceed no further since 2 and 5 are relatively prime. We often call this process **cancellation** since the 3 and the 7 on the top and bottom have in effect cancelled themselves out. Since $42 = 14 \times 3$ and $105 = 35 \times 3$, we could write

$$\frac{\cancel{42}^{\,14}}{\cancel{105}^{\,35}}$$

to cancel 3 top and bottom. Since $14 = 2 \times 7$ and $35 = 5 \times 7$ we write

$$\frac{\cancel{14}^{\,2}}{\cancel{35}^{\,5}}$$

by cancelling 7.

Arithmetic with fractions

The examples below illustrate the basic processes of addition, subtraction, multiplication and division.

To add two fractions with the same denominator we add the numerators and place the result over the common denominator; a similar result holds for subtraction.

To add two fractions having different denominators we first find the LCM of these denominators then replace each fraction by its equivalent having the LCM as its denominator.

To multiply two fractions put the product of their numerators as the numerator of the result, and the product of their denominators as the denominator of the result.

To divide one fraction by another, invert the second fraction and multiply.

Examples

1. $\dfrac{1}{4} + \dfrac{2}{4} = \dfrac{3}{4}$

2. $\dfrac{1}{4} + \dfrac{2}{5} = \dfrac{5}{20} + \dfrac{8}{20} = \dfrac{13}{20}$ (20 is the LCM of 4 and 5)

3. $\dfrac{1}{4} + \dfrac{3}{10} = \dfrac{5}{20} + \dfrac{6}{20} = \dfrac{11}{20}$ (20 is the LCM of 4 and 10)

$\left(= \dfrac{10}{40} + \dfrac{12}{40} = \dfrac{22}{40} = \dfrac{11}{20} \right)$ (40 is the product of 4 and 10)

4. $\dfrac{4}{5} - \dfrac{2}{3} = \dfrac{12}{15} - \dfrac{10}{15} = \dfrac{2}{15}$

5. $\dfrac{4}{5} \times \dfrac{2}{3} = \dfrac{4 \times 2}{5 \times 3} = \dfrac{8}{15}$

6. $\dfrac{4}{15} \times \dfrac{5}{6} = \dfrac{4 \times 5}{15 \times 6} = \dfrac{2 \times 2 \times 5}{5 \times 3 \times 2 \times 3} = \dfrac{2}{3 \times 3} = \dfrac{2}{9}$

$$\left(\text{We could write } \dfrac{\overset{2}{\cancel{4}}}{\underset{3}{\cancel{15}}} \times \dfrac{\overset{1}{\cancel{5}}}{\underset{3}{\cancel{6}}} = \dfrac{2}{9} \right)$$

7. $\dfrac{4}{15} \div \dfrac{5}{6} = \dfrac{4}{15} \times \dfrac{6}{5} = \dfrac{4 \times 6}{15 \times 5} = \dfrac{2 \times 2 \times 2 \times 3}{3 \times 5 \times 5} = \dfrac{8}{25}$

8. $\dfrac{2}{3} \div \dfrac{3}{4} = \dfrac{2}{3} \times \dfrac{4}{3} = \dfrac{8}{9}$

9. $\dfrac{5^{-2}}{5^{-6}} = \dfrac{1}{5^2} \div \dfrac{1}{5^6} = \dfrac{1}{5^2} \times \dfrac{5^6}{1} = 5^4 = \dfrac{1}{5^{-4}}$

10. $\left(\dfrac{1}{3} + \dfrac{1}{4} \right) \div \left(\dfrac{5}{7} - \dfrac{3}{8} \right) = \left(\dfrac{4}{12} + \dfrac{3}{12} \right) \div \left(\dfrac{40}{56} - \dfrac{21}{56} \right)$

$$= \dfrac{7}{12} \div \dfrac{19}{56} = \dfrac{7}{12} \times \dfrac{56}{19} = \dfrac{7 \times 7 \times 8}{3 \times 4 \times 19}$$

$$= \dfrac{7 \times 7 \times 2}{3 \times 19} = \dfrac{98}{57} = 1\dfrac{41}{57} \qquad \blacksquare$$

Exercise 1.3

1 Express the following as fractions in their lowest terms:

(a) $\dfrac{5040}{40\,320}$

(b) $\dfrac{6182}{-616}$

(c) $\dfrac{2^5 \times 3^{-2} \times 17}{2^3 \times 3^{-5} \times 5 \times 11}$

(d) $\dfrac{(-42) \times (308) \times (230)}{(-60) \times (121) \times (-69)}.$

2 Separate out the integral part (i.e. whole number) and the fractional part for the improper fractions in Question 1 $\left(\text{e.g. } \dfrac{7}{3} = 2\dfrac{1}{3} \right).$

3 Carry out the following additions, expressing the improper fraction results as mixed fractions.

(a) $\dfrac{1}{4} + \dfrac{3}{5}$

(b) $\dfrac{2}{7} + \dfrac{1}{5} + \dfrac{5}{9}$

(c) $\dfrac{11}{19} + \dfrac{1}{2} + \dfrac{5}{11} + \dfrac{7}{38}$

(d) $3\dfrac{1}{4} + 2\dfrac{5}{6} + 4\dfrac{3}{8}$

(e) $10\dfrac{5}{8} + 21\dfrac{3}{7} + 36\dfrac{3}{11}$

(f) $101\dfrac{8}{9} + 94\dfrac{7}{8} + 61\dfrac{61}{72}$.

4   Evaluate the following, expressing the improper fraction results as mixed fractions, noting that some may be negative.

(a) $1\dfrac{1}{5} - \dfrac{2}{3}$

(b) $3\dfrac{1}{11} - 4\dfrac{1}{6} - 6\dfrac{1}{4}$

(c) $2\dfrac{15}{17} - 3\dfrac{1}{5} + 1\dfrac{2}{3}$

(d) $26\dfrac{5}{6} - 13\dfrac{1}{12} - 11\dfrac{17}{24} + 2\dfrac{1}{4}$

(e) $\dfrac{75}{8} - \dfrac{61}{6} + \dfrac{11}{4} - \dfrac{131}{12}$

(f) $\left(2\dfrac{1}{4} - 3\dfrac{1}{5}\right) - \left(4\dfrac{11}{21} - 3\dfrac{5}{7}\right) + \left(3\dfrac{1}{5} - 1\dfrac{7}{15}\right)$.

5   Evaluate the following:

(a) $\dfrac{1}{7} \div \dfrac{4}{7}$

(b) $\dfrac{6}{11} \times \dfrac{11}{24} \times \dfrac{3}{19} \times \dfrac{38}{17}$

(c) $2\dfrac{1}{6} \times 3\dfrac{3}{4} \div 4\dfrac{7}{12}$

(d) $\left(-11\dfrac{1}{3}\right) \times 2\dfrac{2}{5} \div \left(-1\dfrac{2}{15}\right)$

(e) $\left(\dfrac{2}{5} \times \dfrac{3}{20}\right) \div \left(\dfrac{9}{100} \times \dfrac{5}{19}\right)$

(f) $\left(\dfrac{77}{8} \div \dfrac{4}{3}\right) \div \left(\dfrac{24}{13} \div \dfrac{6}{11}\right)$.

6*   Express the following as proper fractions or as mixed fractions:

(a) $\dfrac{\dfrac{1}{3} - \dfrac{1}{4}}{\dfrac{5}{7} - \dfrac{1}{2}}$

(b) $\dfrac{3\dfrac{5}{6} - 2\dfrac{7}{8}}{4\dfrac{5}{11} - 3\dfrac{1}{6}}$

(c) $\dfrac{4\dfrac{1}{4} - 7\dfrac{3}{8} + 2\dfrac{1}{3}}{11\dfrac{2}{5} - 14\dfrac{1}{4} + 1\dfrac{1}{6}}$

(d) $\dfrac{\dfrac{2}{5} - \dfrac{1}{6}}{3\dfrac{1}{6} - 2\dfrac{7}{8}} + \dfrac{3\dfrac{1}{8} - 2\dfrac{1}{4}}{\dfrac{5}{24}}$

(e) $\dfrac{\dfrac{2}{5} - \dfrac{1}{6}}{3\dfrac{1}{6} - 2\dfrac{7}{8}} \div \dfrac{3\dfrac{1}{8} - 2\dfrac{1}{4}}{\dfrac{5}{24}}$

(f) $\dfrac{3\dfrac{1}{4} - 2\dfrac{7}{12}}{1\dfrac{1}{5} - \dfrac{2}{7}} - \dfrac{1\dfrac{2}{11} - \dfrac{8}{9}}{\dfrac{23}{99}} \times 1\dfrac{5}{41}$

7* Simplify the following expression:

$$\left(\frac{2\frac{1}{3} - 1\frac{5}{6} + \frac{1}{12}}{\frac{11}{24}} \right)^2 - \left(1\frac{1}{2} \right)^{-3}.$$

Decimals

Fractions whose denominators are powers of 10 are called decimal fractions, or **decimals** for short. They have a special notation. For example, the fractions

$$\frac{2}{10}, \frac{31}{100}, \frac{3}{100}, \frac{407}{1000}, \frac{21}{1000}, \frac{6}{1000}$$

are written respectively as

$$0.2, 0.31, 0.03, 0.407, 0.021, 0.006.$$

Note that 0.2 could be written as 0.20, 0.200, etc., but it is not usual to include these trailing zeros unless there is good reason, e.g. to emphasise the precision of a result.
  Improper fractions can be dealt with similarly; for example,

$$\frac{24\ 357}{1000} = 24 + \frac{357}{1000} = 24.357.$$

Each digit has its special place; the last number above is 2 tens + 4 units + 3 tenths + 5 hundredths + 7 thousandths, i.e. $2 \times 10 + 4 \times 1 + \frac{3}{10} + \frac{5}{100} + \frac{7}{1000}$.
  These place values are important when carrying out arithmetic. For example,

$$2.46 + 13.1 = 15.56 \qquad 12.46 + 0.328 = 12.788$$
$$22.46 - 13.1 = 9.36 \qquad 1.2 \times 1.3 = 1.56.$$

Multiplying by 10 has the effect of moving the decimal point one place to the right:

$$24.357 \times 10 = 243.57.$$

Multiplying by 10^2 moves the decimal point two places to the right, i.e. 2435.7, and so on.

Dividing by 10 moves the decimal point one place to the left, and so on; for example, $24.357 \div 10 = 2.4357$.

In order to cope with very large or very small numbers, it is useful to employ **scientific notation**, i.e. to write a number in the form $a \times 10^r$ where r is the **exponent** and a is a number between 1 and 10 which is called the **mantissa**. Examples are:

$$243.57 = 2.4357 \times 10^2$$

and $$0.002\ 435\ 7 = 2.4357 \times 10^{-3}.$$

We can convert decimals to fractions and sometimes vice versa. To convert a fraction to a decimal, we first replace the fraction by its equivalent using a denominator that is a power of 10.

Examples

$$2.31 = \frac{231}{100}$$ $$2.25 = \frac{225}{100} = \frac{9 \times 25}{4 \times 25} = \frac{9}{4}$$

$$\frac{4}{5} = \frac{4 \times 2}{5 \times 2} = \frac{8}{10} = 0.8$$ $$\frac{31}{40} = \frac{31 \times 25}{40 \times 25} = \frac{775}{1000} = 0.775$$ ■

It is not possible to convert some fractions to an *exact* decimal equivalent. For example, $\frac{1}{3}, \frac{2}{7}, \frac{5}{9}, \frac{4}{11}$ have no exact equivalent in decimals. In such cases the decimal form contains a single repeated digit or a repeated group of digits.

Examples

$$\frac{1}{3} = 0.333\ 333\ldots \qquad = 0.\dot{3}$$

$$\frac{1}{9} = 0.111\ 111\ldots \qquad = 0.\dot{1}$$

$$\frac{1}{11} = 0.090\ 909\ldots \qquad = 0.\dot{0}\dot{9}$$

$$\frac{2}{7} = 0.285\ 714\ 28\ldots = 0.\dot{2}85\ 71\dot{4}$$

where the dots placed over the digits indicate the sequence to be repeated.   ■

We can approximate a decimal by rounding it off. The number 2.346 is said to contain three decimal places (3 d.p.) or (3D); of the two nearest numbers with two decimal places, namely 2.34 and 2.35; 2.35 is the closer, so 2.346 is **rounded off** to 2.35 (2 d.p.).

To go one step further, 2.35 lies between 2.3 and 2.4; hence 2.35 is rounded off to 2.4 (1 d.p.). The convention is that if the last digit is 5, 6, 7, 8 or 9 we **round up**, otherwise we **round down**.

The number 2.35 has 3 **significant figures** (3 s.f.); the number 0.235 has 3 significant figures, 0.0235 also has three. The number 2350 has 4 significant figures if we are sure that the number in question is closer to 2350 than to either 2349 or 2351. However, when we say that the distance of the Earth from the Sun is 93 million miles (i.e. 93 000 000 miles) we mean that the distance is nearer to 93 million miles than to either 92 million miles or to 94 million miles, so there are really only two significant figures, namely 9 and 3.

We can round off a number to a specified significant figure accuracy using the convention mentioned earlier. Hence 12.3456 is rounded to 12.346 to 5 s.f., to 12.35 to 4 s.f., 12.4 (3 s.f.), 12 to 2 s.f. and 10 to 1 s.f.

Exercise 1.4

1 Convert the following decimals to fractions in their lowest form:

(a) 0.0625 (b) −2.262 (c) 3.375 (d) −0.8125.

2 Express the following fractions as decimals and display repeated digits:

(a) $\dfrac{3}{80}$ (b) $\dfrac{1}{15}$ (c) $\dfrac{10}{11}$ (d) $\dfrac{3}{7}$.

3 Express the following decimal quantities in scientific notation and give the number of significant figures:

(a) −0.3152 (b) 2579.31 (c) −0.000 343 4

(d) 8 575 000.

4 Convert the following numbers expressed in scientific notation back into conventional decimals:

(a) 2.412×10^8 (b) -3.142×10^0 (c) 6.2484×10^{-3}

(d) -6.7×10^{-1}.

5 Explain the difference between 0.4696 and 0.469 600.

6 Express the following in the reduced number of significant figures stated:

(a) 6.28345 (5 s.f., 4 s.f., 3 s.f., 2 s.f.)

(b) -3.2925×10^3 (4 s.f., 3 s.f., 2 s.f.).

7 (a) Round up the number 0.444 45 progressively, digit by digit until only one significant figure is left.

(b)   Repeat the process using $\dfrac{4}{9} = 0.\dot{4}$ and 5 s.f. to start.

8   If we wish to divide the number 60 in the ratio 5 to 7, written $5:7$, we first add 5 to 7 to get 12 then $60 \times \dfrac{5}{12} = 25$ and $60 \times \dfrac{7}{12} = 35$. The required ratio is therefore $25:35$.

(a)   Divide 80 in the ratio $2:3$.

(b)   Divide 80 in the ratio $1:4:5$.

9   Numbers can be expressed as **percentages**, i.e. fractions whose denominators are 100, for example,

$$\frac{7}{10} = \frac{70}{100} = 70\%$$

where the % symbol is read 'percent'.

(a)   Express the following fractions or decimals as percentages:

$\dfrac{1}{5}, \dfrac{3}{8}, \dfrac{2}{3}, \dfrac{5}{4}; \; 0.4, 0.835, 1.12.$

(b)   Express the following percentages as fractions in their lowest terms:

23%, 4.12%, 62.5%.

1.3 FRACTIONAL POWERS AND LOGARITHMS

Fractional powers and roots

It is easy to calculate that $3^5 = 243$. Then we say that 3 is the fifth root of 243, written $(243)^{1/5}$ or $\sqrt[5]{243}$. Note that

$$(243)^{1/5}(243)^{1/5}(243)^{1/5}(243)^{1/5}(243)^{1/5} = (253)^{(1/5)\times 5} = 243.$$

Similarly, $128 = 2^7$ so that $2 = (128)^{1/7}$ is the seventh root of 128, written $\sqrt[7]{128}$.

We call the third root of a number the **cube root**; hence 2 is the cube root of 8, since $2^3 = 8$. The second root is called the **square root**; $2 = (4)^{1/2}$ (or $4^{1/2}$) and written $\sqrt{4}$. We do not write $\sqrt[2]{4}$.

You may remember that $(-2) \times (-2) = 4$, i.e. $(-2)^2 = 4$, so we *could* say $-2 = 4^{1/2}$. Indeed, every positive number has in fact got *two* square roots (and two fourth roots, two sixth roots, and so on). The convention is that $4^{1/2}$ means the *positive* square root of 4.

Note that

$$2^{1/5} \times 4^{1/5} = (2 \times 4)^{1/5} = 8^{1/5}$$

that

$$2^{1/3}(16)^{-1/3} = \left(\frac{2}{16}\right)^{1/3} = \left(\frac{1}{8}\right)^{1/3} = \frac{1}{2}$$

and that

$$(32)^{1/3} = (8 \times 4)^{1/3} = 8^{1/3} \times 4^{1/3} = 2(2^2)^{1/3} = 2 \times 2^{2/3} = 2^{5/3}.$$

Powers which are fractions greater than 1 can be treated as follows:

$$3^{5/2} = 3^{2+1/2} = 3^2 \times 3^{1/2} = 9\sqrt{3}.$$

Finally, $2^{3/4} = (2^3)^{1/4} = (2^{1/4})^3$.

For more complicated cases we can use a pocket calculator. Thus to find $(1.2)^3$ we enter 1.2, press the x^y button on the calculator then enter 3 and press the button again.

Similarly, to find $(1.2)^{0.4}$ we put $y = 0.4$ and repeat the procedure. It does not matter whether or not x is a whole number. If y is not a whole number then x must be positive.

Exercise 1.5

1 Determine the following roots exactly without resorting to a calculator:

(a) $9^{3/2}$ (b) $64^{2/3}$ (c) $(125)^{-1/3}$

(d) $(729)^{-5/6}$ (e) $(343/8)^{-2/3}$ (f) $(11/5)^0$

(g) $(0.0016)^{1/4}$ (h) $(6.25)^{-1/2}$.

2 Again, without using a calculator, simplify the following. Note that 4/7 is equivalent to $\frac{4}{7}$, etc.

(a) $(9 \times 10^{-8})^{1/2}(20)^4$ (b) $((16)^{3/2})^{-1/2}$

(c) $(2.5 \times 10^{-3})^{1/2}$ (d) $((4/7)^{1/3})^{-6} \times 10^{-2}$

(e) $6 \times 10^{-2} \div (4 \times 10^{-4})$ (f) $(361/289)^{-1/2} \times 4/17$.

3 Given that $2 = \sqrt{2} \times \sqrt{2}$, $3 = \sqrt{3} \times \sqrt{3}$, etc. show that

(a) $\dfrac{1}{\sqrt{2}} = \dfrac{1}{2}\sqrt{2}$ (b) $\dfrac{1}{\sqrt{3}} = \dfrac{1}{3}\sqrt{3}$

(c) $(\sqrt{3} - 1)(\sqrt{3} + 1) = 2$ (d) $\dfrac{\sqrt{3} - 1}{\sqrt{3} + 1} = 2 - \sqrt{3}$

(e) $\dfrac{2\sqrt{3}+1}{\sqrt{2}} = \sqrt{6} + \dfrac{1}{2}\sqrt{2}$

(f) $\dfrac{1}{\sqrt{3}}\left(\dfrac{\sqrt{2}+1}{\sqrt{2}-1}\right) = \sqrt{3} + \dfrac{2}{3}\sqrt{6}.$

4 Use the x^y button on your calculator to find $(5.236)^y$ where $y = 0.4$ (0.2) 1.4, i.e. 0.4 to 1.4 in steps of 0.2. Repeat the same process for $x = 0.1417$. What do you observe?

5 Retaining 4 s.f., find

(a) $(1.9)^{0.7} \times (0.6)^{-2.4}$

(b) $\dfrac{(101.39)^{0.5} \times (6.237)^{-1.2}}{(11.574)^{2.23}}$

(c) $\dfrac{(19 \times 10^3)^{0.5} \times (3.1412)^{1/3}}{10^{5/2}\left\{3.7 - \left(\dfrac{6.1 \times 1.89}{(7.3)^2}\right)\right\}^{3/2}}.$

6 Given the formula $r = \left(\dfrac{4950\eta^2}{\sigma(\rho - \sigma)g}\right)^{1/3}$ which relates to a sphere falling in a viscous liquid, determine r to 4 s.f. if $\eta = 3.12 \times 10^{-5}$, $g = 9.81$, $\sigma = 1.051$ and $\rho = 11.972$.

7 Determine π on your calculator to 7 d.p. How good are the approximations

(a) $22/7$

(b) 3.142

in absolute terms and as a percentage of the calculator value?

8 Determine which of the following exist and evaluate them when appropriate.

(a) $\sqrt{-4} = (-4)^{1/2}$

(b) $\sqrt[3]{-362} = (-362)^{1/3}$

(c) $\sqrt[4]{-652} = (-652)^{1/4}$

(d) $\sqrt[5]{\dfrac{-1}{9865}} = \left(\dfrac{-1}{9865}\right)^{1/5}$

(e) $(-0.321)^{0.6}.$

Logarithms

In the relationship $10^3 = 1000$ we see that 10 must be raised to the power 3 to equal 1000; we say that 3 is the **logarithm** of 1000 to the **base** 10 or, symbolically, $\log_{10} 1000 = 3$.

Similarly, since $2^5 = 32$, we can say that 5 is the logarithm of 32 to the base 2 and write $\log_2 32 = 5$. Note the following six results which are demonstrated with logarithms to the base 2.

1. The logarithm of the product is the sum of the logarithms.

For example, $2^3 \times 2^4 = 2^{3+4} = 2^7$, i.e. $8 \times 16 = 128$.

Now $\log_2 128 = 7$, $\log_2 8 = 3$ and $\log_2 16 = 4$,

so that $\log_2 128 = \log_2 (8 \times 16) = \log_2 8 + \log_2 16$.

2. The logarithm of the quotient is the difference of the logarithms.

For example, $2^5 \div 2^2 = 2^{5-2} = 2^3$, i.e. $32 \div 4 = 8$.

Now $\log_2 8 = 3$, $\log_2 32 = 5$ and $\log_2 4 = 2$,

so that $\log_2 8 = \log_2(32 \div 4) = \log_2 32 - \log_2 4$.

3. The logarithm of the power of a number is that power multiplied by the logarithm.

For example, $(2^4)^3 = 2^{4 \times 3} = 2^{12}$, i.e. $(16)^3 = 4096$.

Now $\log_2 4096 = 12$ and $\log_2 16 = 4$,

so that $\log_2 4096 = \log_2 (16)^3 = 3 \log_2 16$.

4. The logarithm of 1 is zero.

For example, $2^0 = 1$,

so that $\log_2 1 = 0$.

5. The logarithm of the base is 1.

For example, $2^1 = 2$,

so that $\log_2 2 = 1$.

6. The logarithm of a reciprocal is (-1) multiplied by the logarithm.

For example, $\dfrac{1}{8} = \dfrac{1}{2^3} = 2^{-3}$.

Now $\log_2 8 = 3$,

so that $\log_2\left(\dfrac{1}{8}\right) = \log_2\left(\dfrac{1}{2^3}\right) = \log_2(2^{-3}) = -3 = -\log_2 8$.

In Chapter 2 these results are generalized for any base.

Exercise 1.6

1 Write the following results using logarithms:

(a) $3^4 = 81$

(b) $5^3 = 125$

(c) $7^2 = 49$

(d) $4^3 = 64$

(e) $6^0 = 1$

(f) $8^1 = 8$

(g) $10^{-2} = \dfrac{1}{100}$

(h) $4^{1/2} = 2$

(i) $4^{-1/2} = \dfrac{1}{2}$.

2 Write the following results without using logarithms:

(a) $\log_7 343 = 3$

(b) $\log_5 25 = 2$

(c) $\log_3 3 = 1$

(d) $\log_4 1 = 0$

(e) $\log_6\left(\dfrac{1}{36}\right) = -2$

(f) $\log_{25} 5 = \dfrac{1}{2}$

(g) $\log_8\left(\dfrac{1}{2}\right) = -\dfrac{1}{3}$

(h) $\log_{81}\left(\dfrac{1}{3}\right) = -\dfrac{1}{4}$.

3 Illustrate the six results on page 18 using logarithms to the base 10.

Surds

As we have seen, decimals are particular examples of fractions and some fractions, e.g. $\dfrac{1}{3}$ and $\dfrac{2}{7}$, cannot be expressed exactly as terminating decimals. Not every number can be expressed exactly as a fraction. Consider the number $\sqrt{2}$; a calculator with a ten-digit display will produce the result 1.414 213 562, which can be expressed as the fraction

$$\frac{1\ 414\ 213\ 562}{1\ 000\ 000\ 000}.$$

This is not an exact answer, however; if this number is entered into your calculator and squared, the result is displayed as 1.999 999 999. To quote the original number exactly, we leave it as $\sqrt{2}$ and it is perfectly acceptable to write it in this form unless an approximate answer is wanted, such as an answer correct to 3 s.f. or 2 d.p. Numbers such as $\sqrt{2}, \sqrt{3}, \sqrt{5}, \sqrt{6}$ are called **surds**; they are examples of **irrational numbers,** a wider class of numbers which cannot be expressed as fractions.

Example   Express the following numbers in their simplest form:

(a) $\sqrt{24}$ (b) $\sqrt{8}$ (c) $\sqrt{324}$

(d) $\sqrt{30}$.

Solution:

(a) Since $24 = 4 \times 6$ then $\sqrt{24} = \sqrt{4} \times \sqrt{6} = 2\sqrt{6}$

(b) $8 = 2 \times 2 \times 2$ so that $\sqrt{8} = 2\sqrt{2}$

(c) $324 = 18 \times 18$ and $\sqrt{324} = 18$

(d) $30 = 2 \times 3 \times 5$ and $\sqrt{30}$ is therefore already in its simplest form.

Multiplication of surds

The product $(2 + \sqrt{3})(5 - \sqrt{2})$ can be evaluated by the usual method:

$$(2 + \sqrt{3})(5 - \sqrt{2}) = 2 \times 5 + \sqrt{3} \times 5 - 2 \times \sqrt{2} - \sqrt{3} \times \sqrt{2}$$
$$= 10 + 5\sqrt{3} - 2\sqrt{2} - \sqrt{6}.$$

Sometimes it is possible to simplify the resulting expression:

$$(3 + 2\sqrt{5})(4 - \sqrt{5}) = 3 \times 4 + 2\sqrt{5} \times 4 - 3 \times \sqrt{5} - 2\sqrt{5} \times \sqrt{5}$$
$$= 12 + 8\sqrt{5} - 3\sqrt{5} - 2 \times 5$$
$$= 2 + 5\sqrt{5}.$$

In some special instances the surd parts cancel out:

$$(3 + \sqrt{2})(3 - \sqrt{2}) = 3 \times 3 + \sqrt{2} \times 3 - 3 \times \sqrt{2} - \sqrt{2} \times \sqrt{2}$$
$$= 9 + 3\sqrt{2} - 3\sqrt{2} - 2$$
$$= 7.$$

Rationalising the denominator

It is often the case that a surd in the denominator of a fraction is unwelcome; by a simple manipulation it is possible in effect to transfer the surd to the numerator.

As an example, since $\sqrt{5} \times \sqrt{5} = 5$ then

$$\frac{3}{\sqrt{5}} = \frac{3\sqrt{5}}{\sqrt{5} \times \sqrt{5}} = \frac{3\sqrt{5}}{5}.$$

In a similar way, since

$$(3 - \sqrt{2})(3 + \sqrt{2}) = 7$$

then

$$\frac{2 + \sqrt{3}}{3 - \sqrt{2}} = \frac{2 + \sqrt{3}}{3 - \sqrt{2}} \times \frac{3 + \sqrt{2}}{3 + \sqrt{2}}$$

$$= \frac{(2 + \sqrt{3})(3 + \sqrt{2})}{(3 - \sqrt{2})(3 + \sqrt{2})} = \frac{6 + 3\sqrt{3} + 2\sqrt{2} + \sqrt{6}}{7}.$$

Exercise 1.7

1 Simplify the following as far as possible:

(a) $\sqrt{45}$ (b) $\sqrt{125}$ (c) $\sqrt{48}$

(d) $\sqrt{300}$ (e) $\sqrt{108}$ (f) $\sqrt{1728}$

(g) $\sqrt{252}$ (h) $\sqrt{10\,935}.$

2 Evaluate the following products, simplifying where possible:

(a) $(3 - \sqrt{2})\sqrt{2}$ (b) $\sqrt{3}(4 + 2\sqrt{3})$

(c) $(\sqrt{2} + 3)(\sqrt{2} + 4)$ (d) $(\sqrt{2} + 3)(2\sqrt{2} - 5)$

(e) $(3\sqrt{2} - 4)(5\sqrt{2} + 6)$ (f) $(2 - \sqrt{5})^2$

(g) $(2 + 3\sqrt{5})^2$ (h) $(6 + 2\sqrt{5})(6 - 2\sqrt{5}).$

3 Multiply each of the following by a similar expression to make the product free from surds; find the product in each case.

(a) $(3 - \sqrt{6})$ (b) $(\sqrt{7} + 5)$ (c) $(3\sqrt{3} - 2\sqrt{2}).$

4 Rationalise the denominator and as far as possible simplify each of the following.

(a) $\dfrac{5}{\sqrt{3}}$ (b) $\dfrac{1}{\sqrt{5}}$ (c) $\dfrac{7}{\sqrt{13}}$

(d) $\dfrac{4\sqrt{3}}{\sqrt{2}}$ (e) $\dfrac{1}{\sqrt{64}}$ (f) $\dfrac{\sqrt{3}}{\sqrt{6}}$

(g) $\dfrac{2}{3-\sqrt{3}}$ (h) $\dfrac{4\sqrt{3}}{\sqrt{5}+\sqrt{3}}$ (i) $\dfrac{3}{3\sqrt{5}-5}$

(j) $\dfrac{2+\sqrt{3}}{\sqrt{3}+1}$ (k) $\dfrac{3\sqrt{2}-1}{5-2\sqrt{2}}$ (l) $\dfrac{\sqrt{3}+1}{\sqrt{3}+2}$

(m) $\dfrac{2\sqrt{6}}{\sqrt{6}+5}$ (n) $\dfrac{\sqrt{5}+2}{\sqrt{5}-2}$ (o) $\dfrac{1}{(\sqrt{5}+2)^2}$

(p) $\dfrac{1}{(\sqrt{3}+2)(\sqrt{2}-1)}$.

1.4 ERRORS AND THE MODULUS SIGN

Errors

When measurements are made there is a limit on their accuracy. For example, when measuring a length using a ruler graduated in millimetres we are unable to be more accurate than the nearest 0.5 mm.

If a length has a true value of 127.8 mm and we measure it as 128 mm the **actual error** is $(128 - 127.8)$ mm $= 0.2$ mm. In general,

actual error = measured value − true value.

If the measurement is greater than the true value then the actual error is positive; if the measurement is smaller than the true value then the actual error is negative.

The **fractional error** or **relative error** is found by dividing the actual error by the true value; if expressed as a percentage then it is called the **percentage error**. In the example above the fractional error is $\dfrac{0.2}{127.8}$ and the percentage error is $\left(\dfrac{0.2}{127.8} \times 100\right)\%$.

In practice, of course, we do not know the true value, we merely estimate it via the measured value. When there is uncertainty in a measured quantity it is customary to quote it in the form,

measured value ± maximum error.

Hence, if we measured a temperature as 100.3 °C and we can rely on the thermometer reading to the nearest 0.1 °C then we can quote the result as

$$100.3 \pm 0.1 \text{ °C}.$$

The magnitude of the actual error is called the **absolute error**. In this example the maximum actual error is ± 0.1 °C and the maximum absolute error is 0.1 °C.

Modulus sign

The absolute value of a number ignores any sign. Hence the absolute value of -3.8 is 3.8 and the absolute value of 6.3 is 6.3. We denote the absolute value of a number by placing it between two parallel vertical lines, $|\ |$. Hence $|-3.8| = 3.8$ and $|6.3| = 6.3$. These lines constitute the **modulus sign**. It is a very common symbol in mathematics.

Strictly speaking, we should define the relative error as

$$\left| \frac{\text{actual error}}{\text{exact value}} \right|$$

and the percentage error as $(100 \times \text{relative error})$ %.

One important result involving absolute values is the **triangle inequality**, so called because it corresponds to the geometrical statement that the sum of the lengths of any two sides of a triangle is greater than or equal to the length of the third side. It can be stated as follows:

> The absolute value of the sum of two numbers is less than or equal to the sum of their absolute values, or in symbols
>
> $$|a + b| \leq |a| + |b|$$
>
> where a and b are the two numbers.

Example

a	b	$a+b$	$|a+b|$	$|a|+|b|$
4.2	3.5	7.7	7.7	7.7
4.2	-3.5	0.7	0.7	7.7
-4.2	3.5	-0.7	0.7	7.7
-4.2	-3.5	-7.7	7.7	7.7

■

In the measurement of the temperature quoted earlier, the maximum absolute error is $|\pm 0.1|$ °C $= 0.1$ °C; the maximum relative error is approximately

$\left|\dfrac{0.1}{100.3}\right| = 0.000997$ (6 d.p.). Given the uncertainty in the measurement, it is not sensible to be too precise about the relative error and we would more reasonably quote it as 0.001. The maximum percentage error is approximately 0.01%. Note that we have had to replace the true value by the measured value to calculate these errors, hence they are approximations only. When the actual error is small in size compared to the measured value, these approximations will be quite close to the truth.

Arithmetic with measured values

Example

Let one measured quantity be 10.0 ± 0.2 and a second be 40.0 ± 0.5.

The largest value that the sum can be is $10.2 + 40.5 = 50.7$ and the smallest value is $9.8 + 39.5 = 49.3$. Given this range of possible values we quote the sum as 50.0 ± 0.7. (Note that $10.0 + 40.0 = 50.0$). The absolute error is $0.7 = 0.2 + 0.5$.

If we subtracted the smaller quantity from the larger quantity, we could get a result as large as $40.5 - 9.8 = 30.7$ or a result as small as $39.5 - 10.2 = 29.3$. We quote the difference as 30.0 ± 0.7. (Note that $40.0 - 10.0 = 30.0$. The absolute error is $0.7 = 0.2 + 0.5$.) ∎

The general result of addition and subtraction of two measured values is as follows:

> *The absolute error in the sum or the difference of two measured quantities is the* **sum** *of the individual errors.*

The general result for multiplication and division of two measured values is as follows:

> *The relative error in the product or the quotient of two measured quantities is* **approximately** *the* **sum** *of the individual relative errors.*

A similar result is true for percentage errors.

Example

The largest value that the product of the two measured quantities above can take is $10.2 \times 40.5 = 413.1$ and the smallest is $9.8 \times 39.5 = 387.1$; the value calculated from the measurements is 400.

The maximum relative error in the product is

$$\left|\frac{413.1 - 400}{400}\right| = 0.032\,75 \text{ or } \left|\frac{387.1 - 400}{400}\right| = 0.032\,25.$$

Now the sum of the individual relative errors is

$$\left|\frac{0.2}{10}\right| + \left|\frac{0.5}{40}\right| = 0.02 + 0.0125 = 0.0325.$$

Furthermore, if we divide the larger quantity by the smaller, the largest value that the quotient can be is $\dfrac{40.5}{9.8} \simeq 4.1327$ and the smallest value is $\dfrac{39.5}{10.2} \simeq 3.8725$, whereas the value calculated from the measurements is 4. The symbol $\simeq$ means 'is approximately equal to'. The maximum approximate relative error is

$$\left| \frac{4.1327 - 4}{4} \right| = \frac{0.1327}{4} \simeq 0.0332$$

$$\text{or} \quad \left| \frac{3.8725 - 4}{4} \right| = \frac{0.1275}{4} \simeq 0.0319. \qquad \blacksquare$$

Exercise 1.8

1   Two lengths of 20 cm and 50 cm respectively are measured respectively as 21 cm and 52 cm. Calculate the actual and fractional errors in each measurement. Find the actual errors in the sum and difference and the relative errors in the product and quotient. Verify the general rules on page 24.

2   The square root of 5 is quoted as 2.24 (2 d.p.). Estimate the absolute and relative errors, quoting both to one significant figure.

3   Calculate the absolute error in the product, the relative error in each measurement and the relative error in the product for the following pairs of numbers:

(a)   10 cm measured as 9.8 cm, and 25 cm measured as 25.1 cm

(b)   40 cm measured as 41 cm, and 12.5 cm measured as 12 cm

(c)   30 cm measured as 29 cm, and 20 cm measured as 19.5 cm.

4   (a)   Verify the triangle inequality for the numbers $a = \pm 7, b = \pm 20$.

(b)   By putting $b = -c$ we can obtain the result $|a - c| \le |a| + |c|$. Verify this result in the cases $a = \pm 9, c = \pm 14$.

5   A cube has sides of 2 cm, measured as 2.01 cm. What are the actual and relative errors in the measured volume? How does the relative error in volume relate to the relative error in length?

6   The radius of a sphere is measured as 10 cm but the measurement could be subject to an error of ± 0.5 cm. What are the actual and relative errors in its surface area and its volume?

7 The measurement of the pressure of a gas is subject to a percentage error of 2%, the volume is measured to an accuracy of 3% and its absolute temperature to an accuracy of 5%. Find the approximate percentage error in the gas constant, found by multiplying the pressure by the volume and dividing by the absolute temperature.

8* Evaluate $\dfrac{0.814\,862 - 0.814\,860}{0.629\,322}$. If the numbers quoted are correct to the number of decimal places shown, estimate the maximum relative error in the result. Comment on your answers.

SUMMARY

- **Fundamental theorem of arithmetic:** any integer can be uniquely expressed as the product of prime factors.
- **Highest common factor (HCF):** the HCF of two integers is the largest integer which divides exactly into them both.
- **Lowest common multiple (LCM):** the LCM of two integers is equal to their product divided by their HCF.
- **Modulus or absolute value:** the magnitude of a number; the modulus of x is written $|x|$.
- **Powers**

  0^0 cannot be found $\qquad a^0 = 1$

  $a^x \times a^y = a^{(x+y)} \qquad a^x \div a^y = a^{(x-y)}$

  $(a^x)^y = a^{x \times y} \qquad\qquad a^{-y} = 1/a^y.$

- **Fractions:**

  A fraction is a ratio of integers N/M where $M \neq 0$.
  If $|N| \leq |M|$ then the fraction is **proper**.
  If $|N| > |M|$ then the fraction is **vulgar**.
  If all common factors of N and M are cancelled then the fraction is in its **lowest form**.

- **Decimal places (d.p.):** the number of digits (or figures) after the decimal point.
- **Significant figures (s.f.):** the number of digits in a number which are relevant to its required level of accuracy.
- **Scientific notation:** a number in the form $a \times 10^r$ where r is the exponent and a, the mantissa, is a number between 1 and 10.
- **Surds:** expressions involving the sum or difference of rational numbers and their square roots.
- **Logarithms:**

  $\log_b n = x$ if $b^x = n \qquad \log n^m = m \log n$

  $\log(nm) = \log n + \log m \qquad \log(n/m) = \log n - \log m.$

- **Error:** in an approximation to a number error is defined by the relationship

  error = approximate value − true value.

Answers

Exercise 1.1

1 (a) $2 \times 2 \times 2 \times 3 \times 5 \times 19$ (b) $2 \times 5 \times 11 \times 11$

(c) 41×101 (d) $3 \times 3 \times 7 \times 7 \times 7$

(e) $3 \times 13 \times 13 \times 29$ (f) $11 \times 23 \times 71$

(g) $2 \times 2 \times 5 \times 5 \times 11 \times 59$ (h) $5 \times 5 \times 11 \times 17 \times 17 \times 41$

2 *2280* Sum of digits $= 12$; divisible by 3; $2 + 8 - 2 - 0 = 8$, not divisible by 11.

1210 Sum of digits $= 4$, not divisible by 3; $1 + 1 - 2 - 0 = 0$, divisible by 11.

4141 Sum of digits $= 10$, not divisible by 3; $4 + 4 - 1 - 1 = 6$, not divisible by 11.

3087 Sum of digits $= 18$, divisible by 9; $3 + 8 - 0 - 7 = 4$, not divisible by 11.

14703 Sum of digits $= 15$, divisible by 3; $1 + 7 + 3 - 4 - 0 = 7$, not divisible by 11.

17963 Sum of digits $= 26$, not divisible by 3; $1 + 9 + 3 - 7 - 6 = 0$, divisible by 11.

64900 Sum of digits $= 19$, not divisible by 3; $6 + 9 - 4 = 11$, divisible by 11.

3258475 Sum of digits $= 34$, not divisible by 3; $3 + 5 + 4 + 5 - 2 - 8 - 7 = 0$, divisible by 11.

3 (a) LCM $= 5040$, HCF $= 30$ (b) LCM $= 275\,880$, HCF $= 10$

(c) LCM $= 5832$, HCF $= 9$ (d) LCM $= 3360$, HCF $= 24$

(e) LCM $= 5040$, HCF $= 12$

4 (a) $210 \times 720 = 151\,200 = 5040 \times 30$

(b) $1210 \times 2280 = 2\,758\,800 = 275\,880 \times 10$ (c) $72 \times 729 = 52\,488 = 5832 \times 9$

5 1993, 1997, 1999 are all prime

Exercise 1.2

1 (a) $17\,640$ (b) $231\,525$ (c) $1\,500\,750$

(d) $19\,380$ (e) $123/40$ (f) $25/147\,456$

(g) 1

2 (a) 1 in each case

(b) 1 in each case. $x^0 = 1$ for x as close to zero as one wishes but 0^0 is not defined.

3 (a) 49 (b) 25 (c) 161 (d) 558

Exercise 1.3

1 (a) $\dfrac{1}{8}$ (b) $-\dfrac{281}{28}$ (c) $\dfrac{1836}{55}$ (d) $-\dfrac{196}{33}$

2 (b) $-10\dfrac{1}{28}$ (c) $33\dfrac{21}{55}$ (d) $-5\dfrac{31}{33}$

3 (a) $\dfrac{17}{20}$ (b) $1\dfrac{13}{315}$ (c) $1\dfrac{150}{209}$

    (d) $10\dfrac{11}{24}$ (e) $68\dfrac{201}{616}$ (f) $258\dfrac{11}{18}$

4 (a) $\dfrac{8}{15}$ (b) $-7\dfrac{43}{132}$ (c) $\dfrac{194}{255}$

    (d) $4\dfrac{7}{24}$ (e) $-\dfrac{215}{24} = -8\dfrac{23}{24}$ (f) $-\dfrac{11}{420}$

5 (a) $\dfrac{1}{4}$ (b) $\dfrac{3}{34}$ (c) $1\dfrac{17}{22}$

    (d) 24 (e) $2\dfrac{8}{15}$ (f) $\dfrac{273}{128} = 2\dfrac{17}{128}$

6 (a) $\dfrac{7}{18}$ (b) $\dfrac{253}{340}$ (c) $\dfrac{95}{202}$

    (d) 5 (e) $\dfrac{4}{21}$ (f) $-\dfrac{1349}{1968}$

7 $-\dfrac{1699}{968} = -1\dfrac{731}{968}$

Exercise 1.4

1 (a) $\dfrac{1}{16}$ (b) $-2\dfrac{131}{500}$ (c) $3\dfrac{3}{8}$

    (d) $-\dfrac{13}{16}$

2 (a) 0.0375 (b) $0.0666\ldots$ or $0.0\dot{6}$

    (c) 0.909090909 or $0.\dot{9}\dot{0}$ (d) $0.4285714286\ldots$ or $0.\dot{4}2857\dot{1}$

3 (a) -3.152×10^{-1} (4 s.f.)   (b) 2.57931×10^{3} (6 s.f.)

    (c) -3.434×10^{-4} (4 s.f.)   (d) 8.575×10^{6} (4 s.f.)

4 (a) $241\,200\,000$ (b) -3.142

    (c) $0.006\,248\,4$ (d) -0.67

5   The '00' in $0.469\,600$ are significant digits which means that $0.469\,600$ is accurate to within $\dfrac{1}{2} \times 10^{-6}$.

6 (a) 6.2835, 6.284, 6.28, 6.3

(b) $-3.293 \times 10^3, \ -3.29 \times 10^3, \ -3.3 \times 10^3$

7 (a) 0.4445, 0.445, 0.45, 0.5

(b) 0.4444, 0.444, 0.44, 0.4

8 (a) 32:48 (b) 8:32:40.

9 (a) 20%, 37.5%, $66\frac{2}{3}\%$, 125%; 40%, 83.5%, 112%

(b) $\dfrac{23}{100}, \dfrac{412}{10000} = \dfrac{103}{2500}, \dfrac{625}{1000} = \dfrac{5}{8}$

Exercise 1.5

1 (a) 27 (b) 16 (c) 1/5 (d) 1/243

(e) 4/49 (f) 1 (g) 0.2 (h) 0.04

2 (a) 48 (b) 1/8 (c) 0.05

(d) 49/1600 or 0.030 625 (e) 150 (f) 4/19

4 x^y increases with y for $x > 1$, e.g. $x = 5.236$ and decreases with y for $x < 1$, e.g. $x = 0.1417$. (For $x > 1, x^y < x$ for $y < 1$ and $x^y > x, y > 1; x < 1, x^y > x$ for $y < 1$ and $x^y < x, y > 1$.)

5 (a) 5.340 (b) 4.758×10^{-3} (c) 9.818×10^{-2}

6 $r = 3.498 \times 10^{-3}$

7 $\pi = 3.141\,592\,7$ to 7 d.p.

$22/7 = 3.142\,857\,1$ to 7 d.p.

$22/7 - \pi = 1.264 \times 10^{-3}$, 0.040% of calculator value

$3.142 - \pi = 0.407 \times 10^{-3}$, 0.013% of calculator value

8 (a) No (b) -7.127 (c) No (d) -0.1589 (e) No

If $x > 0$ then $\sqrt{-x}, \sqrt[4]{-x}, \sqrt[6]{-x}$, etc., do not exist but $\sqrt[3]{-x} = -\sqrt[3]{x}, \sqrt[5]{-x} = -\sqrt[5]{x}$, etc. do. Generally $(-x)^{1/y}$ cannot be found unless y is an odd integer.

Exercise 1.6

1 (a) $\log_3 81 = 4$ (b) $\log_5 125 = 3$ (c) $\log_7 49 = 2$

(d) $\log_4 64 = 3$ (e) $\log_6 1 = 0$ (f) $\log_8 8 = 1$

(g) $\log_{10}\left(\dfrac{1}{100}\right) = -2$ (h) $\log_4 2 = \dfrac{1}{2}$ (i) $\log_4\left(\dfrac{1}{2}\right) = -\dfrac{1}{2}$

2 (a) $7^3 = 343$ (b) $5^2 = 25$ (c) $3^1 = 3$

(d) $4^0 = 1$ (e) $6^{-2} = \dfrac{1}{36}$ (f) $25^{1/2} = 5$

(g) $8^{-1/3} = \dfrac{1}{2}$ (h) $81^{-1/4} = \dfrac{1}{3}$

Exercise 1.7

1 (a) $3\sqrt{5}$ (b) $5\sqrt{5}$ (c) $4\sqrt{3}$

(d) $10\sqrt{3}$ (e) $6\sqrt{3}$ (f) $24\sqrt{3}$

(g) $6\sqrt{7}$ (h) $27\sqrt{15}$

2 (a) $3\sqrt{2} - 2$ (b) $4\sqrt{3} + 6$ (c) $14 + 7\sqrt{2}$

(d) $\sqrt{2} - 11$ (e) $6 - 2\sqrt{2}$ (f) $9 - 4\sqrt{5}$

(g) $49 + 13\sqrt{5}$ (h) 16

3 (a) $(3 - \sqrt{6})(3 + \sqrt{6}) = 9 - 3 = 3$

(b) $(\sqrt{7} + 5)(-\sqrt{7} + 5) = -7 + 25 = 18$

(c) $(3\sqrt{3} - 2\sqrt{2})(3\sqrt{3} + 2\sqrt{2}) = 27 - 8 = 19$

4 (a) $\dfrac{5\sqrt{3}}{3}$ (b) $\dfrac{\sqrt{5}}{5}$

(c) $\dfrac{7\sqrt{13}}{13}$ (d) $\dfrac{4\sqrt{6}}{2} = 2\sqrt{6}$

(e) $\dfrac{1}{\sqrt{64}} = \dfrac{1}{8}$ (f) $\dfrac{\sqrt{18}}{6} = \dfrac{\sqrt{2}}{2}$

(g) $\dfrac{6 + 2\sqrt{3}}{6} - \dfrac{3 + \sqrt{3}}{3}$ (h) $\dfrac{4\sqrt{3}(\sqrt{5} - \sqrt{3})}{5 - 3} = 2\sqrt{15} - 6$

(i) $\dfrac{3(3\sqrt{5} + 5)}{45 - 25} = \dfrac{9\sqrt{5} + 15}{20}$ (j) $\dfrac{(2 + \sqrt{3})(\sqrt{3} - 1)}{2} = \dfrac{1 + \sqrt{3}}{2}$

(k) $\dfrac{(3\sqrt{2} - 1)(5 + 2\sqrt{2})}{17} = \dfrac{7 + 13\sqrt{2}}{17}$

(l) $\dfrac{(\sqrt{3} + 1)(-\sqrt{3} + 2)}{1} = -1 + \sqrt{3}$ (m) $\dfrac{2\sqrt{6}(-\sqrt{6} + 5)}{-6 + 25} = \dfrac{10\sqrt{6} - 12}{19}$

(n) $\dfrac{(\sqrt{5} + 2)(\sqrt{5} + 2)}{1} = (\sqrt{5} + 2)^2 = 9 + 4\sqrt{5}$

(o) $\dfrac{(\sqrt{5} - 2)^2}{\{(\sqrt{5} + 2)(\sqrt{5} - 2)\}^2} = \dfrac{9 - 4\sqrt{5}}{(5 - 4)^2} = 9 - 4\sqrt{5}$

(p) $\quad \dfrac{(\sqrt{3}-2)(\sqrt{2}+1)}{(3-4)(2-1)} = \dfrac{\sqrt{6}-2\sqrt{2}+\sqrt{3}-2}{-1} = 2\sqrt{2}-\sqrt{6}+2-\sqrt{3}$

Exercise 1.8

1 Actual errors are 1 cm, 2 cm; fractional errors are 0.05, 0.04. Absolute error in sum is 3 cm, in difference 3 cm. Relative error in product is 0.092, in quotient 0.094.

2 Absolute error is 0.004, Relative error is 0.002.

3

	Absolute error in product	Relative errors in the measurements	Relative error in product
(a)	$-4.02\,\text{cm}^2$	$-0.02,\ 0.004$	$-0.0161*$
(b)	$-8\,\text{cm}^2$	$0.025,\ -0.04$	-0.016
(c)	$-34.5\,\text{cm}^2$	$-0.03,\ -0.025$	-0.0575

* This number is approximate.

5 Actual error in length is 0.1 cm. Actual error in volume is 0.120 601 cm^3.
Relative error in length $=0.05$ and in volume $\simeq 0.015075 \simeq 3 \times$ relative error in length.

6 9.5 cm $<$ radius $<$ 10.5 cm. $1134 <$ surface area of $1257 < 1386$.

$3591 <$ volume of $4189 < 4849$.

$-0.098 <$ relative error in surface area $< 0.102 \simeq 2 \times$ relative area in radius.

$-0.143 <$ relative error in volume $< 0.158 \simeq 3 \times$ relative error in radius.

7 Percentage error is approximately 1%.

8 Calculated value $\simeq 0.000\,003\,178$

Relative error in numerator is $\dfrac{0.000\,000\,5 + 0.000\,000\,5}{0.000\,002} = 0.5$

Maximum absolute error in result is $0.000\,003\,178 \times \left(0.5 + \dfrac{0.000\,000\,5}{0.629\,322}\right)$

$\leq 0.000\,001\,6$

and the result can be quoted as $0.000\,003 \pm 0.000\,002$. This error is approximately two-thirds of the quoted result, which renders the result worthless. This illustrates the danger in quoting too many figures in a calculated result.

2 BASIC ALGEBRA

Introduction

At the simplest level, algebra can be regarded as generalised arithmetic, replacing numbers by letters as variables and allowing us to make general statements about the relationships between them as formulae or in the form of equations or inequalities which have to be solved. Formulae are used in every walk of life: the calculation of interest on a loan, the area of shapes to be cut from sheets of metal or the period of an oscillation. Ohm's law states that the current flowing through a resistor is proportional to the potential difference across its ends and this can be concisely restated as a formula. A thorough understanding of the simpler algebraic processes is vital for later work.

Objectives

After working through this chapter you should be able to

- evaluate algebraic expressions using the BEDMAS convention
- expand expressions using the laws of algebra
- interpret formulae
- rearrange a formula to make another of the variables the subject
- distinguish between an identity and an equation
- obtain the solution of a linear equation in one unknown
- recognise the kinds of solution for two linear simultaneous equations
- understand the terms *direct proportion*, *inverse proportion* and *joint proportion*
- solve simple problems involving proportion
- interpret simple inequalities in terms of intervals on the real line
- solve simple inequalities
- interpret inequalities involving the modulus of a variable

2.1 MANIPULATION OF SYMBOLS

In Section 1.3 we stated in words six laws of logarithms. A more compact statement employs the language of **algebra**. The laws are restated in words and the algebraic equivalent placed underneath. The base of the logarithms is a. If $z = a^x$ then $x = \log_a z$.

1. The logarithm of a product is the sum of the logarithms.

$$\log_a(xy) = \log_a x + \log_a y$$

2. The logarithm of a quotient is the difference of the logarithms.

$$\log_a(x/y) = \log_a x - \log_a y$$

3. The logarithm of the power of a number is that power multiplied by the logarithm.

$$\log_a(x^n) = n \log_a x$$

4. The logarithm of 1 is zero.

$$\log_a 1 = 0$$

5. The logarithm of the base is 1.

$$\log_a a = 1$$

6. The logarithm of a reciprocal of a number is (-1) multiplied by the logarithm of that number.

$$\log_a\left(\frac{1}{x}\right) = \log_a 1 - \log_a x = -\log_a x$$

Not only is the algebraic statement more compact than its counterpart in words, but it is also a general statement from which particular results may be deduced. For example, if we consider logarithms with base 10, and have $x = 3$, $y = 2$, the the second law becomes

$$\log_{10}(3/2) = \log_{10} 3 - \log_{10} 2$$

In this chapter we lay the foundations for a systematic use of the algebraic method.

Algebraic quantities are denoted symbolically by letters. The quantity may have a fixed value; for example, the acceleration due to gravity is denoted by g; it is customary to use letters at the start of the alphabet for fixed values. The quantity may be able to take several values; for example, the time that elapses from a given starting point is denoted by t; it is usual to employ letters at the end of the alphabet for quantities that take several values. Some special numbers are denoted by letters; for example, you will have met the number π and in Chapter 10 we introduce the number e.

Sometimes we have several related quantities or we wish to allow the possibility of generalising our results to situations with more quantities. Then we use the idea of a **subscript**, e.g. x_1, x_2, x_3, where 1, 2 and 3 are the subscripts.

Algebraic expressions

An algebraic expression is a combination of numbers, algebraic quantities and arithmetic symbols; for example, $2x + 5x^2 - 15xy$. The conventions we use are illustrated by the following examples.

Convention	Meaning
$2x$	2 multiplied by x or 2 times x
$5x^2$	5 times x squared
$-2x^3$	minus 2 times x cubed
$15xy$	15 times x times y
$\dfrac{a}{b}$ or a/b	alternative ways of writing $a \div b$
$a/(3b)$	a divided by the product of 3 and b.

Notice how we omit the multiplication sign.

In the examples above, the numbers 2, 5, -2, 15 and 3 are called the **coefficients** of x, x^2, x^3, xy and ab respectively. Conventionally they are placed in front of the algebraic quantities.

It is useful to simplify algebraic expressions; the following examples illustrate the rules.

Examples

1. Where the terms are similar they are collected by adding and subtracting the coefficients. Consider the expression $14x - 3x + 5x$. Now $14 - 3 + 5 = 16$, so the expression can be simplified to $16x$.

2. Where the terms are not similar they cannot be combined by simply adding or subtracting the coefficients. The expression $14x - 3y + 5z$ cannot be simplified further. But partial simplification is sometimes possible. In the expression

$14x - 3y + 5y - 8x$ we can treat the terms in x and the terms in y separately to arrive at $6x + 2y$. (You might find it helpful to rewrite the expression as $14x - 8x + 5y - 3y$.)

3. Different powers of a variable cannot be treated as like terms. The expression $3x + 4x^2$ cannot be simplified since the terms in x and x^2 are not considered as like terms. The expression can be rewritten in a different form which has some benefits; see Section 2.4.

4. Multiplication and division follow the rules for manipulating numbers. For example,

$$a \times b = ab = ba = b \times a$$

$$3a \times 7b = 3 \times 7 \times a \times b = 21ab$$

$$a \times (-b) = -ab, \quad (-a) \times b = -ab, \quad (-a) \times (-b) = ab$$

$$\frac{a}{-b} = -\frac{a}{b} = \frac{-a}{b}, \qquad \frac{-a}{-b} = \frac{a}{b}$$

$$\frac{2 \times 6xy}{z} \times \frac{2 \times xz}{3y} = \frac{\cancel{2}xy}{\cancel{z}}^{4} \times \frac{2y\cancel{z}}{\cancel{3}\cancel{y}}^{1} = 4x \times 2x = 8x^2$$

$$\frac{7 \times ab}{c^2} \div \frac{11 \times ab}{c} = \frac{7ab}{c^2} \times \frac{c}{11ab} = \frac{7\cancel{ab}}{11\cancel{ab}c} = \frac{7}{11c}.$$
∎

Note that cancellation follows the principles laid down for numbers in Chapter 1.

Powers

Just as 5^3 means $5 \times 5 \times 5$, so we write a^3 as a shorthand for $a \times a \times a$. The rules we illustrated in Chapter 1 for powers of numbers can be generalised and expressed compactly using algebraic notation. Note that a^1 is simply a.

1. $\quad a^m \times a^n = a^{m+n} \qquad a^m \div a^n = a^{m-n}$
   $\quad (a^m)^n = a^{mn} \qquad\quad a^0 = 1, a \neq 0$

2. $\quad a^{-n} = \dfrac{1}{a^n}$

3. $\quad (ab)^n = a^n b^n$

Examples

$$a^2 \times a^3 = a^5 \qquad a^6 \div a^2 = a^4$$
$$(a^2)^5 = a^{10} \qquad (3x^2y)^2 = 9x^4y^2$$ ■

Fractional powers are also a generalisation of the case for numbers. The square roots of x are written $\pm\sqrt{x}$ or $\pm x^{1/2}$ and, in general, the nth root of a is written $a^{1/n}$. When $n = 1$, $(a^{1/n})^n = a^1 = a$. The notation $a^{p/q}$ can mean the pth power of the qth root of a, i.e. $(a^{1/q})^p$ or the qth root of the pth power of a, i.e. $(a^p)^{1/q}$, provided the roots exist.

An algebraic expression can be **evaluated** by **substituting** specific numerical values for the algebraic quantities. For example, the value of a^3 when $a = 2$ is $2^3 = 8$, the value of the expression $3x - 2y + 4z$ when $x = 1$, $y = -2$, $z = -3$ is given by $3 \times 1 - 2 \times (-2) + 4 \times (-3) = 3 + 4 - 12 = -5$.

Exercise 2.1

1 Simplify the following to a single logarithm:

(a) $\quad \log_{10} 100 + \log_{10}(1/20)$

(b) $\quad \log_5 6 + \log_5 4 - 3 \log_5 2$

(c) $\quad \log_a 20 + \log_a 3 - 2 \log_a 2$

(d) $\quad -4 \log_a(1/3)$

(e) $\quad \dfrac{1}{2}\log_a(4/9)$

(f) $\quad \dfrac{1}{3}\log_a(27/8) + \dfrac{1}{4}\log_a 16.$

2 Simplify the following by removing the logarithms:

(a) $\quad \log_{10} 1000$

(b) $\quad \dfrac{1}{3}\log_{10} 1000 + 2 \log_{10} 100 - 6 \log_{10} \sqrt{10}$

(c) $\quad \log_a a^4 - 6 \log_a a^{3/2} + 2 \log_a a.$

3* On your calculator find e^x when $x = 1$. This should give $e^1 = 2.718\,281\,8$, i.e. e, the base of natural logarithms. Now find log e, which is in fact $\log_{10} e$. Store this. Find also ln 10, i.e. the natural logarithm of 10 or $\log_e 10$. Hence find $\log_{10} e \times \log_e 10$.

4 Simplify the following algebraic expressions into their simplest form. You may always assume that the contents of a bracket must be evaluated first of all. A full set of rules follows later in the chapter.

(a) $5ab \times 6a^2b$

(b) $3a^2b \times 2ab$

(c) $(4 \times 3ab) \div (6ab^2)$

(d) $(a/b) \div (b/c)$

(e) $(9ab/c) \times (2c/a)$

(f) $(2ab/c) \times (6a^2c/b) \times (bc/8a)$

(g) $(6\sqrt{yz}/p) \times (19p/y^{5/2}) \div (38z^{-3/2}p^{-1})$

(h) $(a^3)^{-1/2}$

(i) $\{(ab)^2\}^{-1/3}$

(j) $(abc)^{1/3} \div (a^2b^2c^2)^{2/3}$.

5 With $a = 1, b = 4, c = 27$, evaluate the following:

(a) $6a^2 + 2bc - 2c^{1/3}$

(b) $3\sqrt{ab} - (3c)^{1/4} + (2b^2)^{1/5}$

(c) $\dfrac{b^{3/2} + c^{2/3}}{\sqrt{a} + \sqrt{b}}$

(d) $(16bc)^{-2/3}$

(e) $\left\{\dfrac{(a + 2b)}{c/3}\right\}^{-1/7}$

(f) $\{(c - 11) \times b\}^{1/3}$.

6 Write down algebraic expressions for the following quantities:

(a) Minus one-third of the square root of x plus y^3, the whole expression divided by the square root of z cubed.

(b) The ratio a to b multiplied by the fourth power of c with 1 added to the total.

(c) The expression in (b) with 5 taken away from it and the result squared.

(d) The sum of x and y, all squared, divided by the sum of the squares of x and y.

(e) One minus the expression in (d).

7 Reduce the following expressions to their simplest form:

(a) $\left(\dfrac{16}{81}\right)^{3/4}$

(b) $\left(\dfrac{25}{9}\right)^{-1/2}$

(c) $b^2x^2y \div bxy$

(d) $108x^3z^2 \div 9xz$

(e) $x^2y^2 \div 15(xy)^{-1/2}$

(f) $(y^3)^{1/4} \times (y^{-1/2})^{1/3}$

(g) $(x^2yz^{-1})^2 \times (x^2y^{-2}z^2)^{-1/2} \div (xy)^{9/2}$

(h) $\dfrac{2^{n+1} - 2^n}{2^n - 2^{n-1}}$.

Expanding bracketed terms

To **expand** expressions involving brackets, each term inside the brackets is operated on by each term outside the brackets. For example,

$$5(7x - 3y) = 5 \times 7x - 5 \times 3y = 35x - 15y$$
$$- 2(x + z) = -2x - 2z$$
$$- 3(a - b) = -3 \times a - 3 \times (-b) = -3a + 3b$$

We can symbolically represent the key rules as

$$a(b + c) = ab + ac \qquad a(b - c) = ab - ac.$$

Note the example

$$c(x + a) = cx + ca = xc + ac = (x + a)c$$

Example

Simplify the expression $2(x + 3y) - 3(2x - y)$.
The expression becomes

$$2x + 6y - 6x + 3y$$

and then

$$-4x + 9y. \qquad \blacksquare$$

When two bracketed expressions are placed next to each other they are assumed to multiply together.

Examples

1. $(x - 3)(3x - 4) = x \times (3x) - 3 \times (3x) + x \times (-4) + (-3) \times (-4)$
$$= 3x^2 - 9x - 4x + 12 = 3x^2 - 13x + 12$$

2. $(2a + 3b)(3a - b) = 6a^2 + 9ba - 2ab - 3b^2$
$$= 6a^2 + 9ab - 2ab - 3b^2$$
$$= 6a^2 + 7ab - 3b^2 \qquad \blacksquare$$

Recall that the product ab is the same as ba, just as 2×3 and 3×2 have the same value.

Convention often puts the letters in alphabetical order, but we sometimes do not follow it until we have obtained the final result.

We may generalise the ideas by producing statements such as

$$(a - b)(c - d) = ac - bc - ad + bd$$

but it is not a useful exercise to try to *remember* such statements. It is far better to *understand* how they are obtained so that you can carry out such manipulations in specific cases.

Three kinds of product which are widely met are the following. These *are* worth remembering.

$$(a + b)^2 = (a + b)(a + b) = a^2 + 2ab + b^2$$
$$(a - b)^2 = (a - b)(a - b) = a^2 - 2ab + b^2$$
$$(a + b)(a - b) = (a - b)(a + b) = a^2 - b^2$$

You may verify them for yourself.

Note that the order of precedence of operations is the same as for numbers, so the acronym BEDMAS applies. For example,

$$5 + 3x^2 y$$

means that x is squared then multiplied by 3 and by y, the result being added to 5. On the other hand $(5 + 3x^2)y$ means $5y + 3x^2 y$.

Adding and subtracting fractions

The rules are parallel to those for numerical fractions.

Examples

1. $\dfrac{a}{c} + \dfrac{b}{c} = \dfrac{a + b}{c}$

2. $\dfrac{a}{2c} + \dfrac{b}{3c} = \dfrac{3a}{6c} + \dfrac{2b}{6c} = \dfrac{3a + 2b}{6c}$

   > Each fraction is replaced by an equivalent fraction whose denominator is the lowest common multiple of the original denominators.

3. $\dfrac{a}{c} + \dfrac{b}{d} = \dfrac{ad}{cd} + \dfrac{cb}{cd} = \dfrac{ad + cb}{cd}$

4. $\dfrac{a}{ce} + \dfrac{b}{de} = \dfrac{ad}{cde} + \dfrac{cb}{cde} = \dfrac{ad + cb}{cde}$

5. $\dfrac{a}{x} + \dfrac{b}{x^2} = \dfrac{ax}{x^2} + \dfrac{b}{x^2} = \dfrac{ax+b}{x^2}$

6. $\dfrac{a}{2x^3} + \dfrac{b}{3x^2} = \dfrac{3a}{6x^3} + \dfrac{2bx}{6x^3} = \dfrac{3a+2bx}{6x^3}$

7. $\dfrac{a}{x^2y} + \dfrac{b}{xy^3} = \dfrac{ay^2}{x^2y^3} + \dfrac{bx}{x^2y^3} = \dfrac{ay^2+bx}{x^2y^3}.$ ∎

Exercise 2.2

1 Multiply out the following algebraic products in full, collecting together all like terms.

(a) $(2x-3)(x-1)$ (b) $(3x+4)(3x-1)(2x+5)$

(c) $(2x+y)(3x-y)$ (d) $(x+z)(x^2+z)$

(e) $(2x+y)(3x-1)$ (f) $(2x+y)(2x-y)(x-y)$

(g) $(z+a)(z+b)$ (h) $(x+1)(x+2)+(x+3)(x+4)$

(i) $(3x-y)(2x-y)-3(x-y)$ (j) $z(x+y)+4x(y-z)+3y(x+2z).$

 2 Expand the following expressions:

(a) $6(x-1)+3(x+4)+5x$ (b) $2x-5+9(3-5x)+11$

(c) $19x+6(4-23x)+74$

(d) $2.18y-5.65-6.14(9.51y-2.11)$ (retain 2 d.p.)

(e) $\dfrac{1}{7}\left(\dfrac{H}{3}-\dfrac{1}{4}\right)-\dfrac{1}{11}\left(3-\dfrac{H}{4}\right)$ (as a fraction)

(f) $14\left(\dfrac{5}{x}-\dfrac{1}{2}\right)+3\left(\dfrac{20}{x}+\dfrac{7}{2}\right)$ (g) $4(x^2-5)+2(x-x^2)$

(h) $2xy-6+x(4-y)+14y(2x-17)$

(i) $x^2(2+x)+5(6-x^2)$

(j) $x_1(2-z_1^2)+z_1(5+2x_1z_1)-2(x_1-z_1^2)$

(k) $x(2+x)+x^2(3-x)+x^3(5+11x-x^2)$

(l) $\pi\left(\dfrac{9.72}{x}-4.18x\right)+\dfrac{1}{e}\left(x-\dfrac{5.18}{x}\right)$ (express this as a decimal to 2 d.p. with $\pi=3.14$ and $e=2.72$).

3 Prove that $a^3+b^3=(a+b)(a^2-ab+b^2)$. Now replace b by $-b$ and find an expression for a^3-b^3.

4 Fully expand the following expressions:

(a) $(x - 1)(x - 3)$ (b) $(2a - 1)(4a - 2)$

(c) $(s + 3t)(s - 2t)$ (d) $(xy - 1)(xy + 1)$

(e) $(3 - x)(5x + 2)$ (f) $(x - 1)(x - 2)(x - 3)$

(g) $(x - 1)(x - 2)(x + 3) + 3(x - 4)(x - 5) + x - 6$

(h) $(p - 2q)(p - 5q)(p - q)$ (i) $(x + y)^4$

(j) $(x + 1)^5$ and identify the coefficient of x^3.

5 Write down single-fraction expressions for

(a) $\dfrac{a}{c} - \dfrac{b}{c}$ (b) $\dfrac{a}{2c} - \dfrac{b}{3c}$ (c) $\dfrac{a}{c} - \dfrac{b}{d}$.

6 Write as single fractions

(a) $\dfrac{a}{c} + \dfrac{b}{c} + 1$ (b) $\dfrac{a}{2c} + \dfrac{b}{5c}$

(c) $\dfrac{a}{3c} + \dfrac{b}{4c} - 2$ (d) $\dfrac{a}{3d} + \dfrac{b}{5d} + \dfrac{c}{6d}$

(e) $\dfrac{a}{x} + \dfrac{b}{x^2} + \dfrac{c}{x^3}$ (f) $\dfrac{a}{3x} + \dfrac{b}{4x^2}$

(g) $\dfrac{a}{2x} + \dfrac{b}{3x^2} - \dfrac{c}{4x^3}$.

7 Write in its simplest form the expression $\dfrac{a + b}{c}$ with

(a) $a = x, b = x^2, c = 1$ (b) $a = 3x, b = 5x^2, c = x$

(c) $a = x, b = y, c = 3y$ (d) $a = -x, b = -2x^2, c = -x^2$

(e) $a = z, b = z^{-1}, c = z^{-1}$ (f) $a = x^{1/2}, b = x^{3/2}, c = x^{3/2}$

(g) $a = x^{1/3}, b = x^{4/3}, c = x^{7/3}$ (h) $a = x - 1, b = x + 1, c = x$

(i) $a = xy, b = xy^2, c - y$ (j) $a = 3x + 1, b = -2x + 1, c = x + 1$.

8 Given that

$$\frac{a}{b} + \frac{c}{d} = \frac{ad + cb}{bd}$$

simplify the following sums of fractions, cancelling where necessary. For example, with $a = x, b = x^2, c = 1, d = x$ the formula becomes

$$\frac{1\,\cancel{x}}{x\cancel{x^2}} + \frac{1}{x} = \frac{2}{x}.$$

(a) $a = x^2, b = x, c = x^3, d = 3x^2$

(b) $a = x - 1, b = x, c = 2x - 1, d = 2x$

(c) $a = x + 1, b = c = x - 1, d = x + 1$

(d) $a = 2x + 1, b = 3x - 1, c = x + 1, d = 2x - 1$

(e) $a = 1 - 3x, b = 3 - 4x, c = x + 6, d = 9 - x$

(f) $a = a_1 + a_2 x, b = b_1 + b_2 x, c = c_1 + c_2 x, d = d_1 + d_2 x$

(g) $a = 1 + x, b = 1 - x, c = 1 + x^2, d = (1 + x)(1 - x)$

(h) $a = 1 - x^2, b = 2 + x^2, c = 1 - x, d = 1 + x$

(i) $a = x, b = y^2, c = x^2, d = y$

(j) $a = \sqrt{x}, b = 1/\sqrt{x}, c = x^{3/2}, d = x^{1/2}.$

9 Using a common denominator, simplify

$$\frac{x}{(x - 1)(x^2 + 1)} + \frac{3x + 5}{(x - 2)(x^2 + 1)}$$

leaving the numerator as an expanded expression.

10 Multiplying top and bottom is a common practice with algebraic fractions, e.g.

$$\frac{1}{ax + b} = \frac{1}{(ax + b)} \times \frac{(ax - b)}{(ax - b)} = \frac{ax - b}{a^2 x^2 - b^2}.$$

Now find the expressions for the following unknowns in brackets

(a) $\dfrac{1}{x - 1} = \dfrac{A}{x^2 - 1}$ (A)

(b) $\dfrac{3x + 2}{2x - 1} = \dfrac{A}{(2x - 1)(x - 1)}$ (A)

(c) $\dfrac{1}{(x - 1)(x - 2)} = \dfrac{A}{(x - 1)(x - 2)(x - 3)}$ (A)

(d) $\dfrac{b}{a + b} = \dfrac{A}{a^2 - b^2}$ (A)

(e)* $\dfrac{\sqrt{x}}{\sqrt{x} + y} = \dfrac{A}{x - y^2} = \dfrac{B}{x^2 - xy}$ (A, B)

(f) $\dfrac{1}{3\sqrt{x} - 2\sqrt{y}} = \dfrac{A}{9x - 4y}$ $(A).$

11 Express the following as a single fraction over a common denominator:

(a) $\quad \dfrac{3}{x} + \dfrac{5x+1}{x-3} + 1$

(b) $\quad \dfrac{x}{y} + \dfrac{x^2}{y^2} - 1$

(c) $\quad 2x + \dfrac{3x-1}{x+1}$

(d) $\quad 3\sqrt{x} + \dfrac{2x-1}{(\sqrt{x} - 1/\sqrt{x})}$

(e) $\quad \dfrac{x^2-1}{x} - 2 + \dfrac{1}{x^2}$

(f) $\quad \dfrac{ax+b}{cx+d} - \dfrac{bx+a}{dx+c}.$

12* Prove the following results:

(a) $\quad (x + z^{-1})(z^{4/3} - z^{-2/3}) = \dfrac{z^2 - z^{-2}}{z^{-1/3}}$

(b) $\quad \left(\dfrac{y^{3/2} + yz}{yz - z^3}\right) - \dfrac{\sqrt{y}}{\sqrt{y} + z} = \dfrac{\sqrt{y}(y + z^2)}{z(y - z^2)}.$

2.2 FORMULAE, IDENTITIES, EQUATIONS

Formulae

A simple formula such as

$$T = 2\pi\sqrt{l/g}$$

shows how the periodic time of a simple pendulum, T, is given in terms of its length, l, and the (constant) acceleration due to gravity, g. Another example is Einstein's famous formula:

$$E = mc^2$$

where E represents energy, m mass and c the velocity of light. A **formula** relates algebraic quantities to each other. Some of these quantities are called **variables** because they can take a range of values. In the pendulum formula T is typically measured in seconds (s) and can in theory have any positive value. Likewise l is often measured in metres (m) and again can have any positive value. The acceleration $g = 9.81\,\mathrm{m\,s^{-2}}$ is fixed, or **constant**, and this being so means that knowing the value of T we can find the value of l or vice versa. Variables with values that are known are called **knowns** and those whose values are not known are called **unknowns**.

Example The area of a circle is given by

$$A = \pi r^2$$

where A is the area and r is the radius of the circle. If the value of π is taken to be 3.14, find the areas of the circles of radius

(a) 4 cm   (b) 2 m.

Solution

(a) When $r = 4$ cm then $A = 3.142 \times 4^2 = 50.272 \, \text{cm}^2$.

(b) When $r = 2$ m then $A = 3.142 \times 2^2 = 12.568 \, \text{m}^2$.

Note the units of the area of the circle.   ■

It is important when writing down a formula to realise that the units in which the values of the variables are measured must be compatible. If we wished to measure the radius of the circle in centimetres and the area in square metres, the formula would have to be modified to read $A = (0.01)^2 \pi r^2$.

Exercise 2.3

 1 (a)   Calculate the period time T of a simple pendulum of length

        (i)   1 m   (ii) 1.3 m

        (iii)   1.4 m. $(T = 2\pi\sqrt{l/g})$, $(g = 9.81 \text{ m s}^{-2})$

    (b)   Older clocks and chronometers were based on the simple pendulum principle. As the temperature of the environment increases, the metal in the pendulum expands and its length increases. T correspondingly changes. In opposite seasons, i.e. summer and winter, how would clocks tend to run (i.e. fast/slow)?

2 Given that $c = 3 \times 10^8 \text{ m s}^{-1}$, and using Einstein's formula $E = mc^2$, calculate the energy in joules stored in 1 kg of material. (For mass, length and time the basic units are kilogram, metre and second.)

3 The surface area (S) and volume (V) of a sphere are given by

    (i) $S = 4\pi r^2$ (ii) $V = \dfrac{4}{3}\pi r^3$.

    (a)   When the radius $r = 10.5$ m, determine S and V.

    (b)   If r is reduced by 1%, by approximately what percentage do S and V reduce?

    (c)   If $S = 100 \text{ m}^2$, find r and hence V. (Retain 4 s.f.)

4   A right circular closed cylinder of radius r and height h has total surface area given by $S = 2\pi r(r + h)$. Find S when $r = 10\,\text{cm}$, $h = 93\,\text{cm}$.

 5   The energy stored in a capacitor, capacitance C farads, charged to a potential of V volts is $\frac{1}{2}CV^2$ joules. When $C = 3.58 \times 10^{-6}$ and $V = 898$ determine the energy stored.

 6   For a lens of focal length f, together with object and image distances u and v, the variables are connected by the formula

$$\frac{1}{f} = \frac{1}{u} + \frac{1}{v}$$

(a)   Find f when $v = 30\,\text{m}$ and $u = 20\,\text{m}$.

(b)   Find v when $f = 10\,\text{m}$, $u = 40\,\text{m}$.

Transposing formulae

The need to rearrange formulae is particularly important when it comes to making one of the variables in an algebraic expression or formula into the subject of that formula. For example, T is the **subject** of the formula

$$T = 2\pi\sqrt{l/g}.$$

This form of manipulation or rearrangement is called **transposing** a formula.

For example, the area A of a rectangle of length l and breadth b is given by the formula $A = lb$. We have measured the area to be $4\,\text{m}^2$ and the length to be 2.5 m; to find a value for the breadth we need to transpose the formula to $b = A/l$, from which we readily find that $b = 4/2.5 = 1.6\,\text{m}$.

The following procedures may be used for transposing formulae:

(i)   The same quantity may be added to or subtracted from both sides.
(ii) Both sides may be multiplied by or divided by the same quantity.

Note that the same operation must be carried out on both sides.

But this must be treated with caution. If we have formulae in which all quantities have positive values only, there should be no problem; this is usually the case with physical quantities. However, since both $(-x)^2$ and $(+x)^2 = x^2$ then squaring may cause difficulties. We shall meet this situation later.

The operations can become second nature after regular practice. Then short cuts are often taken.

Given the formula $a = b - c$ we add c to both sides to obtain $a + c = b$. Alternatively, given $a = b + c$ we subtract c from both sides to obtain $a - c = b$. In effect, as c is transferred to the other side it changes sign.

Given $a = \dfrac{b}{c}$ we multiply both sides by c to obtain $ac = b$. Alternatrively, given $a = bc$ we divide both sides by c to obtain $\dfrac{a}{c} = b$. In effect, as c is transferred to the other side it changes position from top to bottom or vice versa.

Examples

1. Recall that the focal length f of a lens is given by the formula

$$\frac{1}{f} = \frac{1}{u} + \frac{1}{v}$$

where u is the distance of the object from the lens and v is the distance of its image from the lens. First we combine the fractions on the right-hand side so that

$$\frac{1}{f} = \frac{v}{uv} + \frac{u}{uv} = \frac{v + u}{uv}$$

then we invert both sides to obtain

$$f = \frac{uv}{v + u}.$$

(We could test the validity of the lens formula by measuring the object distance with a lens of known focal length, calculating the image distance from the formula and comparing the result with the measured image distance.) First we return to the original formula, subtract $\dfrac{1}{u}$ from both sides to obtain

$$\frac{1}{f} - \frac{1}{u} = \frac{1}{v}$$

which can be restated as

$$\frac{1}{v} = \frac{1}{f} - \frac{1}{u} = \frac{u - f}{fu} = \frac{u - f}{uf}$$

then by inverting both sides we arrive at

$$v = \frac{uf}{u - f}.$$

2. Return to the pendulum formula. To make l the subject, square both sides to obtain

$$T^2 = 4\pi^2 \frac{l}{g}$$

Now multiply by $\frac{g}{4\pi^2}$ to remove the terms multiplying l, i.e.

$$T^2 \times \left(\frac{g}{4\pi^2}\right) = 4\pi^2 \times \frac{l}{g} \times \left(\frac{g}{4\pi^2}\right) = l, \text{ so that } l = g\frac{T^2}{4\pi^2}.$$

3. We take a formula from electrostatics:

$$E = \frac{q_1 q_2}{4\pi\varepsilon_0 r^2}, \quad r > 0.$$

To change the subject of the formula to r, multiply by the r^2 factor in the denominator on the right to obtain

$$r^2 E = \frac{q_1 q_2}{4\pi\varepsilon_0}$$

then divide by E to obtain

$$r^2 = \frac{q_1 q_2}{4\pi\varepsilon_0 E}$$

or

$$r = +\sqrt{\left(\frac{q_1 q_2}{4\pi\varepsilon_0 E}\right)} = \left(\frac{q_1 q_2}{4\pi\varepsilon_0 E}\right)^{1/2}$$

taking the positive square root. We can do this because r is positive, since it is a distance. ∎

Sometimes it is not possible to transpose a formula to obtain the required subject. Consider the formula

$$c = \frac{1 - T(a + 2b)}{T^2}.$$

We wish to make T the subject. First we multiply by T^2 to obtain

$$cT^2 = 1 - T(a + 2b)$$

then we add $T(a + 2b)$ to both sides and produce

$$cT^2 + T(a + 2b) = 1.$$

It is not possible to separate T from the left-hand side to obtain a formula with T only on the left-hand side and a, b and c on the right-hand side. Deeper mathematics is required to isolate T.

Exercise 2.4

1   Transpose the following formula to make the specified variable (or variables) the subject, or maybe conclude that such a transposition is not possible.

(a) $PV = RT$ (V)

(b) $E = mc^2$ (m, then c)

(c)   Power $=$ resistance $\times$ (current)^2 (current)

(d) $3(2x - 1) + 5(6 - 3x) = 5x - 11$ (x)

(e) $3.18\,(9.2 - x) + 2.64\,(3.03 - 5.5x) = 7.5$ (x, to 2D)

(f) $\dfrac{1}{5}\left(\dfrac{x}{3} - \dfrac{1}{9}\right) + \dfrac{1}{7}\left(\dfrac{x}{11} - 21\right) = 0$ (x, as a fraction)

(g) $9\left(6 - \dfrac{1}{x}\right) + 4\left(\dfrac{3}{x} - 2\right) = \dfrac{17}{x}$ (x)

(h) $(1 + t)(2a + 3b) = 5ct$ (t)

(i) $(\lambda + \mu)(t - 2) = \lambda t - 10$ (t)

(j) $F = \dfrac{GmM}{r^2}$ (G, then r)

(k) $at = (2bt^2) \div ct^3$ (t)

(l) $\sqrt{x} = \dfrac{\sqrt{(1 + x)}}{2}$ (x)

(m) $\dfrac{1 - x}{1 + x} = 3$ (x)

(n) $5 = \sqrt{\dfrac{4 - x}{4 + x}}$ (x)

(o) $\left(\dfrac{a + p}{a - p}\right)^{-1/3} = b^{2/3}$ (a, then p)

(p) $\left(\dfrac{1-x}{3}\right)^{-1/4} = \left(\dfrac{4}{2-x}\right)^{1/4}$ (x)

(q) $T(T+a) = (b-T)(c-T)$ (a, then T)

(r) $T(T+a) = (b-T)(c-T^2)$ (a, then T)

(s) $(1-x^2)(2-x^2) = x^4 - 3x^2 - x$ (x)

(t) $(1-t)(2-t)(3-t) = 6t^3 - t^2 + 7$ (t)

(u) $(1-t)(2-t)(3-t) = 11t(1-t) + 7$ (t)

(v) $\dfrac{a(b+p)}{p} = (b^2 + 2bq + p^2) - p(b+p)$ (p)

(w)* $x^3 + y^3 = x^2 - xy + y^2$ (x)

2   Determine the height h of a closed right circular cylinder whose total surface area S is $100\,\text{m}^2$ and whose radius r is $2\,\text{m}$. Use $S = 2\pi r(r+h)$.

3   Power measured in watts is dissipated in an electric circuit according to the formula $P = VI$ where V is the electric potential difference (volts) and I the current (amps).

(a)   Determine the current flowing through a 60 watt light bulb given $V = 240$ volts.

(b)   If $V = RI$, where R is the electrical resistance of the bulb coil (ohms), determine R.

4   Electrical resistors, R_1, R_2, R_3, connected in parallel have a composite resistance R where

$$\frac{1}{R} = \frac{1}{R_1} + \frac{1}{R_2} + \frac{1}{R_3}$$

(a)   Determine R where $R_1 = 10\,\Omega$ (ohms), $R_2 = 20\,\Omega$, $R_3 = 30\,\Omega$.

(b)   If $R = 12\,\Omega$ and $R_2 = R_3 = 40\,\Omega$ what must be the value of R_1?

5   In the motion of a train along a straight track with uniform acceleration a, the distance s covered by the train obeys the law

$$s - ut + \frac{1}{2}at^2$$

where u is the initial speed.

(a)   Obtain a in terms of the other variables.

(b)   Working in metres and seconds, obtain a (m s^{-2}) when $u = 10\,\text{km h}^{-1}$, $s = 100\,\text{m}$, $t = 100\,\text{s}$.

6   If an explosive force W makes a crater in the ground of radius r and ρ is the density of the soil medium, then $r\left(\dfrac{\rho}{W}\right)^{1/3} = C$, a constant.

(a)   Express W in terms of the other variables.

(b)   If $C = \rho = 1$ and $r = 10$, find W.

 7   The stopping distance of a car measured in feet when its speed is measured as v in mph, is given by $d = 1.08v + 0.057v^2$.

(a)   Find d when $v = 60$ mph.

(b)*   Use trial and error, incrementing upwards one unit at a time, to find the maximum speed of a car to the nearest 1 mph if it has to stop in 300 feet.

8   Celsius and Fahrenheit temperatures are related by the formula

$$\frac{C}{5} = \frac{F - 32}{9}$$

(a)   What Fahrenheit temperature corresponds to $100\,^\circ$C?

(b)   What is the Celsius equivalent of $80\,^\circ$F?

(c)   When do the Celsius and Fahrenheit temperatures coincide?

Identities

An **identity** is a statement that two algebraic expressions are absolutely identical. Whatever values are given to the variable quantities involved, the resulting numerical values of the expressions are equal.

Example

$$(x + 2)(x + 3) \equiv x^2 + 5x + 6$$

is an identity. If, for instance, we put $x = 3$ we obtain

$$5 \times 6 \text{ on the left-hand side and } 9 + 15 + 6 \text{ on the right.}$$

Both numbers are 30 and the result is true. Similarly, putting $x = -4$ gives $(-2) \times (-1)$ and $(-4)^2 + 5 \times (-4) + 6$; since both numbers are equal to 2 the result is again true. (You can verify the identity by multiplying out the left-hand side in its general form.)   ■

Notice the use of the $\equiv$ sign to emphasise equivalence.

Strictly speaking, we should rewrite results like $(a + b)(a - b) = a^2 - b^2$ as identities, namely $(a + b)(a - b) \equiv a^2 - b^2$, since whatever values of a and b are used, the left-hand side and the right-hand side are equal.

Sometimes we can use an identity to determine unknown coefficients. For example, suppose that

$$x^2 + 2x + 1 \equiv x^2 + 2ax + 2a + b$$

then the coefficients of x on each side must be equal. Hence 2 must equal $2a$, from which we deduce that a is equal to 1. The constant terms on each side must be equal so that $2a + b$ must equal 1, and since a has the value 1 it follows that b has the value -1.

Equations

An **equation** is a statement that two quantities are equal. It is usual for such an equation to contain at least one unknown quantity. In the case of an equation in one unknown the equation will only be valid if the unknown is assigned one or more of certain special values, which are called **solutions** of the equation; sometimes they are called **roots** of the equation.

For example, the equation

$$x + 2 = 5$$

has only one solution, or root, namely $x = 3$. For no other value of x will the left-hand side of the equation be equal to 5. The value for x is said to **satisfy** the equation. The process of finding all the solutions of the equation is called **solving** the equation.

The most simple form of equation in one variable or unknown is the **linear equation**. In its standard form it is written $ax = b$. Provided a is non-zero, the solution can be obtained from the formula $x = b/a$, produced by dividing both sides of the equation by a. This means there is only one solution. Notice that the manipulation of equations follows the same rules as transposition of formulae. You should be clear what distinguishes a formula from an equation.

Example

Solve the equation

$$4x + 5 = 7x - 4.$$

Transfer $4x$ to the right-hand side of the equation (by subtracting $4x$ from both sides) to obtain

$$5 = 7x - 4x - 4.$$

Now add 4 to both sides; this transfers -4 to the left-hand side, so that

$$5 + 4 = 7x - 4x$$

i.e. $9 = 3x$

or $3x = 9$

Hence $x = 3$. ■

Simultaneous equations

If we have an equation of the form

$$x + 2y = 3,$$

neither x nor y can be found without knowledge of the other; for example, if $x = -5$, $-5 + 2y = 3$, so $2y = 8$ and $y = 4$.

But if a second *simultaneous* equation $4x - 3y = 1$ is introduced, a solution can be found.

We write

$$x + 2y = 3 \tag{1}$$

$$4x - 3y = 1. \tag{2}$$

Multiply equation (1) by 3 and equation (2) by 2 then add them to give

$$3x + 6y + 8x - 6y = 3x + 8x = 9 + 2$$

hence $11x = 11$ and $x = 1$.

Substituting into either (1) or (2) gives $y = 1$.

It is not surprising that two simultaneous equations are needed to find two unknowns uniquely, but this information may not be sufficient, as the following examples will show.

Example

Consider the sets of equations

(a) $x + y = 1$
$\quad\;\; x + y = 2$

(b) $x + y = 1$
$\quad\;\; 2x + 2y = 2$

The first set of equations is algebraic nonsense and the second set gives repeated information. We say that set (a) is **inconsistent**; the equations can never be solved. On the other hand, set (b) has infinitely many 'solutions', for example, $x = 0, y = 1$, or $x = 1, y = 0$, or $x = 2, y = -1$. ∎

In general, we require an equation for each unknown. Hence with three unknowns we require three simultaneous equations, with the proviso that we may find no solution or infinitely many possible solutions.

Example

Consider the system of equations

$$3x + y + z = 14 \tag{1}$$

$$2x - y + 2z = 15 \tag{2}$$

$$x + 2y - 3z = -17. \tag{3}$$

We can, for example, eliminate z temporarily by multiplying equation (1) by 2 and subtracting it from equation (2). More concisely, $(2) - 2 \times (1)$ gives

$$(2x - y + 2z) - 2(3x + y + z) = -4x - 3y = 15 - 2 \times 14 = -13.$$

Changing sign throughout, i.e. multiplying both sides by -1, we obtain

$$4x + 3y = 13. \tag{4}$$

We can also eliminate z by adding $3 \times$ equation (2) to $2 \times$ equation (3), or more concisely $3 \times (2) + 2 \times (3)$, to give

$$(6x - 3y + 6z) + (2x + 4y - 6z) = 8x + y = 3 \times 15 - 2 \times 17 = 45 - 34 = 11.$$

So we have

$$8x + y = 11. \tag{5}$$

Equations (4) and (5) can be solved to give $x = 1, y = 3$ and any of the equations (1), (2) or (3) shows that $z = 8$. It is usually advisable to check your results in the two equations other than the one just used to find z. ■

This is a good illustration of a wider principle in that three linear equations in three unknowns can be solved, provided each of the three equations represents a totally independent and consistent piece of information. In fact this result can be extended upwards to greater numbers of linear equations in the same number of unknowns.

Exercise 2.5

1   Establish whether the following are identities or equations. Solve any equations for x.

(a) $(1 - x)(1 + x + x^2 + x^3) = 1 - x^4$

(b) $(1 - x)(1 + x + x^2 + x^3) = 1 - x^5$

(c) $x^3 - 2x^2 = 14 - 7x$

(d) $x(x^2 + 11) + 6(x^2 + 1) - x^2(x + 6) - (6x + 11) = 0$

(e) $x(x^2 + 11) + 6(x^2 + 1) - x^2(x + 6) - (11x + 6) = 0$

2   In the following identities compare coefficients so that the unknown constants can be found. For example,

$$x^2 + 2x + 1 \equiv x^2 + 2ax + a + b \text{ gives}$$
$$a = 1 \text{ and } 2a + b = 1, \text{ so that } b = -1.$$

(a) $x^3 + ax^2 + (a + b)x + a + b + c \equiv x^3 + 3x^2 + 2x + 1$

(b) $\left(x - \dfrac{1}{x}\right)^2 \equiv x^2 + k + \dfrac{m}{x^2} + \dfrac{n}{x^4}$

(c) $x^2 \equiv (x - 1)^2 + a(x - 1) + b$

(d) $2(x + y)^2 - (x - y)^2 + 3(x + y) - (x - y) \equiv x^2 + axy + y^2 + bc + cy$

(e) $(x + 1)^3 - x^3 \equiv 3x^2 + px + q$

3   The identity $a(x + 1)(x + 2) + b(x + 1) + c \equiv 2x^2 - 5x + 7$ is given.

(a)   Expand the brackets and determine a, b, c (the coefficients of x^2, x and the constant term).

(b)   Alternatively, let $x = -1$ then $x = -2$ in order to read off c, b and a in turn.

4   Given that

$$x^3 + 3px^2 + qx + r \equiv (x + a)^3$$

equate the coefficients to show that

(a) $q = 3a^2$ \hspace{3cm} (b) $r = a^3$.

How are q and r related?

5   Determine a, given that

$$4x^4 - 41x^3 - 28x^2 - 8x + 1 \equiv (4x^2 + 3x + 1)(x^2 + ax + 1).$$

6   Identities involving powers can be compared as follows. If, for example,

$$(xy)^a (x/y)^b \equiv x^2, \text{ then } x^a y^a x^b y^{-b} \equiv x^2.$$

Then

$$x^{a+b} \equiv x^2 \quad \text{and} \quad y^{a-b} \equiv 1, \quad \text{so that } a = b = 1.$$

Determine the powers in the identities that follow:

(a) $(x^2y^3)^a\left(\dfrac{x}{y^3}\right)^b \equiv x^3$

(b) $x^a y^{-b} \equiv 1 \quad (x \neq y)$

(c) $(x^a y^{-b})^2 x^3 y^{-2} \equiv x^5 y^{-3}$

(d) $(xy)^a (x^2/y)^b \left(\dfrac{1}{xy}\right)^c \equiv 1$

(In (d) determine the relationship between a and c.)

(e) $(xyz)^a (x^2/y)^b \left(\dfrac{1}{xy}\right)^c \equiv z^4$

(f) $(xy^a z^2)^2 (z/x^2)^b (yz^{-3})^c \equiv z^{-1}$

(g) $(x^{2/3}\sqrt{yz^a})^2 (x^{-a}y^{-1/3}z^b) \equiv \dfrac{y^{2/3}}{z}$

(h) $(x^a yz^{-1})^b (2yz^2)^c (x/y)^a \equiv 4(x^6 z)^2$.

7 In physics use is made of **dimensional analysis** so that physical quantities are expressible in terms of powers of the basic quantities, mass $[M]$, length $[L]$ and time $[T]$. By equating identical powers on both sides of the equation, find x, y and z to satisfy

(a) $\dfrac{[L]^3}{[T]} = \dfrac{[M]^x}{[L]^x [T]^x} \cdot [L]^y \cdot \dfrac{[M]^2}{[L]^{2z}[T]^{2z}}$

i.e. note that

$$3 = -x + y - 2z \quad [L]$$
$$-1 = -x \qquad\;\; - 2z \quad [T]$$
$$0 = \quad x \qquad + z \quad [M]$$

(b) $\dfrac{[M][L]}{[T]^2} = [L]^x \cdot \dfrac{[M]^y}{[L]^y [T]^y} \cdot \dfrac{[L]^z}{[T]^z}$.

8* The wind force F acting upon a large high-sided vehicle is believed to be proportional to its exposed area A, the relative velocity of the wind V and the density of the air ρ, i.e. $F \propto A^p V^a \rho^r$.

In dimensional analysis

$$[F] = [MLT^{-2}] \qquad [V] = [LT^{-1}] \qquad [A] = [L^2] \qquad [\rho] = [ML^{-3}].$$

By writing

$$[MLT^{-2}] = [LT^{-1}]^p [L^2]^q [ML^{-3}]^r$$

obtain values for p and q and r.

9   Determine the unknown variables, i.e. solve the simultaneous equations, in the following cases. If the equations cannot be solved, state the reason why.

(a) $3x + 2y = 5$ given $y = 1$

(b) $6x + 2y + 3z = 19$ given $x = 1, y = 2$

(c) $4x - 5y = 4$
       $6x + 10y = 5$

(d) $9z - 11b = 4a$ find $\dfrac{z}{a}, \dfrac{b}{a}$
       $3z + 5b = 2a$

(e) $3.52y + 4.61z = 19.81$ (retain 3 d.p.)
       $-2.56y + 7.81z = 25.02$

(f) $x - 3y = 2$ (g) $x - 3y = 2$
       $-2x + 6y = -4$ $-2x + 6y = -2$

(h) $3x + 2y + 5z = 22$ (i) $6p + 5q - 14r = -12$
       $-2x - 6y + 7z = 7$ $p + q + r = 15$
       $5x + 3y - 5z = -4$ $p - q + 2r = 16$

(j) $\dfrac{3}{x} + \dfrac{2}{y} = 5$ (k) $x + y = 3$
                                                       $x^2 - y^2 = -3$
       $\dfrac{5}{y} = 6 - \dfrac{1}{y}$

(l) $3x + 2y + 4z = 8$ (m)* $xy + 2xz + 3yz = 6$
           $x + y + z = 3$ $-3xy + 4xz + 11yz = 12$
           $x + 2z = 2$ $-2xy - 5xz + 15yz = 8$
                                                         $xyz = 1$

Hint: in (m) divide by xyz throughout and work with $\dfrac{1}{x}, \dfrac{1}{y}, \dfrac{1}{z}$.

10*   Solve the sets of equations given below by using one equation to reduce the number of variables in the other equations.

(a) $ab + c = 11$ (b) $a^2 + b + c = 4$
              $b - c = 1$ $a(a + b) + c = 2$
           $2b + 3c = 12$ $b - 1 = 0$

(c) $x^2 + 2y^2 = 9$
          $13x^2 - 3y^2 = 1$
          $x(x - 2y) = 5.$

In the following M, N and P are integers lying between 1 and 9.

(d) $2M + N = 10$
       $MN = 12$

(e) $M + 2N + 3P = 10$
       $MN = 6$

(f) $M + N + P = 6$
       $MNP = 6.$

2.3 PROPORTIONALITY

The simplest relationship between two variables is one in which one of the variables is proportional to the other.

The variable y is said to be **directly proportional to** (or sometimes simply **proportional to**) the variable x if $y = Cx$ where C is a constant, called the **constant of proportionality**. It is also possible to write the direct proportion relationship as $y \propto x$ but we shall not use that notation here.

It follows from this definition that if the value of x is doubled then the value of y is doubled, if the value x is divided by 5 then the value of y is divided by 5 and so on. The converse is also true.

As an example, suppose that $y = 3x$. When $x = 2, y = 6$; when $x = 4, y = 12$; when $x = \frac{2}{5}, y = \frac{6}{5}$. If we halve the value of y to 3 then the value of x is also halved, to 1.

Two examples of direct proportion are the potential difference across the ends of an electrical conductor and the current flowing through the conductor, and the load placed on a spring suspended vertically and the extension produced in the spring.

Other forms of proportionality exist. The variable y is said to be **inversely proportional to** the variable x if

$$y = \frac{C}{x}$$

where C is a constant. Here, for example, if the value of x is doubled then the value of y is halved and vice versa. An example of inverse proportion is the pressure of a fixed mass of an ideal gas and its volume, when the temperature of the gas is held constant.

Examples

1. The potential difference across the ends of a conductor is 6 volts when the current flowing through it is 0.5 amp. Find the potential difference across the ends when the current is 0.35 amp then find the current flowing through the conductor when the potential difference is 0.8 volt.

   Let the potential difference be V (volts) when the current is i (amps); then $V = Ri$ where R is a constant (called the **resistance** and measured in ohms). Then

$6 = R \times 0.5$, so $R = 12$ and the relationship is $V = 12i$. If $i = 0.35$ then $V = 12 \times 0.35 = 4.2$, so the potential difference is 4.2 volts.

When $V = 0.8$ we have the result that $0.8 = Ri = 12i$, so

$$i = \frac{0.8}{12} = \frac{8}{120} = \frac{1}{15} \text{ amp.}$$

It is possible to use an alternative method if the value of R is not required. Let the potential difference be V_1 (volts) when the current is i_1 (amps) and be V_2 (volts) when the current is i_2 (amps). Then $V_1 = Ri_1$ and $V_2 = Ri_2$ so that

$$R = \frac{V_1}{i_1} = \frac{V_2}{i_2}$$

Hence

$$V_2 = \frac{V_1 \times i_2}{i_1} \quad \text{and} \quad i_2 = \frac{V_2 \times i_1}{V_1}.$$

Check that these two formulae give the same values for the problems we have just solved.

2. The pressure of an ideal gas is 1×10^5 pascals (Pa) when its volume is 0.8 litre. The gas is compressed to a volume of 0.5 litre; what is its pressure? At what volume would the pressure be 2.5×10^4 Pa? Assume constant temperature.

Let the pressure of the gas be p (Pa) and its volume be V (litres). Then $p = \dfrac{k}{V}$ where k is a constant. Therefore

$$1 \times 10^5 = \frac{k}{0.8}$$

so $k = 0.8 \times 10^5 = 8 \times 10^4$. When

$$V = 0.5, p = \frac{8 \times 10^4}{0.5} = 16 \times 10^4 = 1.6 \times 10^5 \text{ Pa}$$

and when

$$p = 2.5 \times 10^4, 2.5 \times 10^4 = \frac{8 \times 10^4}{V}$$

so

$$V = \frac{8 \times 10^4}{2.5 \times 10^4} = \frac{8}{2.5} = \frac{32}{10} = 3.2 \text{ litres.}$$

Again, we note an alternative method. If

$$p_1 = \frac{k}{V_1} \quad \text{and} \quad p_2 = \frac{k}{V_2}$$

then

$$k = p_1 V_1 = p_2 V_2.$$

Hence

$$V_2 = \frac{p_1 V_1}{p_2} \quad \text{and} \quad p_2 = \frac{p_1 V_1}{V_2}. \qquad\blacksquare$$

Other forms of proportionality

For the simple pendulum we had the formula

$$T = 2\pi \sqrt{\frac{l}{g}}$$

where T is the period (in seconds), l is the length of the pendulum (in metres) and g is the constant acceleration due to gravity ($9.81 \, \text{m s}^{-2}$).

We may write this relationship as

$$T = C\sqrt{l} \quad \text{or} \quad T = Cl^{1/2}$$

where C is the constant $\dfrac{2\pi}{\sqrt{g}} = 0.641$ (3 d.p.). We say that T is proportional to the square root of l.

The volume of a sphere of radius a is given by $V = \dfrac{4}{3}\pi a^3$. We say that V is proportional to the cube of a. If a is measured in metres (m) then V is measured in cubic metres (m^3).

In both these examples we can work out values directly from the formula since the constant has the same value for all cases. Whatever sphere we deal with, the constant of proportionality is always the same, namely $4\pi/3$. However, in the case of the pressure and volume of an ideal gas the constant of proportionality depends upon temperature and upon the gas under consideration. In the case of the electrical conductor the constant of proportionality depends, among other factors, on the material from which the conductor is made.

Example   When an object falls through the air, the upwards resistance to its motion is proportional to the square of its speed. On measuring the speed on two occasions separated by 10 s it is observed that the speed has doubled. By how much has the resistance increased?

Let v be the speed of the object in metres per second (m s^{-1}) and R the resistance in kiloNewtons (kN). Then $R = cv^2$ where c is a constant. Hence $R_1 = cv_1^2$ and $R_2 = cv_2^2$. Therefore

$$\frac{R_1}{v_1^2} = \frac{R_2}{v_2^2} \quad \text{and} \quad \frac{R_2}{R_1} = \frac{v_2^2}{v_1^2}.$$

Since $v_2/v_1 = 2$, we get $R_2/R_1 = 4$, so the resistance has quadrupled. It is worth noting that if such an object were allowed to fall for long enough (by being dropped from a sufficiently great height) then its speed would not continue to increase without limit. At some point the resistance would reach a value equal to the weight of the object; then the speed of the object would remain constant at the **terminal speed**.   ■

Simultaneous proportionality

In physics it is known that the gravitational force of attraction between two objects varies directly as the mass of the first object, directly as the mass of the second object and inversely as the square of the distance between the objects.

The phrase 'varies directly as' is another way of saying 'is directly proportional to' and the phrase 'varies inversely as' means the same as 'is inversely proportional to'.

Hence if the mass of the first object is m_1 (kg), the mass of the second object is m_2 (kg), the distance between them is r (m) and the gravitational force of attraction is F (kN) then we may say that

(i) $F = k_1 m_1$ where k_1 is a constant
(ii) $F = k_2 m_2$ where k_2 is a constant
(iii) $F = k_3/r^2$ where k_3 is a constant.

We can combine these three relationships into a single statement:

$$F = \frac{Gm_1 m_2}{r^2}$$

where G is a constant. G is known as the **universal constant of gravitation**; its value is approximately $6.67 \times 10^{-11}\,\text{N m}^2\,\text{kg}^{-2}$. We can reassure ourselves that the combined formula is correct. For example, statement (i) really means that m_2 and r do not vary. In the combined formula, if m_2 and r are constants,

$$F = \frac{\text{constant} \times m_1 \times \text{constant}}{(\text{constant})^2}$$

i.e. $F = \text{constant} \times m_1$

which is statement (i).

Check that (ii) and (iii) are also embraced in the combined formula.

Example   Find the force of attraction between two masses, one of 10^4 kg and the other of 5×10^4 kg, which are approaching each other, at the instant when they are 10^4 km apart. If the distance between them is halved, what happens to the force of attraction?

$$F = \frac{6.67 \times 10^{-11} \times 10^4 \times 5 \times 10^4}{10^4 \times 10^4} = \frac{6.67 \times 5 \times 10^{-3}}{10^8}$$

$$= 6.67 \times 5 \times 10^{-11} = 3.335 \times 10^{-10} \text{ N}$$

Since the force of attraction is inversely proportional to the square of the distance apart, halving the distance results in a fourfold increase in the force, which will then be

$$4 \times 3.335 \times 10^{-10} \text{ N} = 1.334 \times 10^{-9} \text{ N.}$$ ■

Exercise 2.6

 1   The pressure P in a tank of water is proportional to depth h below the surface. Find the formula relating the pressure in pascals and the depth in millimetres, and complete the table below.

Pressure p (Pa)	2452.5		5886
Depth h (mm)	250	400	

 2   The pressure of a fixed mass of an ideal gas expanding at constant temperature is directly proportional to its density. Find the formula relating the pressure in pascals to the density in kilograms per cubic metre, and complete the table below.

Pressure p (Pa)		10^4		10^5
Density ρ (kg m^{-3})			0.09	0.18

3   The frequency f of vibration of a stretched string is inversely proportional to its length h. Find the formula relating the frequency (Hz) to the length (m), and complete the table below.

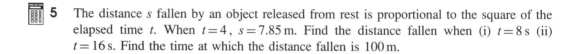

Frequency f (Hz)		250	320
Length h (m)	1.5	0.6	

4   Using the data of Question 3 find the effect of increasing the length by 25%, 50%, 150%. Also find the effects of increasing the length from 300 m to 400 m, and of decreasing the length from 200 m to 150 m.

5   The distance s fallen by an object released from rest is proportional to the square of the elapsed time t. When $t = 4$, $s = 7.85$ m. Find the distance fallen when (i) $t = 8$ s (ii) $t = 16$ s. Find the time at which the distance fallen is 100 m.

6   The power P watts (W) in an electrical circuit is proportional to the square of the current i amps (A) flowing through it. When $i = 3$ A, $P = 90$ W. What is the effect of increasing the current to 4 A? If the power is 50 W, what is the current in the circuit?

7   The velocity of sound v in a gas is proportional to the square root of the pressure p in the gas and inversely proportional to the square root of its density ρ. Write down a formula relating v to p and ρ. Given that $v = 328$ when $p = 10^5$ and $\rho = 1.3$, find the value of p when $v = 300$ and $\rho = 1.25$. Also find the value of ρ when $p = 3 \times 10^4$ and $v = 350$.

8   A satellite is in geostationary orbit above the Earth, i.e. it is always directly above the same point on the Earth's surface, hence its period of orbit is the same as that of the Earth, namely 86 400 s. The formula relating the period T in seconds to the radius r of the orbit in kilometres (relative to the centre of the Earth) is

$$T = 2\pi\sqrt{\frac{r^3}{GM}}$$

where the universal gravitational constant, $G = 6.67 \times 10^{-11}\,\mathrm{N\,m^2\,kg^{-2}}$ and the mass of the Earth, $M = 6 \times 10^{24}$ kg. What is the value of r? If the satellite is moved into an orbit where $r = 40\,000$ km, find the effect on T.

9   The velocity v of sound in a solid is related to the Young's modulus E and the density ρ of the solid by the formula

$$v = \sqrt{\frac{E}{\rho}}$$

If v is measured in metres per second ($\mathrm{m\,s^{-1}}$) and ρ in kilograms per cubic metre ($\mathrm{kg\,m^{-3}}$) what are the units of E?

10*   Gravitational acceleration g at a point is inversely proportional to the square of the distance r of that point from the centre of the Earth. If the gravitational acceleration at the surface of the Earth is g_0, write down a formula for g in terms of r. The Earth may be considered as a sphere of radius R.

What is the effect on the period of a simple pendulum if it is moved from the Earth's surface to a point at altitude 1 km? You may assume that $R = 6400$ km and that $g_0 = 9.81\,\mathrm{m\,s^{-2}}$.

2.4 FACTORISATION

Algebraic expressions often need to be factorised so they can be simplified, perhaps to solve equations. Manipulative skill and experience play an important part. Looking for **factors** is often based on trial and error, and we must always remember that certain expressions do not **factorise**, i.e. they do not decompose into multiplied components. Even when factors can be found it may be very difficult to discover them.

From earlier work we can see that

$$(a + b)(a - 2b) \equiv a^2 + ba - 2ab - 2b^2$$
$$\equiv a^2 - ab - 2b^2.$$

It is a much harder task, starting from the expression $a^2 - ab - 2b^2$ to deduce that it can be written as the product $(a + b)(a - 2b)$. This is the art of factorisation.

In the following examples we look essentially for the highest common factors of the terms in the expression.

Examples
1. Factorise $27x^2 - 18x^3$.
   The HCF of the numbers 27 and 18 is 9; the HCF of x^2 and x^3 is x^2. Hence

$$27x^2 - 18x^3 \equiv 9x^2(3) - 9x^2(2x)$$
$$\equiv 9x^2(3 - 2x).$$

2. Factorise $20x + 4x^3$.
   The HCF of 20 and 4 is 4 and of x and x^3 is x. Hence

$$20x + 4x^3 \equiv 4x(5) + 4x(x^2)$$
$$\equiv 4x(5 + x^2).$$

3. Factorise $4xy + 8y^2 + 10x^2y$.
   The HCF of 4, 8 and 10 is 2 and of xy, y^2 and x^2y is y. Hence

$$4xy + 8y^2 + 10x^2y \equiv 2y(2x + 4y + 5x^2).$$ ■

Quadratic expressions

An expression of the type $ax^2 + bx + c$ is called a **quadratic expression** (in x). The only requirement is that a must be non-zero. Examples are $-2x^2 + 5x + 6$, $2x^2 - 7$, $x^2 + 3x$. If possible, a quadratic expression can be factorised as the product of linear factors.

First we consider the case where the coefficient of x^2 is 1. The process is one of systematic trial and error.

Example
Consider the expression $x^2 - 5x + 6$. Try $(x - \alpha)(x - \beta)$ as a possible factorisation. Multiplying this out gives $(x - \alpha)(x - \beta) = x^2 - (\alpha + \beta)x + \alpha\beta$.

Comparison with the original expression implies that the numbers α and β must obey the equations

$$\alpha + \beta \ = \text{minus the coefficient of } x \ = 5$$
$$\alpha\beta \quad = \text{the constant term} \qquad = 6.$$

By trial and error we see that the numbers 2 and 3 clearly satisfy these equations. Hence

$$x^2 - 5x + 6 \equiv (x - 2)(x - 3) \equiv (x - 3)(x - 2).$$ ■

It is worth remarking that if the constant term is positive then α and β have the same sign, positive if the coefficient of x is positive and negative if it is negative. If the constant term is negative then α and β have opposite signs; if the coefficient of x is positive then the larger-sized number is positive and vice versa.

Example

$$x^2 + 4x + 3 \equiv (x + 1)(x + 3) \qquad x^2 + x - 6 \equiv (x + 3)(x - 2)$$

$$x^2 - 4x + 3 \equiv (x - 1)(x - 3) \qquad x^2 - x - 6 \equiv (x + 2)(x - 3). \qquad \blacksquare$$

Certain forms of the quadratic expression have simple factorisations which are worth learning:

$$x^2 + 2ax + a^2 \equiv (x + a)^2$$
$$x - 2ax + a^2 \equiv (x - a)^2$$
$$x^2 - a^2 \equiv (x + a)(x - a).$$

Examples

$$x^2 + 6x + 9 \equiv (x + 3)^2$$
$$x^2 - 4x + 4 \equiv (x - 2)^2$$
$$x^2 - 25 \equiv (x + 5)(x - 5). \qquad \blacksquare$$

If you can 'spot' a factor then it allows a simplification of the factorisation process. In general, if $(x - a)$ is a factor of an expression then putting $x = a$ makes the value of the expression equal to zero. For example, $x^2 - 5x + 6 \equiv (x - 2)(x - 3)$. If we put either $x = 2$ or $x = 3$ the expression $x^2 - 5x + 6$ is zero.

Example

Consider the expression $x^3 - 12x^2 + 44x - 48$. We see that $x = 2$ makes the value of the expression $8 - 12 \times 4 + 88 - 48 = 0$.

Hence $x^3 - 12x^2 + 44x - 48 \equiv (x - 2) \times$ (quadratic).

In the second pair of brackets we can fill in some gaps. $x^3 = x \times x^2$ and $-48 = -2 \times 24$. Then $x^3 - 12x^2 + 44x - 48 \equiv (x - 2)(x^2 + \alpha x + 24)$ where α has to be determined.

Looking at the terms in x^2 on the right-hand side, we obtain $-2x^2 + \alpha x^2$; on the left-hand side we have $-12x^2$, hence $\alpha = -10$ and $x^3 - 12x^2 + 44x - 48 \equiv (x - 2)(x^2 - 10x + 24)$.

The quadratic expression can be factorised by the method outlined earlier as $(x - 4)(x - 6)$, so $x^2 - 12x^2 + 44x - 48 \equiv (x - 2)(x - 4)(x - 6)$.   ■

The situation where the coefficient x^2 is not 1 is more tedious and requires greater care.

Example   Consider the expression $14x^2 + 65x - 25$. We could try dividing by 14 and proceeding as before but this leads to difficult arithmetic. It is better to note that 7 and 2, and 14 and 1 are factor combinations of 14 and that 5 and 5, and 25 and 1 are factor combinations of 25. The factors to try are

$$(7x + 5), (7x - 5), (2x + 5), (2x - 5) \text{ and so on}$$

until the combinations of (7, 2) or (14, 1), (5, 5) or (25, 1) are exhausted. Eventually we see that the factors are

$$(14x - 5) \quad \text{and} \quad (x + 5).$$ ■

But suppose we have the expression $14x^2 + x + 25$; none of the possible combinations will work and we must accept that there are no factors. In Section 4.2 we shall see how to determine this possibility in advance of any trial and error process for the factors.

Algebraic fractions

The simplification of algebraic fractions is often helped by factorisation. For example,

$$\frac{6x}{2x^3 + 4x} \equiv \frac{6x}{2x(x^2 + 2)} \equiv \frac{3}{x^2 + 2}.$$

Examples   1. Simplify the fraction $\dfrac{x - 2}{x^2 - x - 2}$.
   The denominator can be factorised into the product $(x - 2)(x + 1)$. Hence, cancelling the factor $(x - 2)$ top and bottom, we obtain the result $\dfrac{1}{x + 1}$.

2. Simplify $\dfrac{x^2 - 5x + 6}{x^2 - 6x + 9}$.
   Factorising the numerator and denominator and cancelling we obtain

$$\frac{(x - 2)(x - 3)}{(x - 3)^2} = \frac{x - 2}{x - 3}.$$

3. Simplify $\dfrac{5xy + x}{2x + x^2y}$.

The fraction can be written to allow cancellation:

$$\frac{5xy + x}{2x + x^2y} = \frac{x(5y + 1)}{x(2 + xy)} = \frac{5y + 1}{2 + xy}.$$

4. Simplify $\dfrac{(x^2 + y^2)(x + y)}{(x - y)^2}$.

There are no common factors between the denominator and the numerator, indeed the numerator cannot be factorised further. The fraction is in its simplest form.

5. Write the following expression as a single fraction:

$$\frac{2}{x^2 + 4x + 3} + \frac{3}{5x^2 - 5x - 10}.$$

First we factorise the denominators to obtain

$$\frac{2}{(x + 3)(x + 1)} + \frac{3}{5(x + 1)(x - 2)}.$$

The HCF of the denominators is $5(x + 3)(x + 1)(x - 2)$. Hence the expression becomes

$$\frac{2 \times 5(x - 2)}{5(x + 3)(x + 1)(x - 2)} + \frac{3(x + 3)}{5(x + 1)(x - 2)(x + 3)} \equiv \frac{10(x - 2) + 3(x + 3)}{5(x + 3)(x + 1)(x - 2)}$$

$$\equiv \frac{10x - 20 + 3x + 9}{5(x + 3)(x + 1)(x - 2)}$$

$$\equiv \frac{13x - 11}{5(x + 3)(x + 1)(x - 2)}.$$

No cancellation is possible between the numerator and the denominator and we cannot simplify the expression further.

6. Write the following expression as a single fraction:

$$E = \frac{3x + 2}{x^2 + x - 2} + \frac{4}{x^2 - 1}.$$

Here is the solution:

$$E \equiv \frac{3x+2}{(x+2)(x-1)} + \frac{4}{(x-1)(x+1)}$$

$$\equiv \frac{(3x+2)(x+1)}{(x+2)(x-1)(x+1)} + \frac{4(x+2)}{(x+2)(x-1)(x+1)}$$

$$\equiv \frac{(3x+2)(x+1) + 4(x+2)}{(x+2)(x-1)(x+1)}$$

$$\equiv \frac{3x^2+2x+3x+2+4x+8}{(x+2)(x-1)(x+1)}$$

$$\equiv \frac{3x^2+9x+10}{(x+2)(x-1)(x+1)}.$$ ∎

Exercise 2.7

1 Factorise the following expressions where possible or conclude that there are no simple factors.

(a) $x^2 - 7x + 12$

(b) $9x^2 - 100$

(c) $49x^2 - 7$

(d) $4x^2 - 4x - 8$

(e) $21x^2 + 13x - 2$

(f) $33x^2 - 35x - 12$

(g) $8p^2 - 27pq + 9q^2$

(h) $x^2 + 2x + 3$

(i) $x^3 - 11x^2 + 26x - 16$

(j) $30x^3 - 127x^2 + 174x - 77$

(k) $6x^3 - 17x^2 - 61x + 132$

(l) $3m(m+2n) - 2mn - 4n^2$

(m) $x - y^2 + x^2 - y$

(n) $(3p+q)(q-p) + q^3 - p^3$

2 Express as a single fraction, factorising and cancelling where possible:

(a) $\dfrac{x+4}{x+3} - 1$

(b) $\dfrac{1}{x+1} - \dfrac{1}{(x+1)^2}$

(c) $\dfrac{3}{2x+1} + \dfrac{5x}{4x^2+4x+1}$

(d) $\dfrac{5x+1}{x^2-x-2} + \dfrac{2}{x^2-4}$

(e) $\dfrac{1}{5}\left(\dfrac{1}{x+1} - \dfrac{1}{x+6}\right)$

(f) $\dfrac{1}{(x-1)(x+1)(x-2)} + \dfrac{2}{(x-1(x+1)(x-3)}$

(g) $\dfrac{x+1}{(x+2(x+3)}+\dfrac{x+2}{(x+1)(x+3)}+\dfrac{x+3}{(x+1)(x+2)}$

(h) $\dfrac{p+2q}{p^2-q^2}-\dfrac{1}{p+q}$ $\qquad$ (i) $\dfrac{\sqrt{a}+3\sqrt{b}}{a-b}-\dfrac{2}{\sqrt{a}+\sqrt{b}}$

(j) $\dfrac{m(1+t^2)}{(1+t^3)}-\dfrac{m^{-5/2}}{m^{-7/2}(1+t)}$.

3 The identity

$$\frac{1}{(x-1)(x-2)}\equiv\frac{A}{x-1}+\frac{B}{x-2}$$

can be written as

$$\frac{1}{(x-1)(x-2)}\equiv\frac{A(x-2)+B(x-1)}{(x-1)(x-2)}$$

i.e. $A(x-2)+B(x-1)\equiv 1$, by comparing numerators.

(a) Comparing coefficients of x and the constant term, determine A and B.

(b) Instead, try setting $x=1$ and $x=2$ in turn, reading off A and B.

4 Repeat the procedures in Question 3 for the following identities finding A, B, C as appropriate.

(a) $\dfrac{3x+1}{(x-1)(x-2)}\equiv\dfrac{A}{x-1}+\dfrac{B}{x-2}$

(b) $\dfrac{1}{(x+1)(2x+5)}\equiv\dfrac{A}{(x+1)}+\dfrac{B}{(2x+5)}$

(c) $\dfrac{5x+6}{(2x+1)(3x+1)}\equiv\dfrac{A}{2x+1}+\dfrac{B}{3x+1}$

(d) $\dfrac{1}{(x-1)(x-2)(x-3)}\equiv\dfrac{A}{x-1}+\dfrac{B}{x-2}+\dfrac{C}{x-3}$

(e) $\dfrac{2x+1}{(x+1)(x+2)(x+3)}\equiv\dfrac{A}{x+1}+\dfrac{B}{x+2}+\dfrac{C}{x+3}$

(f)* $\dfrac{x^2-x+3}{(2x+1)(x+3)(x-1)}\equiv\dfrac{A}{2x+1}+\dfrac{B}{x+3}+\dfrac{C}{x-1}$

(g)* $\dfrac{1}{(x+1)(x+2)^2}\equiv\dfrac{A}{x+1}+\dfrac{B}{x+2}+\dfrac{C}{(x+2)^2}$

In (g) the identity must hold for all x, e.g. $x=0$. Use this fact to find B, knowing A and C.

5* The following expressions all have factors which involve small integers. Step sensibly through the integers between -4 and 4 to identify factors of the following expressions:

(a) $x^3 + 2x^2 - 5x - 6$ (b) $x^3 - 14x^2 + 35x - 6$

(c) $x^3 - 3x + 2$ (d) $x^3 - 3x^2 + 3x - 1$

(e) $x^4 - 2x^3 - 13x^2 + 14x + 24.$

Show that $x + 11$ and $x - 6$ are each factors of the expression

$x^5 + 7x^4 - 61x^3 - 16x^2 + 300x + 396$

and that the other factors are in (a).

6* By letting $a = b$ show that $a - b$ is a factor of

$a^2(b - c) + b^2(c - a) + c^2(a - b).$

Hence show that the expression is equivalent to $-(a - b)(b - c)(c - a).$

2.5 SIMPLE INEQUALITIES

An inequality is a statement about two quantities which are not equal. As an example '5 is greater than 3' which is written $5 > 3$, where the symbol $>$ means 'is greater than'. Numbers can be represented by points on a line called the **real line**. Figure 2.1 shows a section of the line with some numbers indicated on it.

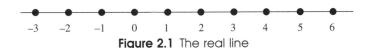

Figure 2.1 The real line

We see that the number 5 lies to the right of the number 3, which is the geometrical equivalent of the statement $5 > 3$. Conversely 3 lies to the left of 5; we say that 3 is less than 5 and we write $3 < 5$. There are two other symbols that we introduce here: $\leq$ means 'is less than or equal to' and $\geq$ means 'is greater than or equal to'. Hence $4 \leq 7$ and $4 \leq 4, 7 \geq 4$ and $4 \geq 4$ are all true statements.

As with equations, inequalities can be unravelled by applying some simple rules:

(i)   If the same quantity is added to or subtracted from both sides of an inequality then the inequality is still true.

For example, given that $5 > 3, 5 + 4 > 3 + 4$ and $5 - 2 > 3 - 2.$

(ii) If both sides of an inequality are multiplied by or divided by the same *positive* quantity (i.e. a quantity > 0) then the inequality is still true.

$$\text{For example, } 5 \times 2 > 3 \times 2 \text{ and } \frac{5}{4} > \frac{3}{4}.$$

(iii) If both sides of an inequality are multiplied or divided by the same *negative* quantity (i.e. a quantity < 0) then the inequality is no longer true, but if the inequality sign is reversed the resulting inequality is true.

For example, $5 \times (-2)$ is not greater than $3 \times (-2)$ since -10 lies to the of left -6.

$$\text{However, } 5 \times (-2) < 3 \times (-2); \text{ similarly, } \frac{5}{-4} < \frac{3}{-4}.$$

When an inequality is multiplied (divided) by a negative quantity, it reverses all three components of the inequality: left-hand side, inequality sign and right-hand side. Hence if $6 > -2$ then $-6 < +2$.

We may summarise the results algebraically as

(i)   if $a > b$ then $a + c > b + c$
(ii) if $a < b$ then $a + c < b + c$
(iii) if $a > b$ and $\lambda > 0$ then $\lambda a > \lambda b$
(iv) if $a > b$ and $\lambda < 0$ then $\lambda a < \lambda b$

Other results which are of use are

(v)   if $a > b$ then $a - b > 0$, and vice versa
(vi) if $ab > 0$ then *either* $a > 0$ and $b > 0$
              or $a < 0$ and $b < 0$
(vii) if $ab < 0$ then *either* $a > 0$ and $b < 0$
              or $a < 0$ and $b > 0$.

Example

(a) Show that if $a > b$ then $a^2 > b^2$ if $b > 0$ but $a^2 < b^2$ if $a < 0$.
(b) If $a^2 > b^2$, is it true that $a > b$?

Solution

(a) If $b > 0$ then it follows that $a > 0$. And if $a > b$,

$$a \times a > a \times b \quad \text{and} \quad b \times a > b \times b$$

so that $a^2 > ab > b^2$, i.e. $a^2 > b^2$. If $a < 0$ then $b < 0$ also. Hence if $a > b$,

$$a \times a < a \times b \quad \text{and} \quad b \times a < b \times b$$

so that $a^2 < ab < b^2$, from which we conclude that $a^2 < b^2$.

(b) If $a = -3$ and $b = 2$ then $a^2 = 9$, $b^2 = 4$ and $a^2 > b^2$. But $a < b$. Hence the claim is false. ∎

Intervals on the real line

Suppose we wish to find the values of x for which the inequality

$$5 - 4x > 3x + 2$$

is true.
We add $4x$ to both sides so that

$$5 > 7x + 2$$

then we subtract 2 from both sides to obtain

$$3 > 7x \text{ or } 7x < 3$$

finally, we divide both sides by 7 giving

$$x < \frac{3}{7}.$$

This represents an **interval** on the real line which is depicted in Figure 2.2(a). Figure 2.2(b) depicts the interval $x \geq -\frac{4}{5}$. If the inequality includes the $=$ option, we say that the endpoint is included and the interval is **closed** at that end; the symbol ● is used to indicate this, whereas if the endpoint is not included we use the symbol ○ and declare the interval **open** at that end.

$$\frac{3}{7} \qquad\qquad\qquad -\frac{4}{5}$$

(a)   (b)

Figure 2.2 Intervals on the real line

(b)

Figure 2.3 Finite intervals on the real line

Figure 2.3(a) depicts an interval open at both ends; it is the interval where $x < 3$ and also $x > -2$; we write these two inequalities as one statement:

$$-2 < x < 3.$$

Figure 2.3(b) depicts an interval closed at both ends, it is the interval where $x \leq 2$ and also $x \geq -4$; we write the inequalities as one statement:

$$-4 \leq x \leq 2.$$

Inequalities and the modulus sign

The **absolute value** of x is written $|x|$. The equation $|x| = 2$ has two solutions $x = 2$ and $x = -2$. What interpretation can we put on the inequality $|x| < 2$? It is, in fact, the interval $-2 < x < 2$. In general the inequalities $|x| < a$ and $-a < x < a$ are equivalent. It follows that the inequality $|x| > a$ represents two intervals: $x > a$ and $x < -a$. Refer to Figure 2.4.

(a)   (b)

Figure 2.4 Intervals described by the modulus sign

We can rewrite inequalities involving the modulus sign as the following example shows.

Example   Rewrite the following inequalities without using the modulus sign:

(a) $|2x - 1| < 4$ (b) $|3x + 2| = 1$ (c) $|3 - x| \geqslant 5$.

Solution

(a) The first step is to write the inequality as

$$-4 < 2x - 1 < 4.$$

If we add 1 to each component we obtain

$$-3 < 2x < 5.$$

Dividing throughout by 2 gives

$$-\frac{3}{2} < x < \frac{5}{2}.$$

(b) Either $3x + 2 = 1$ or $3x + 2 = -1$

so $3x = -1$ or $3x = -3$.

Therefore $x = -\dfrac{1}{3}$ and $x = -1$ are the solutions.

(c) Either $3 - x \geq 5$ or $3 - x \leq -5$

so $-x \geq 2$ or $-x \leq -8$.

We multiply both sides of each inequality by -1; remember to reverse the sense of the inequality signs. Therefore $x \leq -2$ or $x > 8$. ∎

We can also rewrite inequalities to include the modulus sign.

Example   Using the modulus sign rewrite the inequality $4 \leq x \leq 7$.

Solution

We first find the centre of the interval $4 \leq x \leq 7$ then the distance from the centre to either end of the interval.

The centre of the interval is

$$\frac{1}{2}(4 + 7) = 5\frac{1}{2}.$$

The length of the intervals is $(7 - 4) = 3$; the distance to either end is $1\frac{1}{2}$.

Hence $\left| x - 5\frac{1}{2} \right| \leq 1\frac{1}{2}$ is the restatement.

If we wish to avoid fractions, we may multiply all values by 2 to obtain

$$|2x - 11| \leq 3. \quad \blacksquare$$

Exercise 2.8

1   Resolve the following inequalities:

   (a) $x + 7 > -3$

   (b) $5x + 5 < 2$

   (c) $3x - 5 \geq 2$

   (d) $x + 2 < 2x - 4$

   (e) $12 - x \geq 6$

   (f) $7x - 8 \geq 10x + 3$

   (g) $5 + 4x < 3x - 2$

   (h) $2 - x < 5 - 6x$

   (i) $11 - x < 3x + 2$

   (j) $9 - 15x < 6x - 19$

   (k) $3 - 2x < 5x - 4.$

2   Express the following in modulus form:

   (a) $2 < x < 8$

   (b) $3 \leq x \leq 11$

   (c) $-5 < x < 6$

   (d) $6 < 3x < 24$

   (e) $3 \leq -2x \leq 11$

   (f) $9 < 5x + 3 < 15$

   (g) $a < x < b$

   (h) $3 < ax < 5.$

3   Express the following as inequalities for x:

   (a) $|x - 5| < 6$

   (b) $|x + 11| > 10$

   (c) $|2x - 1| < 4$

   (d) $|5 - 3x| < 6$

   (e) $\left| \dfrac{x}{6} + 15 \right| > 15.$

4*   If $c > 0$ write $|ax + b| < c$ as an inequality in x and draw a sketch to represent the result on the real line.

SUMMARY

- **Algebraic expressions** are evaluated in this order: **Brackets, Exponentiation, Division/Multiplication, Addition/Subtraction**

- **Expansion of expressions** follows the Distributive laws:

  $$a(b + c) = ab + ac$$
  $$(a + b)c = ac + bc.$$

  Note that

  $$(a + b)^2 = a^2 + 2ab + b^2, (a - b)^2 = a^2 - 2ab + b^2,$$
  $$(a - b)(a + b) = a^2 - b^2.$$

- **A formula** relates one algebraic quantity, the **subject**, to the others

- **An identity** states that two algebraic expressions are always equal to each other; an **equation** is true for one or more values of the unknown(s)

- **Simultaneous equations**: two linear simultaneous equations in two unknowns are **inconsistent** if they have no solution; if they are consistent they either have a **unique solution** or **infinitely many solutions**

- **Proportionality**: defining C to be a constant, we can say that

  y is **directly proptional to** x if $y = Cx$

  y is **inversely proportional to** x if $y = \dfrac{C}{x}$

  z is **jointly proportional to** x **and** y if $z = Cxy$.

- **Inequalities**

  $a > b$ implies $a + c > b + c$

  $a < b$ implies $a + c < b + c$

  if $a > b$ and if $\lambda > 0$ then $\lambda a > \lambda b$; if $\lambda < 0$ then $\lambda a < \lambda b$

  $ab > 0$ implies either $a > 0, b > 0$ or $a < 0, b < 0$

  $ab < 0$ implies either $a > 0, b < 0$ or $a < 0, b > 0$.

- **Intervals**: inequalities can be represented by intervals on the real line.
- **Inequalities involving $|x|$**

$|x| < a$ is equivalent to $-a < x < a$

$|x| > a$ is equivalent to $x < -a$ or $x > a$

$|x| = a$ is equivalent to $x = -a$ or $x = a$.

Answers

Exercise 2.1

1 (a) $\log_{10} 5$ (b) $\log_5 3$ (c) $\log_a 15$

  (d) $\log_a 81$ (e) $\log_a(2/3)$ (f) $\log_a 3$

2 (a) 3   (b) 2   (c) -3

3 1

4 (a) $30a^3b^2$ (b) $6a^3b^2$ (c) $2/b$

  (d) ac/b^2 (e) $18b$ (f) $3a^2bc/2$

  (g) $3py^{-2}z^2$ (h) $a^{-3/2}$ (i) $(ab)^{-2/3} = a^{-2/3}b^{-2/3}$

  (j) $1/(abc) = (abc)^{-1}$

5 (a) 216   (b) 5   (c) 17/3

  (d) 1/144   (e) 1   (f) 4

6 (a) $\dfrac{-\dfrac{1}{3}\sqrt{x}+y^2}{z^{3/2}}$ (b) $\dfrac{ac^4}{b}+1$

  (c) $\left(\dfrac{ac^4}{b}+1-5\right)^2 = \left(\dfrac{ac^4}{b}-4\right)^2$ (d) $\dfrac{(x+y)^2}{x^2+y^2} = 1 + \dfrac{2xy}{x^2+y^2}$

  (e) $-\dfrac{2xy}{x^2+y^2}$

7 (a) $8/27$ (b) $3/5$ (c) bx

  (d) $12x^2z$ (e) $(xy)^{5/2}/15$ (f) $y^{7/12}$

  (g) $x^{-3/2}y^{-3/2}z^{-3}$ (h) 2

Exercise 2.2

1 (a) $2x^2 - 5x + 3$ (b) $18x^3 + 63x^2 + 37x - 20$

  (c) $6x^2 + xy - y^2$ (d) $x^3 + x^2z + xz + z^2$

  (e) $6x^2 + 3xy - 2x - y$ (f) $4x^3 - 4x^2y - xy^2 + y^3$

  (g) $z^2 + z(a+b) + ab$ (h) $2x^2 + 10x + 14$

  (i) $6x^2 - 5xy - 3x + y^2 + 3y$ (j) $7xy - 3xz + 7yz$

2 (a) $14x + 6$ (b) $33 - 43x$

(c) $98 - 119x$

(d) $60.57y - 18.61$

(e) $65H/924 - 95/308$

(f) $\dfrac{7}{2} + \dfrac{130}{x}$

(g) $2(x^2 + x - 10)$

(h) $4x + 29xy - 238y - 6$

(i) $x^3 - 3x^2 + 30$

(j) $x_1 z_1^2 + 2z_1^2 + 5z_1$

(k) $-x^5 + 11x^4 + 4x^3 + 4x^2 + 2x$

(l) $\dfrac{28.62}{x} - 12.76x$

4 (a) $x^2 - 4x + 3$

(b) $8a^2 - 8a + 2$

(c) $s^2 + st - 6t^2$

(d) $x^2 y^2 - 1$

(e) $-5x^2 + 13x + 6$

(f) $x^3 - 6x^2 + 11x - 6$

(g) $x^3 + 3x^2 - 33x + 60$

(h) $p^3 - 8p^2 q + 17pq^2 - 10q^3$

(i) $x^4 + 4x^3 y + 6x^2 y^2 + 4xy^3 + y^4$

(j) $x^5 + 5x^4 + 10x^3 + 10x^2 + 5x + 1$; the coefficient of x^3 is 10

5 (a) $\dfrac{a - b}{c}$

(b) $\dfrac{3a - 2b}{6c}$

(c) $\dfrac{ad - bc}{cd}$

6 (a) $\dfrac{a + b + c}{c}$

(b) $\dfrac{5a + 2b}{10c}$

(c) $\dfrac{4a + 3b - 24c}{12c}$

(d) $\dfrac{10a + 6b + 5c}{30d}$

(e) $\dfrac{ax^2 + bx + c}{x^3}$

(f) $\dfrac{4ax + 3b}{12x^2}$

(g) $\dfrac{6ax^2 + 4bx - 3c}{12x^3}$

7 (a) $x + x^2$

(b) $3 + 5x$

(c) $\dfrac{x + y}{3y}$

(d) $\dfrac{1 + 2x}{x}$

(e) $1 + z^2$

(f) $\dfrac{1}{x} + 1$

(g) $\dfrac{x + 1}{x^2}$

(h) 2

(i) $x + xy$

(j) $\dfrac{x + 2}{x + 1}$

8 (a) $\dfrac{4x}{3}$

(b) $\dfrac{(4x - 3)}{2x}$

(c) $\dfrac{2x^2 + 2}{(x - 1)(x + 1)}$

(d) $\dfrac{7x^2 + 2x - 2}{(3x - 1)(2x - 1)}$

(e) $\dfrac{x^2 - 49x + 27}{(4x - 3)(9 - x)}$

(f) $\dfrac{x^2(a_2 d_2 + b_2 c_2) + x(a_1 d_2 + a_2 d_1 + b_1 c_2 + b_2 c_1) + a_1 d_1 + b_1 c_1}{(b_1 + b_2 x)(d_1 + d_2 x)}$

(g) $-\dfrac{2(x^2 + x + 1)}{(x + 1)(x - 1)}$

(h) $-\dfrac{(2x^3 + x - 3)}{(x^2 + 2)(x + 1)}$

(i) $\dfrac{x + x^2 y}{y^2}$

(j) $2x$

9 $\dfrac{4x^2 - 5}{(x-1)(x-2)(x^2+1)}$

10   (a) $x+1$ (b) $3x^2 - x - 2 \equiv (3x+2)(x-1)$

(c) $x - 3$ (d) $ab - b^2 \equiv b(a-b)$

(e) $A \equiv \sqrt{x}(\sqrt{x} - y) \equiv x - \sqrt{xy}$ $B \equiv Ax \equiv x^2 - x^{3/2}y$

(f) $3\sqrt{x} + 2\sqrt{y}$

11   (a) $\dfrac{6x^2 + x - 9}{x(x-3)}$ (b) $\dfrac{x^2 + xy - y^2}{y^2}$

(c) $\dfrac{2x^2 + 5x - 1}{x+1}$ (d) $\dfrac{5x - 4}{\sqrt{x}\left(1 - \dfrac{1}{x}\right)} = \dfrac{\sqrt{x}(5x-4)}{x-1}$

(e) $\dfrac{x^3 - 2x^2 - x + 1}{x^3}$ (f) $\dfrac{(ad - bc)(x^2 - 1)}{(cx+d)(dx+c)}$

Exercise 2.3

1   (a)   (i) 2.006 s   (ii) 2.287 s   (iii) 2.374 s

(b)   Faster in winter, slower in summer

2 9×10^{16} joules

3   (a)   (i) $1385 \, \text{m}^2$ (ii) $4849 \, \text{m}^3$

(b) S by 2% approx. V by 3% approx.

(c) $V = 94.03 \, \text{m}^3$

4 $6472 \, \text{cm}^2$

5   1.443 joules

6   (a)   12 m   (b)   13.33 m

Exercise 2.4

1   (a) $V = RT/P$ (b) $m = E/c^2, c = (E/m)^{1/2}$

(c)   current $= (\text{power/resistance})^{1/2}$ (d)   19/7

(e)   1.68   (f)   2618/69 or $37\dfrac{65}{69}$

(g)   7/23   (h) $t = \dfrac{2a + 3b}{2a + 3b - 5c}$

(i) $\quad t = \dfrac{2(\lambda + \mu - 5)}{\mu}$

(j) $\quad G = \dfrac{Fr^2}{Mm}, r = \left(\dfrac{GmM}{F}\right)^{1/2}$

(k) $\quad (2b/ac)^{1/2}$

(l) $\quad 1/3$

(m) $\quad -1/2$

(n) $\quad -48/13$

(o) $\quad a = -p(b^2 + 1)/(b^2 - 1), p = a(1 - b^2)/(1 + b^2)$

(p) $\quad -2$

(q) $\quad a = (b(c - T) - cT)/T, T = \dfrac{bc}{a + b + c}$

(r) $\quad a = \dfrac{(b - T)(c - T^2) - T^2}{T}$, T not possible

(s) $\quad 2$

(t) $\quad -1/11$

(u) $\quad$ not possible

(v) $\quad bp$

(w) $\quad x = 1 - y$

2 5.958 m

3 (a) 0.25 amp

(b) 960 ohm

4 (a) 5.455 Ω

(b) 30 Ω

5 (a) $a = 2(s - ut)/t^2$

(b) -3.556×10^{-2} m s^{-2}

6 (a) $W = \text{constant} \times \rho r^3$

(b) 1000

7 (a) 270 feet

(b) 63 mph (64 mph will not stop in 300 feet)

8 (a) 212 °F

(b) 26.67 °C

(c) $-40°$

Exercise 2.5

1 (a) Identity

(b) Equation: $x = 0$ or 1

(c) Equation; $x = 2$

(d) Equation; $x = 1$

(e) Identity

2 (a) $a = 3, b = -1, c = -1$

(b) $k = -2, m = 1, n = 0$

(c) $a = 2, b = 1$

(d) $a = 6, b = 2, c = 4$

(e) $p = 3, q = 1$

3 $a = 2, b = -11, c = 14$

4 $q^3 = 27r^2$

5 $a = -11$

6   (a) $a = b = 1$ (b) $a = b = 0$
      (c) $a = 1, b = \frac{1}{2}$ (d) $a = c, b = 0$
      (e) $a = c = 4, b = 0$ (f) $a = -1, b = 1, c = 2$
      (g) $z = 2/3, b = -7/3$ (h) $a = 4, b = 2, c = 2$

7   (a) $x = -1, y = 4, z = 1$ (b) $x = y = z = 1$

8 $p = 2, q = 1, r = 1$

9   (a) $x = 1$ (b) $z = 3$
      (c) $x = 13/14, y = -2/35$ (d) $z/a = 7/13, b/a = 1/13$
      (e) $y = 1.002, z = 3.532$
      (f)   The second equation repeats the first, but with a factor of -2.
      (g)   Multiply the first equation by -2 to obtain $-2x + 6y = -4$, which has the same left hand side as the second equation. The equations are therefore inconsistent.
      (h) $x = 1, y = 2, z = 3$ (i) $p = 8, q = 2, r = 5$
      (j) $x = y = 1$ (k) $x = 1, y = 2$
      (l)   Cannot be solved. The first equation is equivalent to the sum of twice the second equation and the third equation.
      (m) $x = y = z = 1$

10 (a) $a = b = 3, c = 2$ (b) $a = -1, b = 1, c = 2$
      (c) $x = 1, y = -2$ or $x = -1, y = 2$ (d) $M = 3, N = 4$
      (e) $M = 3, N = 2, P = 1$
      (f)   Integers 3, 2, 1, M, N, P are all interchangeable, e.g. $M = 3, N = 2, P = 1$ or $M = 3, N = 1, P = 2$

Exercise 2.6

1 $P = 9.81h$

P	3924	5886
h	400	600

2 $P = 5.\dot{5} \times 10^5 \, \rho$

P	10^4	5×10^4
ρ	0.018	0.09

3 $f = 150/h$

f	100	320
h	1.5	0.468 75

4 Reduction in frequency to 80%, $66\frac{2}{3}\%$, 40%, 75% reduction, $33\frac{1}{3}\%$ increase.

5 (i) 31.4 m (ii) 125.6 m; 14.3 s approx.

6 $P = 160$ W, $\sqrt{5}$ amp

7 $v = k\sqrt{\dfrac{p}{\rho}}$ or $v = \sqrt{\dfrac{\gamma p}{\rho}}$, k and γ constants; 8.044×10^{4}, 0.287

8 42 300 km (note units), $T = 79\,500$ s $\simeq 22.1$ h

9 $\text{kg m}^{-1}\text{s}^{-2}$

10 $g = g_0 R^2/r^2$, $T = 2\pi\dfrac{r}{R}\sqrt{\dfrac{l}{g_0}}$, increase by a factor $\dfrac{6401}{6400}$

Exercise 2.7

1
(a) $(x - 4)(x - 3)$
(b) $(3x - 10)(3x + 10)$
(c) $7(x^2 - 1)$
(d) $4(x - 2)(x + 1)$
(e) No factors
(f) $(3x - 4)(11x + 3)$
(g) $(p - 3q)(8p - 3q)$
(h) No factors
(i) $(x - 1)(x - 2)(x - 8)$
(j) $(x - 1)(5x - 7)(6x - 11)$
(k) $(x - 4)(x + 3)(6x - 11)$
(l) $(m + 2n)(3m - 2n)$
(m) $(x - y)(x + y + 1)$
(n) $(q - p)[p^2 + p(q + 3) + q(q + 1)]$

2
(a) $\dfrac{1}{x + 3}$
(b) $\dfrac{x}{(x + 1)^2}$
(c) $\dfrac{11x + 3}{(2x + 1)^2}$
(d) $\dfrac{5x^2 + 13x + 4}{(x + 2)(x + 1)(x - 2)}$
(e) $\dfrac{1}{(x + 1)(x + 6)}$
(f) $\dfrac{3x - 7}{(x + 1)(x - 1)(x - 2)(x - 3)}$
(g) $\dfrac{3x^2 + 2x + 14}{(x + 1)(x + 2)(x + 3)}$
(h) $\dfrac{3q}{(p + q)(p - q)}$
(i) $\dfrac{5\sqrt{b} - \sqrt{a}}{a - b}$
(j) $\dfrac{mt}{(1 - t + t^2)(1 + t)}$ or $\dfrac{mt}{1 + t^3}$

3 $A = -1, B = 1$

4 (a) $A = -4, B = 7$

(b) $A = \dfrac{1}{3}, B = -\dfrac{2}{3}$

(c) $A = -7, B = 13$

(d) $A = \dfrac{1}{2}, B = -1, C = \dfrac{1}{2}$

(e) $A = -\dfrac{1}{2}, B = 3, C = \dfrac{5}{2}$

(f) $A = -1, B = \dfrac{3}{4}, C = \dfrac{1}{4}$

(g) $A = 1, B = C = -1$

5 (a) $(x + 3)(x + 1)(x - 2)$

(b) $(x - 3)(x^2 - 11x + 2)$

(c) $(x - 1)^2(x + 2)$

(d) $(x - 1)^3$

(e) $(x + 3)(x + 1)(x - 2)(x - 4)$

Exercise 2.8

1 (a) $x > -10$

(b) $x < -3/5$

(c) $x \geq 7/3$

(d) $x > 6$

(e) $x \leq 6$

(f) $x \leq -11/3$

(g) $x < -7$

(h) $x < 3/5$

(i) $9 < 4x \text{ or } x > 9/4$

(j) $4 < 3x \text{ or } x > 4/3$

(k) $x > 1$

2 (a) $|x - 5| < 3$

(b) $|x - 7| \leq 4$

(c) $\left| x - \dfrac{1}{2} \right| < \dfrac{11}{2}$

(d) $|x - 5| < 3$

(e) $|2x + 7| \leq 4$

(f) $|5x - 9| < 3$

(g) $\left| x - \left(\dfrac{a + b}{2} \right) \right| < \dfrac{b - a}{2}$

(h) $|ax - 4| < 1$

3 (a) $-1 < x < 11$

(b) $x < -21, x > -1$

(c) $-\dfrac{3}{2} < x < \dfrac{5}{2}$

(d) $-\dfrac{1}{3} < x < \dfrac{11}{3}$

(e) $x < -120, x > -60$

4

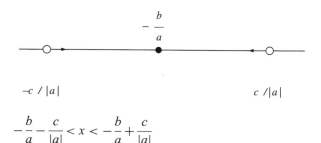

$$-\dfrac{b}{a} - \dfrac{c}{|a|} < x < -\dfrac{b}{a} + \dfrac{c}{|a|}$$

3 STRAIGHT LINES

Introduction

The simplest relationship between two variable quantities is a linear one, which can be represented pictorially by a straight line. Even when the relationship is not linear, we often transform it to a linear version in order to use the straight line representation, from which we obtain information about the relationship by experimental observation of the variables. The relationship between the speed of an object falling under gravity and the time of travel is linear, as is the relationship between the load applied to a spring and the extension it produces, according to Hooke's law.

Objectives

After working through this chapter you should be able to

- understand the Cartesian coordinate system for graphs
- plot points on a graph using Cartesian coordinates
- obtain the gradient and intercept of a straight line from its graph
- find the equation of a straight line which passes through a given point with a known gradient
- find the equation of a straight line through two given points
- recognise when two lines are parallel
- recognise when two lines are perpendicular
- understand the significance of ill-conditioning
- obtain the solution of two simultaneous linear equations in two unknowns
- interpret simultaneous linear inequalities in terms of regions in the plane
- recognise how to reduce a relationship to linear form
- plot a log-linear relationship and a log-log relationship
- use log-linear and log-log graph paper

3.1 GRAPHS AND PLOTTING

The first steps in determining the relationship between two variable quantities are often to carry out experiments and collect data on pairs of values of the two variables, then to put them on a **graph** or scaled picture.

As an example, suppose that we wish to study the relationship between the frequency y of a wave in kiloHertz and the wavelength x in metres. Experiments yield the following results.

x	200	300	400	500	600	750	800
y	1500	1000	750	600	500	400	375

The conventional approach is to choose two **axes**, which are lines intersecting at right angles at a point called the **origin**. Suitable scales are placed on the axes and the points are plotted as in Figure 3.1.

Each pair of values from the set of data corresponds to a point on the graph. To plot the point corresponding to the first pair of values we designate, by convention, the horizontal axis as the **x-axis** and the vertical axis as the **y-axis**. We move along the horizontal axis a distance of 200 units to the right then parallel to the y-axis a distance of 1500 units

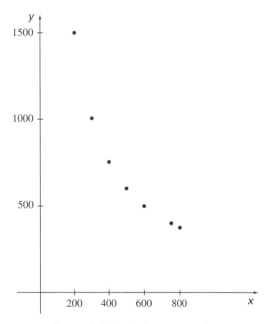

Figure 3.1 Plotting a graph

upwards. At the meeting point we place a mark to represent the point. Similarly for the second point we move along the horizontal axis a distance of 300 units to the right then parallel to the y-axis a distance of 1000 units upwards. Other points are treated in a similar way. In drawing a graph it is wise to choose the scales on the axes sensibly so that they use as full an extent of the graph (squared) paper as is consistent with reading the values easily. The scales for the two axes need not be the same, so be careful when interpreting information from the graph, such as rate of increase.

The value of x for any point is called the **x-coordinate** and the value of y is called the **y-coordinate**. A common notation for a point in general is (x, y); examples from the data are (200, 1500), (400, 750). In older textbooks the x-coordinate is called the *abscissa* and the y-coordinate the *ordinate*.

The axes divide the **coordinate plane**, called the x-y plane, into four **quadrants** as in trigonometry. Figure 3.2 shows the quadrants and a point in each. Note that in the first quadrant both coordinates are positive and in the third they are both negative; in the second the y-coordinate is positive whereas the x-coordinate is negative and the reverse is true in the fourth quadrant.

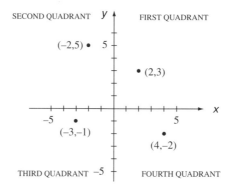

Figure 3.2 The quadrants of the plane

Remember the rule: upwards is positive, to the right is positive. By implication, downwards is negative, to the left is negative.

To obtain a continuous reading we try to draw a smooth curve through or close to the points, as in Figure 3.3.

Here is an example of reading information from graphs. Figure 3.4 is obtained by plotting the points below and drawing a smooth curve through them.

x	1	2	3	4	5
y	2	5	10	17	26

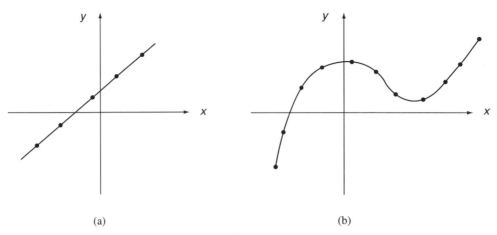

(a)   (b)

Figure 3.3   Drawing a graph

From the graph we can estimate that at $x = 2\frac{1}{2}$, y is approximately 7.5, written $y \simeq 7.5$ at $x = 3\frac{1}{2}$, $y \simeq 13.25$; at $x = 1\frac{1}{2}$, $y \simeq 1.25$; and at $x = 6$, $y \simeq 37$.

The first two estimates were made at values of x within the given range and the last two estimates were made outside. Reading within the range of the tabulated data is known as **interpolation**, a reasonably safe process numerically; reading outside the data is called **extrapolation**. Extrapolation is more suspect but it is sometimes the only available option.

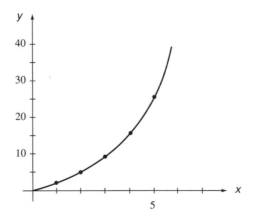

Figure 3.4   Reading values from a graph

Plotting from an equation

Sometimes we are given an equation and in order to draw a graph we generate points by using the equation. For example, consider the equation

$$y = 2 - x^2.$$

We select the values shown below.

x	-2	-1	0	1	2	3
y	-2	1	2	1	-2	-7

The points are plotted in Figure 3.5.

A smooth curve can be drawn through the points. The number of points we choose depends on the nature of the curve and the degree of smoothness sought.

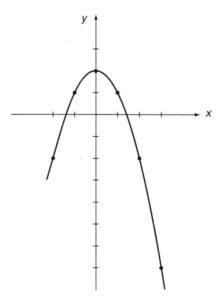

Figure 3.5 Plotting a graph from an equation

Exercise 3.1

1   A beaker of boiling water, initially at a temperature of $100\,°C$, is left to cool in a room whose temperature is $20\,°C$. It is stirred frequently and the times when it reaches certain specified temperatures are recorded below.

$T\,(°C)$	100	80	60	50	40
$t\,(min)$	0	10	24	34	48

Draw the graph of temperature against time and interpolate from it the time when

(a) $T = 90\,°C$ (b) $T = 65\,°C$.

(c)   Similarly, estimate the temperature after 5 min.

(d)   By continuing the graph, extrapolate to the time when the temperature falls to $30\,°C$. Estimate its value in minutes.

2   Mark out a graph scaled from -1 to $+1$ on the y-axis and from 0 to $360°$ on the x-axis.

(a)   Using equally spaced points at $20°$ intervals, plot $\sin x°$. Put your calculator in degrees mode and press the SIN button.

(b)   Complete the picture by noting that $\sin 90° = 1$, $\sin 180° = 0$ and $\sin 270° = -1$.

(c)   Sketch the curve.

(d)   What would happen if you repeated the process as far as $720°$?

3   Repeat Question 2 using your calculator for

(a) $\cos x°$ (b) $\tan x°$.

4   Plot the graph of $y = 2 - x^2$ and, using the same axes, plot the graph of $y = x$.

(a)   Estimate from the graph the x and y coordinates of the two meeting points.

(b)   The exact solution should satisfy the equation $x = 2 - x^2$. Check your estimates in this way.

3.2 THE STRAIGHT LINE

An important class of relationship between two variables is the **linear relationship**, whose graph is a straight line. As an example, consider

$$y = 3x + 4.$$

We chose the values shown below.

x	-2	-1	0	1	2	3	4
y	-2	1	4	7	10	13	16

And we have plotted the results in Figure 3.6(a).

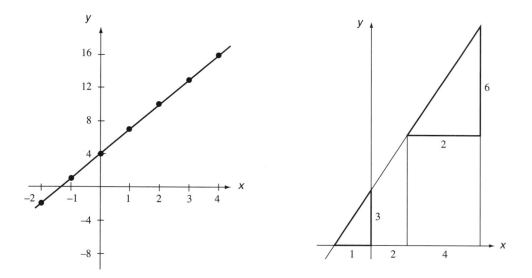

Figure 3.6 Intercept and gradient

The graph is a straight line drawn through the plotted points. The value of the y-coordinate when $x = 0$ is called the y-intercept or, more simply, the **intercept** of the straight line. In this example the intercept is 4.

In Figure 3.6(b) we have constructed some triangles which help to demonstrate that the **gradient** of the straight line is the same, no matter where it is measured (which is why the line is 'straight').

The triangles demonstrate that, if we move from any one point on the line to any other, the increase in the y-value is three times the increase in the x-value. In this example the gradient is 3. Plot the line for yourself, draw a triangle where you wish and verify that the gradient is 3.

As a second example, consider the relationship

$$y = 5 - 2x.$$

The values below are used to plot the graph in Figure 3.7(a).

x	-2	-1	0	1	2	3	4
y	9	7	5	3	1	-1	-3

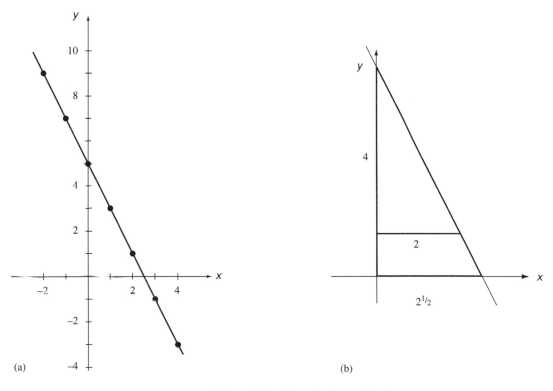

(a)   (b)

Figure 3.7 Graph of $y = 5 - 2x$

The graph is again a straight line. The intercept is 5. As x is increased, y *decreases* at double the rate. The triangle we have drawn in Figure 3.7(b) apparently shows that the gradient is $4/2 = 2$; however, because we wish to demonstrate that y decreases from left to right, we say that the gradient is -2.

In general, the equation of a straight line can be written as

$$y = mx + c$$

where m is the gradient and c is the intercept.

A straight line is uniquely determined by any two points upon it. For example, suppose we seek the equation of the line passing through the points $(-2, 1)$ and $(3, 11)$.

Substituting the first pair of values into the general equation, we obtain

$$1 = -2m + c$$

and substituting the second pair, we obtain

$$11 = 3m + c.$$

Subtracting the first equation from the second, we obtain

$$10 = 3m - (-2m) = 5m.$$

It is straightforward to obtain the solution

$$m = 2 \quad \text{and} \quad c = 5$$

so the required equation is

$$y = 2x + 5.$$

There are two special classes of line. Lines parallel to the x-axis have general equation $y = c$ (gradient zero). Lines parallel to the y-axis have general equation $x = b$ (gradient infinity) see Figure 3.8.

Note in particular that the x-axis has the equation $y = 0$ and the y-axis has the equation $x = 0$.

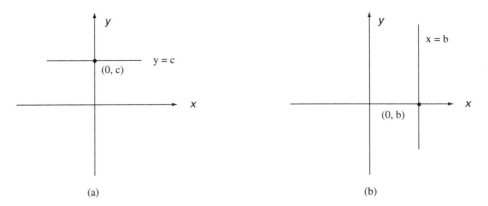

Figure 3.8 Lines parallel to the axes

In Figure 3.9(a) we show the lines $y = 4$ and $y = -2$. Figure 3.9(b) shows the lines $x = -1$ and $x = 3$, which are parallel to the y-axis. It is necessary to take care with such lines; it is not possible to regard their equations as special cases of the general equation of a straight line. (What value to give to m is not clear.)

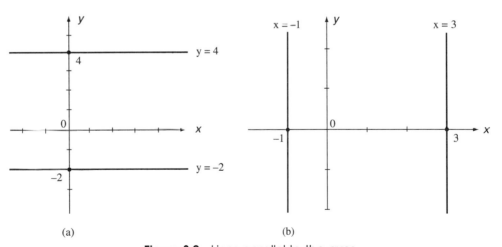

Figure 3.9 Lines parallel to the axes

Figure 3.10 shows four examples of straight lines. In diagrams (a) and (c) the lines have a positive gradient and the general direction is 'south-west' to 'north-east' whereas in diagrams (b) and (d) the gradient is negative and the lines slope in a general 'north-

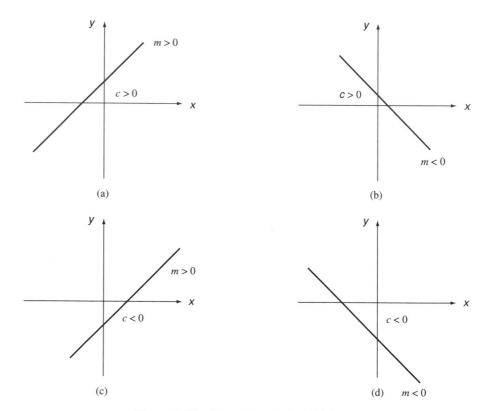

Figure 3.10 Examples of straight lines

west' to 'south-east' direction. You may have deduced that when a straight line passes through the origin, the intercept is zero and its equation is simply $y = mx$.

Exercise 3.2

1 For each of the following straight lines determine the gradient and the intercept (on the y-axis). Draw the graph in each case.

(a) $y = 3x - 5$ (b) $y = -\dfrac{x}{5} + 1$ (c) $2y + 3x + 1 = 0$.

2 Determine the equation of the straight line which

(a)   passes through the points (1, 3) and (2, 6)
(b)   has intercepts on the axes at the points (0, 2) and (5, 0).

3 Which of the following sets of points are collinear, i.e. lie on a straight line?

(a)   (0, 5), (−2, 1), (0.5, 6)   (b)   (0, 2), (2, 1), (4, 3)

(c)   (0, 4), (9, 1), (−6, 6)

First plot a graph to estimate m and c in the equation $y = mx + c$ for the cases where the points are collinear. Obtain the exact values by evaluation.

4* A line perpendicular to the straight line $y = mx + c$ has gradient $-1/m$. Starting with the straight line passing through the points (0, 0) and (3, 4), determine the equation of a line distance 5 units from the origin at its closest point, the point (3, 4). What are the coordinates of its intercepts on the axes?

Forms of the straight line equation

We have stated that the general equation of a straight line is $y = mx + c$. There are in fact several other forms of the equation which can be used in different circumstances.

Given a point on the line and the gradient

Let the given point A be (x_1, y_1) and the gradient of the line be m. Consider a point P on the line; see Figure 3.11(a). Let the coordinates of P be (x, y). The point N is such that APN is a right-angled triangle as shown with AN parallel to the x-axis and NP parallel to the y-axis.

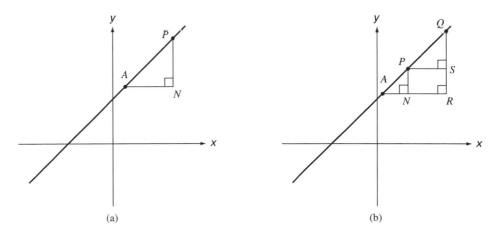

(a)   (b)

Figure 3.11 Finding the equation of a straight line

The y-coordinate of N is y_1, the same as for A; the x-coordinate of N is the same as for P, namely x.

Now the ratio PN/AN is the gradient of the line, namely m. The length AN is the difference between the x-coordinates of N and A, i.e. $x - x_1$, and the length NP is the difference between the y-coordinates, i.e. $y - y_1$. Therefore

$$\frac{y - y_1}{x - x_1} = m.$$

The equation of the line is

$$y - y_1 = m(x - x_1).$$

Example   Find the equation of the line with gradient -2 passing through the point $(1, -3)$. Here $m = -2, x_1 = 1$ and $y_1 = -3$. The equation is

$$y - (-3) = -2(x - 1)$$

i.e. $\qquad y + 3 = -2x + 2 \quad$ or $\quad y = -2x - 1$.   ∎

Given two distinct points on the line

Let the two points be $P(x_1, y_1)$ and $Q(x_2, y_2)$; see Figure 3.11(b). The gradient of the straight line segment PQ is $\dfrac{QS}{PS}$. The point S has the coordinates (x_2, y_1), found in a similar way to the coordinates of N. The gradient of PQ is $\dfrac{(y_2 - y_1)}{(x_2 - x_1)}$. Since this is the same as the gradient of the line segment AP, we obtain the result

$$\frac{y - y_1}{x - x_1} = \frac{y_2 - y_1}{x_2 - x_1}.$$

Therefore

$$y - y_1 = \left(\frac{y_2 - y_1}{x_2 - x_1}\right)(x - x_1).$$

Example

The equation of the line passing through the points $(1, -3)$ and $(-2, 4)$ is

$$\frac{y - (-3)}{x - 1} = \frac{4 - (-3)}{-2 - 1}$$

i.e.

$$\frac{y + 3}{x - 1} = \frac{7}{-3}$$

so

$$y + 3 = -\frac{7}{3}(x - 1)$$

and

$$y = -\frac{7}{3}x + \frac{7}{3} - \frac{9}{3}$$

i.e.

$$y = -\frac{7}{3}x - \frac{2}{3}.$$

∎

Given the intercepts on the axes

Let the intercepts be a and b as shown in Figure 3.12. In the formula for the previous case we have $(x_1, y_1) = (a, 0)$ and $(x_2, y_2) = (0, b)$. Hence

$$y - 0 = \left(\frac{b - 0}{0 - a}\right)(x - a)$$

i.e.

$$y = -\frac{b}{a}(x - a)$$

i.e.

$$y = -\frac{b}{a}x + b.$$

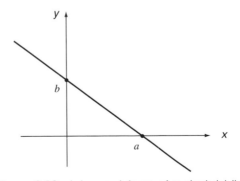

Figure 3.12 Intercept form of a straight line

A common version of this equation is found by writing

$$\frac{y}{b} = -\frac{1}{a}(x - a) = -\frac{x}{a} + 1$$

from which we find that

$$\frac{x}{a} + \frac{y}{b} = 1.$$

If we put $x = 0$ the equation becomes $\frac{y}{b} = 1$, so $y = b$; similarly if $y = 0$, $x = a$.

Example   The line with intercepts of 2 on the x-axis and -3 on the y-axis can be written

$$\frac{x}{2} + \frac{y}{-3} = 1$$

Multiply by -6 to obtain

$$-3x + 2y = -6$$

i.e. $$2y = 3x - 6$$

or $$y = \frac{3}{2}x - 3.$$ ■

'General' equations

The general equation has the form

$$ax + by = c$$

or

$$lx + my + n = 0.$$

Consider the equation $2x + 5y = 10$.

Then

$$5y = -2x + 10$$

or $\qquad y = -\dfrac{2}{5}x + 2.$

This line has gradient $-\dfrac{2}{5}$ and intercept 2.

Example   The line with equation

$$y = \frac{4}{7}x - \frac{3}{7}$$

can be written

$$7y = 4x - 3$$

or $\qquad 4x - 7y - 3 = 0$

i.e. $\qquad 4x - 7y = 3.$ ∎

Exercise 3.3

 1   Determine the equations of the straight lines which pass through the following sets of points:

(a)   (2, 0) and (−1, 5)   (b)   (−11, 4) and (6, 3)

(c)   (9, 6) and (3, 3)

(d)   (2.624, 3.979) and (−15.127, −22.345).

2   In the form $\dfrac{x}{a} + \dfrac{y}{b} = 1$ write down the equations of the straight lines with the following intercepts:

(a)   (6, 0), (0, −5)   (b)   (9, 0), (0, 5)

(c)   (6.3275, 0) (0, −17.613).

In each case rewrite the equation as $lx + my + n = 0$.

3   Determine the following lines:

(a)   perpendicular to the line $x + y = 1$ and passing through the origin

(b)   perpendicular to the line $3x - y = 10$ and passing through (4, −1).

4   Determine the equations of the following lines:

(a)   a line passing through the origin and the point $\left(-\dfrac{4}{5}, \dfrac{7}{5}\right)$

(b)   a line of gradient 2 which passes through the point (2.125, 4.275).

5   Draw on the same axes the following straight lines:

(a) $y = 2x + 1$ (b) $3x + 4y = 5$ (c) $16x - 12y = 17.$

 6   **Hooke's law** states that the extension X of a helical spring beyond its natural length is directly proportional to the tension T in the spring. The following data are obtained.

X (mm)	2	4	6	8	10
T (kg)		15.2		30.4	38

(a)   Fill in the gaps in the table.
(b)   If the length of the spring when unstretched is 130 mm and Y represents the stretched length, determine the constants a and b where

$$Y = aT + b.$$

 7   The cost of a domestic repair facility consists of a call-out charge (£C) and an attendance fee of p pence per minute.

(a)   Write down in a formula the full charge £F made to a customer for an m minute call-out plus attendance.
(b)   If the attendance, but not the call-out, is VAT-rated at $V\%$, write down the modified formula.
(c)   A plumber charges £60 for call-out, 94p per minute attendance and VAT is rated at 17.5%. How much does a customer pay for a 45 minute visit in accordance with (b) above?

 8   **Boyle's law** states that, for a given mass of ideal gas, the product of pressure P and volume V is constant, i.e. $PV = C$. The following table gives pressures in newtons per square metre ($\mathrm{N\,m}^{-2}$) and volumes in cubic metres (m^3).

P	6.000	7.000	8.000	9.000
V	3.194	2.738		2.129

Determine C and find the volume when $P = 8\,\mathrm{N\,m}^{-2}$.

3.3 INTERSECTION OF TWO LINES

If we seek the solution of the simultaneous linear equations

$$2x - y = -4 \quad \text{and} \quad x + 2y = 28$$

then we have two approaches: **graphical** and **algebraic**. The algebraic approach may be subdivided into two methods, variations on essentially the same idea.

Graphical approach

First, we arrange the equations into the forms

$$y = 2x + 4 \quad \text{and} \quad y = -\frac{x}{2} + 14$$

and draw *on the same axes* the graphs of the lines representing these equations. We find, as in Figure 3.13, that the lines intersect at a point given by $x = 4, y = 12$. These values represent the *unique* solution of the equations.

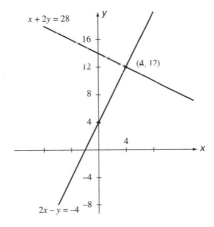

Figure 3.13 Intersection of straight lines

Algebraic approach

Eliminate one variable

Consider again

$$2x - y = -4 \tag{1}$$
$$x + 2y = 28. \tag{2}$$

If we multiply equation (1) by 2 we obtain

$$4x - 2y = -8. \tag{3}$$

Adding equations (2) and (3) gives

$$5x = 20$$

from which we find that $x = 4$. Substituting this into equation (2) gives

$$4 + 2y = 28$$
so that $\qquad 2y = 24$
and therefore $\quad y = 12.$

Note that we could have eliminated x by multiplying equation (2) by 2 and subtracting equation (1) from this new equation. Try it.

Substitute for one variable

If we rewrite (1) as $y = 2x + 4$ and substitute this into equation (2) we obtain

$$x + 2(2x + 4) = 28$$
i.e. $\qquad x + 4x + 8 = 28$
so $\qquad 5x = 20$
therefore $\qquad x = 4$

and we can substitute this into the form $y = 2x + 4$ to find that $y = 12$. Note the similarity between substitution and elimination.

Graphical methods: advantages and disadvantages

A disadvantage of the graphical method for practical solution of simultaneous linear equations is the problem of accuracy. If the coefficients are not so simple and the intersection point does not have integer coordinates, it will be difficult to obtain an accurate solution.

But the graphical representation does have the advantage of pointing out difficulties that may arise.

Ill-conditioning

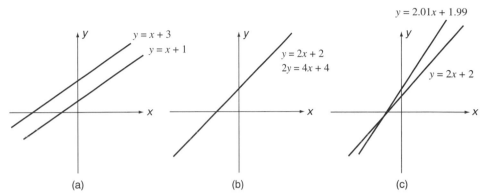

Figure 3.14 Pairs of lines: (a) parallel, (b) coincident and (c) nearly coincident

In Figure 3.14(a) the two lines are **parallel**, i.e. they have the same gradient, so the corresponding equations have no solution. In Figure 3.14(b) the lines are coincident, so each point on the common line represents a possible solution. In Figure 3.14(c) the lines are nearly coincident, the point of intersection is difficult to determine accurately and a slight change in some or all of the coefficients can lead to a completely different solution; this is the case known as **ill-conditioning**.

When the lines are perpendicular (Figure 3.15) the point of intersection is relatively easy to find accurately by graphical means. If the lines are $y = m_1 x + c_1$ and $y = m_2 x + c_2$ then the condition that they are perpendicular is $m_1 m_2 = -1$.

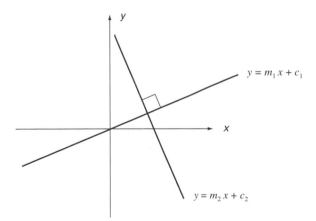

Figure 3.15 Perpendicular lines

Exercise 3.4

1   Given the equation

$$L(x) = \frac{(x - x_2)}{(x_1 - x_2)}y_1 + \frac{(x - x_1)}{(x_2 - x_1)}y_2$$

what does $L(x) = $ a constant represent?

2   A quadrilateral is bounded by the straight lines

$$x + y = 2, \quad x = 0, \quad 3x - 4y = 4, \quad x + y = 8.$$

Draw the graphs of these lines and determine the coordinates of the vertices.

3   Determine the points of intersection of the following pairs of lines:

(a) $y = 2x + 3$ and $y + 3x + 1 = 0$

(b) $11x - 5y = 2$ and $3x - 5y + 15 = 0$.

4*   Determine as best you can the intersection points of the following pairs of lines:

(a) $y = 2.01x + 1.99$ and $y = 2x + 1$

(b) $y = 2.001x + 1.99$ and $y = 2x + 1$.

5*   Draw the graph of an arbitrary line $y = mx + c$ ($c \neq 0$). On the same axes draw the line $y = -x/m$, i.e. a line perpendicular to $y = mx + c$ drawn from the origin. If the two lines meet at (x_1, y_1) determine x_1 and y_1 in terms of m and c and write down the minimum distance of $y = mx + c$ from the origin.

3.4 LINEAR INEQUALITIES

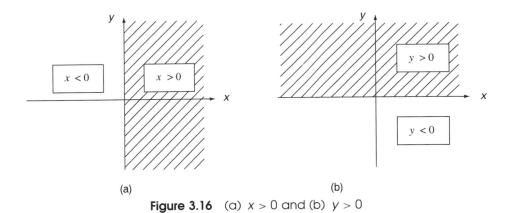

(a)   (b)

Figure 3.16   (a) $x > 0$ and (b) $y > 0$

The y-axis is the line along which $x = 0$. To the right are points (x, y) such as $(2, 1)$, $(3, -4)$ etc. They have the common feature that $x > 0$; the region is shown shaded in Figure 3.16(a). To the left of the y-axis is the region where $x < 0$. Hence the x–y plane is divided into three distinct, non-overlapping regions: $x < 0, x = 0, x > 0$. Similarly the x-axis, where $y = 0$, is the dividing line between the regions $y > 0$ and $y < 0$, as shown in Figure 3.16(b).

Figure 3.17(a) shows that the line $x = 3$ divides the plane into the three regions $x < 3, x = 3$ and $x > 3$. In Figure 3.17(b) the line $y = -2$ is the boundary between the regions $y > -2$ and $y < -2$. Note that these results apply in the context of points in the plane. In general the inequality $y > mx + c$ can be depicted by drawing the line $y = mx + c$ and deciding on which side of it lie those points that satisfy the inequality.

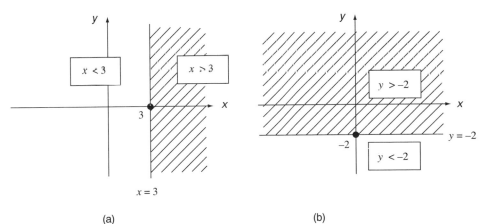

(a)   (b)

Figure 3.17   (a) $x > 3$ and (b) $y > -2$

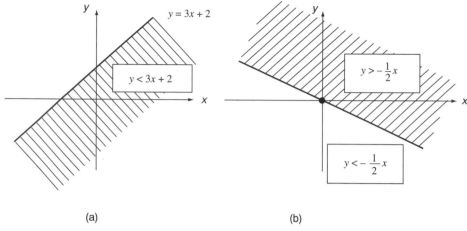

(a)   (b)

Figure 3.18   (a) $y < 3x + 2$ and (b) $y > -\frac{1}{2}x$

Figure 3.18(a) shows the line $y = 3x + 2$. On one side of it lie the points (x, y) for which $y > 3x + 2$, and on the other side lie those points for which $y < 3x + 2$. The origin does not lie on the line and it serves as a simple test case. Since $0 < 3 \times 0 + 2$ the points 'below' the line satisfy the inequality $y < 3x + 2$, shaded in the figure.

In Figure 3.18(b) the line $y = -\frac{1}{2}x$ passes through the origin, so we need another test point; $(1, 1)$ will be suitable. At $(1, 1)$, $1 > \frac{1}{2} \times 1$, so the points which satisfy the inequality $y > \frac{1}{2}x$ lie 'above' the line, as depicted.

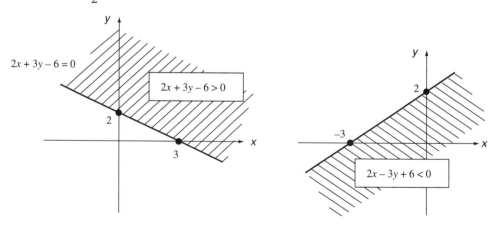

(a)   (b)

Figure 3.19   (a) $2x + 3y - 6 > 0$ and (b) $2x - 3y + 6 < 0$

To find the region for which $ax + by + c > 0$ needs more thought. Figure 3.19(a) shows the region $2x + 3y - 6 > 0$, determined by using the line $2x + 3y - 6 = 0$ and the test point $(0, 0)$. In Figure 3.19(b) we have depicted the region $2x - 3y + 6 < 0$, determined by using the line $2x - 3y + 6 = 0$ and the test point $(0, 0)$.

Inequalities involving modulus signs

The relationship $y = |x|$ depicted in Figure 3.20(a) consists of two straight lines; $y = x$ for $x \geq 0$ and $y = -x$ for $x \leq 0$. Thus $|3| = 3$ and $|-2| = -(-2) = 2$, as we know already. Figure 3.20(b) shows the graph of $y = |x - 1|$. We have used the values below.

x	-2	-1	0	1	2	3
$x - 1$	-3	-2	-1	0	1	2
$|x - 1|$	3	2	1	0	1	2

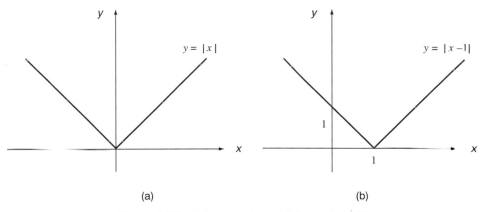

(a)   (b)

Figure 3.20   (a) $y = |x|$ and (b) $y = |x - 1|$

Note that this graph is effectively the graph of $y = |x|$ displaced to the right by 1 unit. To solve the inequality $|x| < 2$ we can use the graph of $y = |x|$.

In Figure 3.21(a) we have superimposed the graphs of $y = 2$ and $y = |x|$. The graphs intersect at two points: when x is positive the intersection occurs at $x = 2$, and when x is negative the intersection occurs at $-x = 2$, i.e. $x = -2$. Hence the regions is $-2 < x < 2$ shown shaded in the figure. Remember that we are in two dimensions and the region where $-2 < x < 2$ looks quite different from the interval $-2 < x < 2$ on the real line.

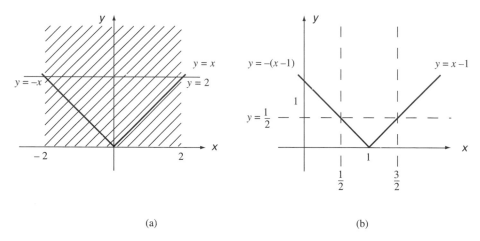

Figure 3.21 (a) $|x| < 2$ and (b) $|x - 1| < 2$

To solve the inequality $|x - 1| > \dfrac{1}{2}$ we refer to Figure 3.21(b). The graph of $y = |x - 1|$ consists of the lines $y = (x - 1)$ for $x \geq 0$ and $y = -(x - 1)$ or $y = 1 - x$ for $x \leq 0$; either line gives $y = 0$ at $x = 1$. Again, there are two intersection points. For x positive we solve $x - 1 + \dfrac{1}{2}$ so that $x = \dfrac{3}{2}$; for x negative we solve $-(x - 1) = \dfrac{1}{2}$ so that $x = \dfrac{1}{2}$. For the inequality $|x - 1| > \dfrac{1}{2}$ to be true, therefore, either $x < \dfrac{1}{2}$ or $x > \dfrac{3}{2}$. Note how the use of the graphs helps us with the algebraic solution.

Since we do not use the graphs to determine the points of intersection, we need not draw them accurately. When the graphs are roughly correct but not as accurate as could be, we say that we *sketch* the graphs rather than plot them. Sketches can be good approximations to the true graph or less accurate, depending on the use to which they are put. We shall return to this idea quite often.

Example   Consider now the inequality $|x| < \dfrac{1}{2}x + 2$. We can solve this by a graphical approach. In Figure 3.22(a) we have plotted, using the same axes, the graphs of $y = |x|$ and $y = \dfrac{1}{2}x + 2$. Where the line $y = \dfrac{1}{2}x + 2$ lies above the graph of $y = |x|$ then $|x| < \dfrac{1}{2}x + 2$; the region is shaded.

The graphs intersect at two points. For positive x the point is where $x = \dfrac{1}{2}x + 2$, i.e. $\dfrac{1}{2}x = 2$ so that $x = 4$. For negative x the point is where $-x = \dfrac{1}{2}x + 2$ so that $-\dfrac{3}{2}x = 2$ or $x = -\dfrac{4}{3}$. We *could* have determined the points graphically but it would have been

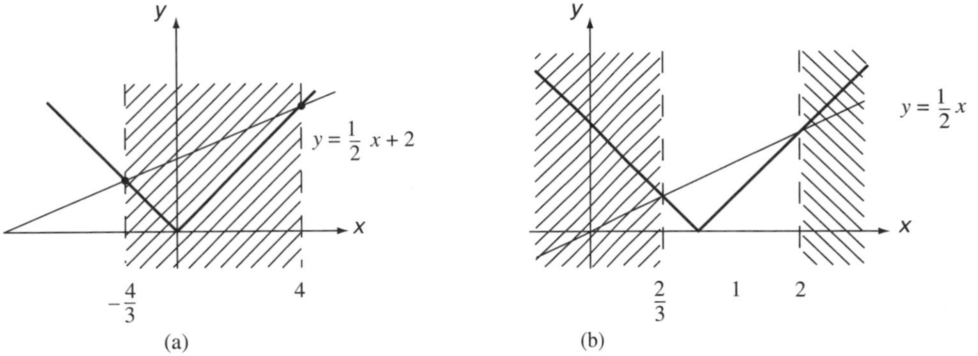

Figure 3.22 (a) $|x - 1| < \frac{1}{2}x + 2$ and (b) $|x - 1| > \frac{1}{2}x$

difficult to obtain accurate values, especially the value $x = -\frac{4}{3}$. Hence if $-\frac{4}{3} < x < 4$ then $|x| < \frac{1}{2}x + 1$. ∎

Example
We now consider the inequality $|x - 1| > \frac{1}{2}x$. In Figure 3.22(b) we have superimposed the graphs of $y = |x - 1|$ and $y = \frac{1}{2}x$. When x is greater than 1 the graphs intersect at the point where $x - 1 = \frac{1}{2}x$, i.e. at $x = 2$. When x is less than 1 the intersection occurs where $-(x - 1) = \frac{1}{2}x$, i.e. $x - 1 = -\frac{1}{2}x$, so $x = \frac{2}{3}$. Then $|x - 1| > \frac{1}{2}x$ if either $x < \frac{2}{3}$ or $x > 2$. These regions are shaded in Figure 3.22(b). ∎

Exercise 3.5

1   Sketch the following regions:

(a) $x > -3$ (b) $y < 4$

(c) $x > -3, y < 4$ (i.e. both are satisfied)

(d) $y > 2x + 1$ (e) $2x + 3y + 2 < 0$

(f) $2 - y + 3x < 2y - x + 5$.

2 In linear programming a **feasible region** is defined as a region in which a number of inequalities hold. Draw the feasible region where the inequalities $x > 0$, $y > 0$, $2x + 5y < 7$, $5x + 2y < 7$ are satisfied simultaneously.

3 Shade in the region where the following inequalities hold simultaneously:

$$x + y + 1 < 0, \quad y - 2x + 1 > 0, \quad y < 0.$$

4 Sketch the following graphs:

(a) $y = -|x + 1|$ (b) $y = |2x - 1|$

(c) $y = |x| + |x - 1|$ (d) $y - |x| - |x - 1|$

(e) $x = |y + 1|$ (f) $y = |x + 1| + |x| + |x - 1|$.

5 Copy Figure 3.21 and shade the regions defined by the following inequalities:

(a) $|x| < y < 2$ (b) $y > |x - 1|, x > 0$

(c) $2 < y < |x|, x < 0$ (d) $0 < y < |x - 1|, x > \dfrac{3}{2}$.

6 Draw sketches like Figure 3.21 to depict

(a) $|2x + 1| > 3$ (b) $\left| x - \dfrac{5}{2} \right| < \dfrac{9}{2}$.

7* For which values of x do the following inequalities hold?

(a) $|x - 1| < 2x - 5$ (b) $|2x| < x + 6$

(c) $|2 - 3x| < 5 - x$ (d) $|x + 1| < |x - 1|$

(e) $|2 - x| < |3 + x| < 6$ (f) $|x| + |x - 5| > 10 - |x - 3|$.

8* Illustrate graphically the inequalities in Question 7.

3.5 REDUCTION TO LINEAR FORM

There are many examples of laws and models in which there is a non-linear relationship between the dependent variable y and the independent variable x. In such cases the graph of y against x is a curve rather than a straight line. However, in some practical cases we can transform the relationship or use special graph paper so that a straight line graph results.

We consider first the idea of a transformation. We shall sometimes use natural logarithms. You may prefer to use logarithms to the base 10. This is unimportant, but you must be consistent.

Examples

Power law, $y = ax^n$.

Here y is assumed to behave like a power of x. The power may be positive or negative and may not be an integer. Taking logarithms we obtain

$$\ln y = \ln a + n \ln x$$

and we plot **ln y** against **ln x** to obtain a straight line graph with slope n and intercept $\ln a$. We can therefore find the value of n directly; the value a can be found using the INV and LN keys.

The power law occurs widely in science and engineering:

(a) $V = K\sqrt{T}$, where V is the velocity of sound in a gas and T is its absolute temperature; K is a constant.

(b) $Q = Cr^4$, where Q is the quantity of viscous fluid passing through a narrow pipe of radius r under constant pressure against gradient; C is a constant.

The power law may be modified to include cases such as $y - b = a(x - d)^n$, i.e. a shift of origin to the point (b, d).

Exponential law, $y = a^x$

Here y behaves like a constant raised to the variable power x. This law occurs in processes involving growth and decay. Taking logarithms gives

$$\ln y = x \ln a.$$

We plot **ln y** against **x** to obtain a straight line graph.

Some examples are

(a) $R = P\left(1 + \dfrac{c}{100}\right)^n$

where R represents the return upon the principal P in £, \$, etc., invested for n years at $c\%$; n need not necessarily be an integer.

(b) $T - T_c = (T_0 - T_c)e^{-kt}$

is Newton's law of cooling for the temperature T of a liquid at time t which cools from an initial temperature T_0 at $t = 0$ in an environment where the temperature is T_c. K is a constant and $e \simeq 2.718$ is the exponential base. By putting $y = T - T_c, a = T_0 - T_c$ the formula becomes $y = ae^{-kt}$. Taking logarithms we obtain $\ln y = \ln a - kt$. Plotting $\ln y$ against t gives a straight line graph.

Quadratic forms

You cannot transform the general quadratic form $y = ax^2 + bx + c$ into a linear form but it is possible to deal with reduced quadratic forms, such as $y = ax^2 + c$ and $y = ax^2 + bx$.

In the first case plot $\ln(y - c)$ against $\ln x$. In the second case observe that $\frac{y}{x} = ax + b$ and plot $\frac{y}{x}$ against x.

Graph paper

You can use conventional graph paper in conjunction with the transformations above to plot straight line graphs. The quantities a, b, $\ln a$, etc., can be estimated from the straight lines and fitted to the data so that an algebraic relationship between x and y can be established.

With special graph paper you can overcome the need for taking logs; it allows you to plot the x and y data directly. The paper is designed to adjust the scale as necessary, in effect taking the logarithm for you. We will describe two types of paper.

Log-linear paper

The scale is linear horizontally and logarithmic vertically. Two cycles of 10 are shown in Figure 3.23, but typically the paper comes in 2, 3 and 4 cycles to represent powers 10^2, 10^3 and 10^4 in the data respectively. To use it for an exponential law plot, observe first the number of powers of 10 through which y varies, so the appropriate number of cycles can be chosen, then just plot directly on the paper. You will see that the plotted points lie on a straight line, but remember the slope of the line is *not* equal to the parameter k in the experimental relationship $y = ae^{-kx}$.

Example

A radioactive element decays into a more stable element by emitting rays and particles. If m is the mass of the element remaining after time t and m_0 is the mass at time $t = 0$, then the law

$$m = m_0 e^{-kt}$$

models the amount m of the element remaining at time t, where k is a positive constant.

The following data represents the remaining mass kilograms of an element at different times recorded in days.

t (day)	0	200	400	1000	1500	2000	3000
m (kg)	10	8.099	6.560	3.485	2.058	1.215	0.423

Depict the data on a log-linear graph.

The points are plotted in Figure 3.23 and a straight line drawn through them. Observe that the straight line fits the middle data points relatively well. ∎

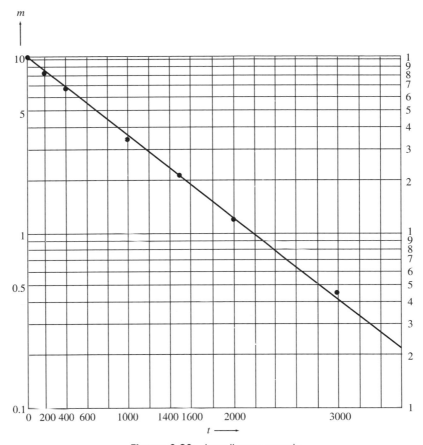

Figure 3.23 Log-linear graph

Example

In an attempt to assess the declining purchasing power of money, we can use log-linear graph paper to gauge unit costs at some future time, provided the rate of inflation is assumed constant. To illustrate this, take 1995 as a baseline, i.e. $P = 1$ in 1995, and an optimal headline inflation rate, e.g. 4%. Using $R = P\left(1 + \dfrac{4}{100}\right)^n$ we can determine (on the calculator using the x^y button) that after 25 years, in 2020, prices will have risen by a factor of 2.666.

We put the points (1995, 1) and (2020, 2.666) on a graph and draw a straight line through them; see Figure 3.24. We can then read off remaining inflation information from the graph, e.g.

        1975   0.45
        2015   2.16

An overall aggregate of 4% inflation means that goods purchased for £1 in 1995 cost 45p in 1975 and will cost £2.16 in 2015. However, we know that the headline inflation rate

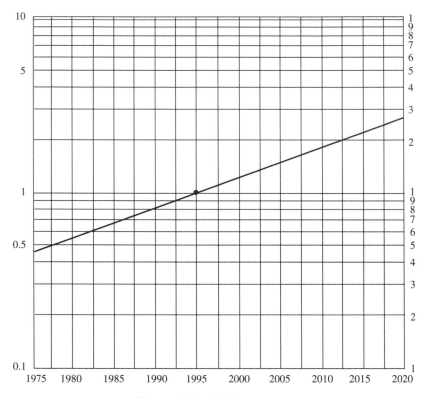

Figure 3.24 Inflation graph

changes all the time, so the figures mean little, but you do have the flexibility to adjust the average headline inflation rate as you choose and you can read off projections from the straight line so drawn. ∎

Log-log paper

To depict general power law relationships we require log-log graph paper. A 3×2 cycle is shown in Figure 3.25, representing a variation of 10^2 in each variable, though the graph paper can be obtained in a variety of combinations, e.g. 2×3, 3×3, 3×4.

Example
A gas expands adiabatically according to the law $PV^\gamma = C$. The following data are obtained, where P is measured in pascals (Pa) and V in cubic metres (m^3). Plot the data on 3 cycle $\times$ 2 cycle log paper and draw a straight line through the points. Estimate γ and C. Estimate from the graph the value of P when $V = 25$ and the value of V when $P = 1$.

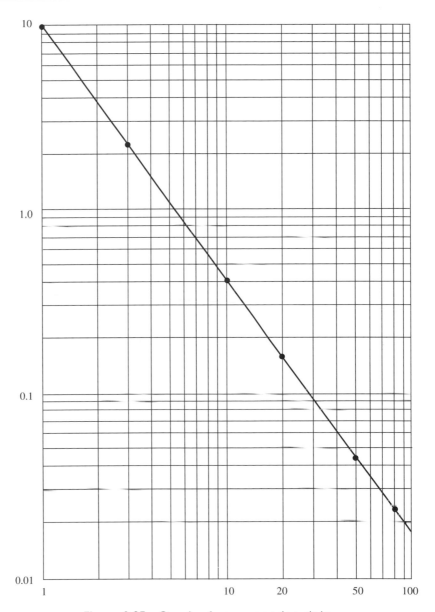

Figure 3.25   Graph of gas expansion data

P (Pa)	10.0	2.20	0.844	0.417	0.160	0.0451	0.0171
V (m^3)	1	3	6	10	20	50	100

Taking logarithms, the relationship becomes $\log P + \gamma \log V = \log C$, i.e.

$$\log P = -\gamma \log V + \log C.$$

The data plotted from bottom right to top left shows that $\gamma \approx 1.38$, $C \approx 10.0$. We did this by taking two well-spaced values (P_1, V_1) and (P_2, V_2) so that

$$\ln P_1 + \gamma \ln V_1 = \ln C$$
$$\ln P_2 + \gamma \ln V_2 = \ln C$$

and eliminated in turn $\ln C$ and γ.

In order to plot the points with reasonable accuracy it is important to know the number of significant figures of each quantity, not the absolute accuracy.

The data are plotted in Figure 3.25 together with the straight line. We shall use logarithms to the base 10, i.e. the LOG button on a calculator.

When $V = 1, P = 10, \log V = 0, \log P = 1$.

When $V = 100, P = 0.0171, \log V = 2, \log P = -1.767$.

The gradient is $(-1.767 - 1)/(2 - 0) = -1.38$ (3 s.f.) so $\gamma = 1.38$ (3 s.f.).

We can find $\log C$ by using the equation $\log P = -\gamma \log V + \log C$ together with a pair of values for P and V and our estimated value of γ. Alternatively, we can note that when $V = 1, PV^\gamma = P = C$ so   that $C = 10$.   From   the   graph   we   find   that   when $V = 25, p \simeq 0.116$ and when $P = 1, V \simeq 5.2$. (The more accurate values, obtained from a calculator, are 0.118 and 5.3 respectively.)   ■

Exercise 3.6

1   In the following examples corresponding data pairs (x, y) are observed where y is a function of x.

(a) $y = ax^2 + bx + c$, c known   (b) $y = \ln kx^2$

(c) $y = cx^2 + dx^4$ (d) $y = axe^{bx}$

(e) $x^2 + y = ax^2 y$

What would you plot against what, in order to obtain a suitable straight line graph?

2   Redraw the log-linear straight line for a mean headline inflation rate which corresponds to unit costs in 1955 being 11% of their 1995 values (see the example on p. 117). What is the mean inflation rate and how will 2040 prices compare to those of 1995 on that basis?

3 The following data represent the density ρ in cm^{-3} of water measured at temperatures between $40\,^{\circ}\text{C}$ and $100\,^{\circ}\text{C}$.

T	40	60	80	100
ρ	0.9922	0.9832	0.9718	0.9584

Assuming that $\rho = \rho_0 + \alpha t + \beta t^2$, where $\rho_0 = 0.9922$ and $t = T - 40$, plot a graph of $\dfrac{\rho - \rho_0}{t}$ against t to obtain estimates for α and β.

4 In the seventeenth century the astronomer Johannes Kepler postulated his famous third law (a result later proved mathematically). This stated that the squares of the periodic times of the planets which orbit the Sun were proportional to the cubes of their mean distance from the Sun. Using data from the table below and assuming that $T \propto r^b$, plot $\ln T$ against $\ln r$ on log-log paper and obtain an estimate for b.

Planet	Orbit period, T (years)	Mean distance from sun, r (km $\times 10^6$)
Mercury	0.241	58
Venus	0.616	108
Earth	1.000	150
Mars	1.881	228
Jupiter	11.85	778
Saturn	29.44	1 427
Uranus	83.94	2 870
Neptune	164.7	4 500
Pluto	248.2	5 900

5 Draw a straight line graph to represent the relationship between object distance u and image distance v of a lens whose focal length f is 20 cm, given the law

$$\frac{1}{f} = \frac{1}{u} + \frac{1}{v}.$$

6* The angular momentum h of a spinning electron takes the value

$$h = \sqrt{\{|q|(|q| + 1)\}}\ b/2\pi$$

where b is a constant to be estimated. If estimates of h and q are known, explain how they can be used to estimate b from a straight line graph.

SUMMARY

- **Graph**: a scaled picture which shows the relationship between two variables
- **Cartesian coordinates**: the point (x, y) is a distance x from the y-axis and distance y from the x-axis
- **Equations of a straight line:**

    A line with gradient m and intercept c has equation $y = mx + c$.

    A line through (x_1, y_1) with gradient m has equation $y - y_1 = m(x - x_1)$

    The general equation is $ax + by + c = 0$

- **Parallel lines** have equal gradients, i.e. $m_1 = m_2$
- **Perpendicular lines** have gradients that multiply to give -1, i.e. $m_1 m_2 = -1$
- **Ill-conditioning**: a system of simultaneous linear equations is ill-conditioned if small changes in the coefficients lead to relatively large changes in the solution
- **Simultaneous inequalities** define regions of the x–y plane where the coordinates of the points satisfy the inequalities
- **Reduction to linear form**: a relationship can be rewritten in a form that allows a straight line graph to be drawn
- **Log-linear relationship**: by taking logarithms the relationship $y = ax^n$ becomes $\log y = \log a + n \log x$ and we may plot $\log y$ against $\log x$. The gradient is n and the intercept is $\log a$
- **Log-log relationship**: the relationship $y = ka^x$ becomes

    $\log y = \log k + x \log a$ and we may plot $\log y$ against x

- **Special graph paper** can be used to make direct plots of log-linear and log-log relationships

Answers

Exercise 3.1

1 (a)   4.6 minutes   (b)   20.0 minutes

   (c)   89.2 °C   (d)   72 minutes

2 (d)   waveform is reproduced; we say that sin x is periodic

3 sin $x°$ and cos $x°$ have a period of 360°; tan $x°$ has a period of 180°

4 Meeting points $(-2, -2)$, $(1, 1)$.

Exercise 3.2

1 (a)

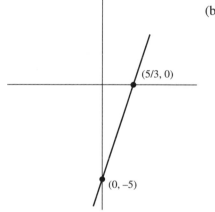

(b)

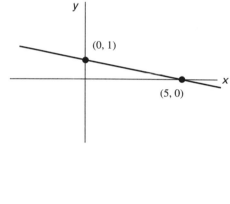

(c)

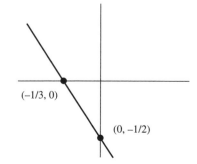

2 (a) $y = 3x$

(b) $2x + 5y = 10$

3 (a) Yes, $y = 2x + 5$

(b) No

(c) Yes, $y = -x/3 + 4$

4

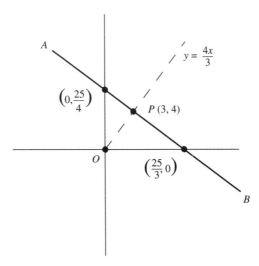

$OP = 5$ units. The line of which OP is a part is $y = \dfrac{4x}{3}$. Line AB is $y = -\dfrac{3x}{4} + \dfrac{25}{4}$. Both lines pass through P (3, 4).

Exercise 3.3

1 (a) $y = -\dfrac{5}{3}(x - 2)$

(b) $y = \dfrac{1}{17}(57 - x)$

(c) $y = \dfrac{1}{2}(x + 3)$

(d) $y = 1.4830x + 0.0876$

2 (a) $\dfrac{x}{6} - \dfrac{y}{5} = 1$ or $-5x + 6y + 30 = 0$

(b) $\dfrac{x}{9} + \dfrac{y}{5} = 1$ or $5x + 9y - 45 = 0$

(c) $\dfrac{x}{6.2375} - \dfrac{y}{17.613} = 1$ or $17.613x - 6.2375y + 109.86 = 0$

3 (a) $y = x$

(b) $x + 3y = 1$

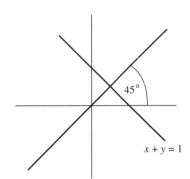

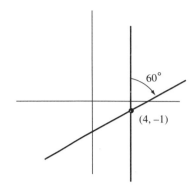

4 (a) $y = -7x/4$

(b) $80x - 40y + 1 = 0$

5

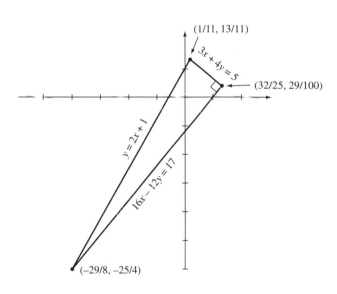

6 (a) 7.6, 22.8

(b) $Y = 3.8T + 130$

7 (a) $F = C + \dfrac{pm}{100}$

(b) $F = C + \dfrac{pm}{100}\left(1 + \dfrac{V}{100}\right)$

(c) £109.70

8 $C = 19.164$ $V = 2.396$ when $P = 8$

Exercise 3.4

1 $L(x)$ is a linear expression and $L(x_1) = y_1$, $L(x_2) = y_2$. $L(x) =$ a constant must therefore be the straight line passing through the points (x_1, y_1) and (x_2, y_2).

2

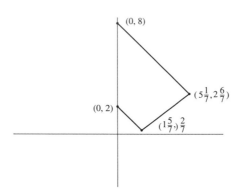

3 (a) $\left(-\dfrac{4}{5}, \dfrac{7}{5} \right)$ (b) $(2.125, 4.275)$ or $\left(\dfrac{17}{8}, \dfrac{171}{40} \right)$

4 (a) $(-99, -197)$ (b) $(-990, -1979)$

5

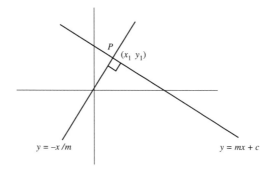

$(x_1, y_1) \equiv \left(\dfrac{-mc}{1 + m^2}, \dfrac{c}{1 + m^2} \right)$ or $\dfrac{c}{1 + m^2}(-m, 1)$

P is $\dfrac{c}{\sqrt{1 + m^2}}$ units distant from the origin

Exercise 3.5

1 (a)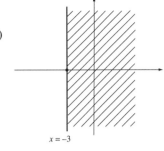

(b)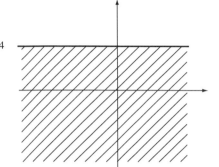

(c)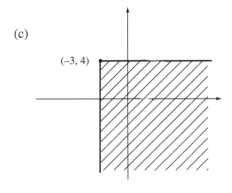

(d)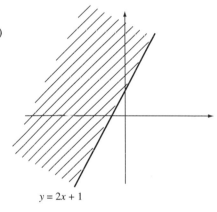

(e)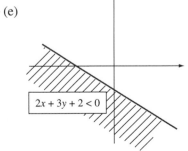

(f)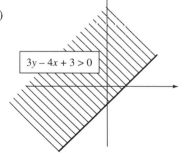

2

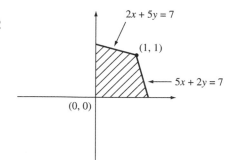

3

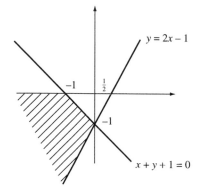

4 (a)

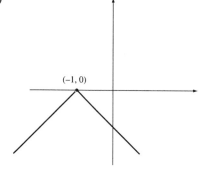

(b)

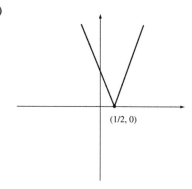

(c) $y = 1 - 2x, x < 0$
      $ = 1, \qquad 0 < x < 1$
      $ = 2x - 1, x > 1$

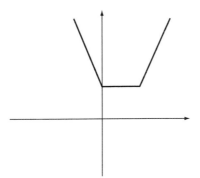

(d) $y = -1, \qquad x < 0$
      $ = 2x - 1, 0 < x < 1$
      $ = 1, \qquad x > 1$

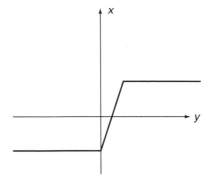

(e)

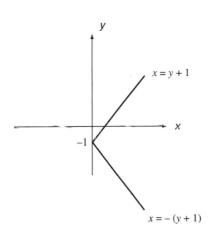

(f) $y = -3x, \qquad x < -1$
      $y = 2 - x, -1 < x < 0$
      $ = 2 + x, \qquad 0 < x < 1$
      $ = 3x, \qquad x > 1$

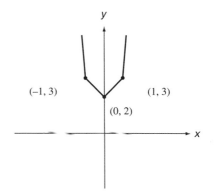

5

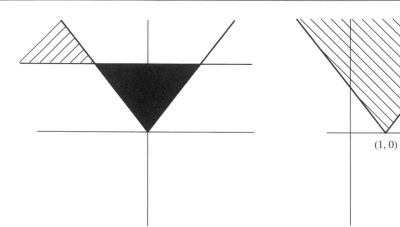

(a)   (b)   (c)   (d)

6   (a)   (b)

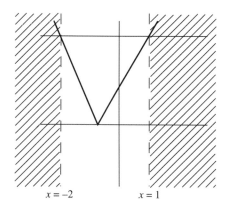 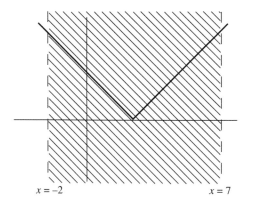

7   (a) $x < 4$ (b) $-2 < x < 6$ (c) $-\dfrac{3}{2} < x < \dfrac{7}{4}$

      (d) $x < 0$ (e) $-\dfrac{1}{2} < x < 3$ (f) $x < -\dfrac{2}{3}, x > 6$

8 Areas are shaded.

(a)

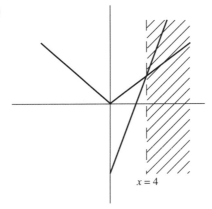

$x = 4$

(b)

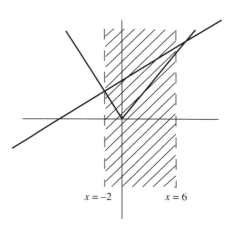

$x = -2$ $x = 6$

(c)

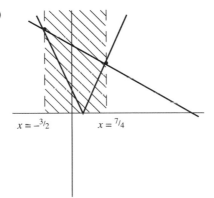

$x = -^3/_2$ $x = ^7/_4$

(d)

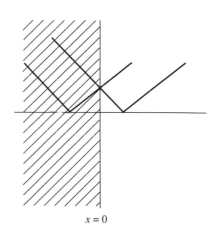

$x = 0$

(e)

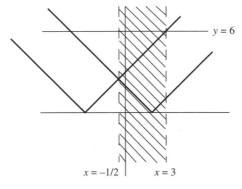

$y = 6$

$x = -1/2$ $x = 3$

(f)

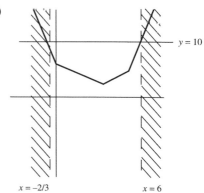

$y = 10$

$x = -2/3$ $x = 6$

Exercise 3.6

1 (a) $\dfrac{y-c}{x}$ against x, $\dfrac{y-c}{x} = ax + b$

   (b) y against $\ln x$, $y = 2\ln x + \ln k$

   (c) $\dfrac{y}{x^2}$ against x^2, $\dfrac{y}{x^2} = a + bx^2$

   (d) $\ln\dfrac{y}{x}$ against x, $\ln\dfrac{y}{x} = bx + \ln a$

   (e) $\dfrac{1}{y}$ against $\dfrac{1}{x^2}$, $\dfrac{1}{y} + \dfrac{1}{x^2} = a$

2 Mean inflation rate $= 5.67\%$; 2040 prices approximately 12 times 1995 prices.

3 $\alpha \simeq -4.0 \times 10^{-4}, \beta \simeq -2.5 \times 10^{-6}$

4 $b \simeq 1.50$

5
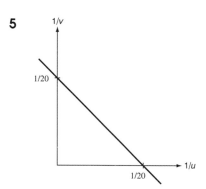

6 Plot $\dfrac{h^2}{|q|}$ against $|q|$.

4 QUADRATICS AND CUBICS

Introduction

There are many relationships in which one variable can be expressed as a power of another. One variable is often expressed as a sum of two or more powers of the other. The simplest examples of this are the quadratic relationship and the cubic relationship. The distance an object travels under a constant acceleration is a quadratic expression in the time of travel. Cubic curves are used in the computer-aided design of roads and three-dimensional structures to achieve a smooth transition between one component and another. The study of quadratics and cubics will provide an introduction to the characteristics of more complicated relationships.

Objectives

After working through this chapter you should be able to

- sketch the graph of a quadratic expression
- relate changes in a quadratic expression to transformations of its graph
- locate the vertex on the graph of a quadratic expression
- state the criterion that determines the number of roots for a quadratic equation
- factorise a quadratic expression and hence solve the related quadratic equation
- express a quadratic expression as the sum or difference of two squares
- use the formula to solve a quadratic equation
- state the formulae for the sum and the product of the roots of a quadratic equation
- sketch the general shape of a cubic expression
- identify local maxima and local minima on the graph of a cubic expression
- recognise a point of inflection on the graph of a cubic expression

4.1 THE QUADRATIC CURVE

The characteristic feature of a straight line is that it has no bending; anyone driving a car in a straight line path does not have to turn the steering wheel. A simple curve which *does* have a sense of bending associated with it is the graph of the relationship $y = x^2$ shown in Figure 4.1(a). The graph can be drawn by using a table of values.

x	-3	-2	-1	0	1	2	3
x^2	9	4	1	0	1	4	9

The data points are plotted and a smooth curve is drawn through them. However, the accuracy achieved by this process depends upon the number and location of the data points and the smoothness of the curve that is drawn.

Two features are noteworthy (refer to Figure 4.1(b)).

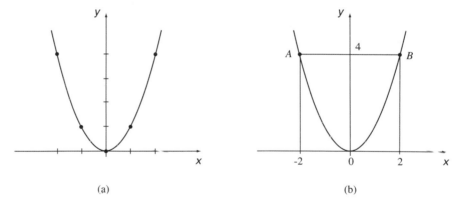

(a)   (b)

Figure 4.1   The graph of $y = x^2$

(i)   The curve has a **vertex** at the origin which is its minimum point.

(ii) The curve is symmetrical about the y-axis; we say it is an **even function**.

Since $(-x_0)^2 = x_0^2$ it follows that the values x_0 and $-x_0$ lead to the same value of y; Figure 4.1(b) shows this with the example $x_0 = 2$. This illustrates the point that a positive number has two square roots; for example, -2 and $+2$ both square to 4, so 4 has the two square roots -2 and 2. The number 0 has the sole square root 0, and since no number squares to a negative number, a negative number has no square roots.

Figure 4.2 shows some examples of the more general relationship $y = ax^2$ where a is a constant. In diagram (a) the constant is positive and the effect of increasing the value of a

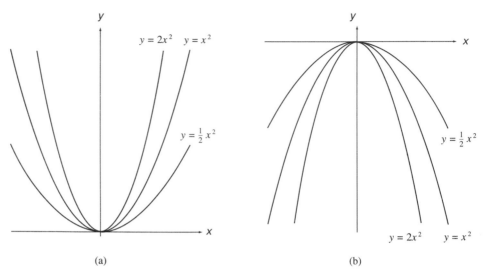

Figure 4.2 The general relationship $y = ax^2$

is to compress or extend the curve horizontally; if $a > 1$ the curve is compressed relative to $y = x^2$, whereas if $a < 1$ the curve is extended relative to $y = x^2$.

In diagram (b) the constant is negative and the curves are 'upside down' with a maximum at the origin. It is generally true that the curve of $y = -ax^2$ is the mirror image in the x-axis of the curve $y = ax^2$.

Figure 4.3 shows two examples of curves having the general form $y = ax^2 + c$ where c is a second constant. Diagram (a) represents the relationship $y = x^2 + 2$ whereas diagram (b) represents $y = x^2 - 1$. Notice that both curves are identical in shape to $y = x^2$ but their

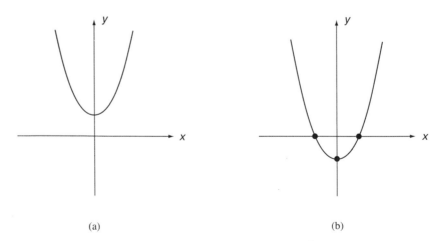

Figure 4.3 The relationship $y = ax^2 + c$

position relative to the x-axis is different. In each case the curves have effectively moved vertically; in (a) the curve has been moved upwards by 2 units whereas in (b) the curve has been moved downwards by 1 unit.

In diagram (a) we see that the curve does not meet the x-axis whereas in diagram (b) the curve **crosses** the x-axis twice, at points A and C. In fact A is at $x = -1$ and C is at $x = 1$; the lowest point B is where $x = 0$ and this value of x is halfway between those at the crossing points. This is always the case where there are two crossing points.

To complete the picture we say that the curve of $y = x^2$ (Figure 4.1(a)) **touches** the x-axis, at the origin.

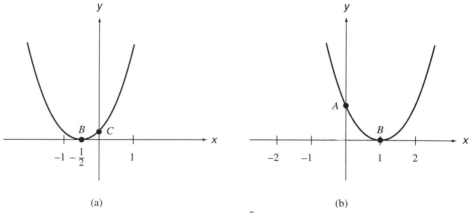

(a)   (b)

Figure 4.4   (a) $y = \left(x + \dfrac{1}{2}\right)^2$ and (b) $y = (x - 1)^2$

Figure 4.4(a) shows the curve of the relationship $y = \left(x + \dfrac{1}{2}\right)^2$. B is the point on the x-axis $(y = 0)$ where $x = -\dfrac{1}{2}$; in effect the curve of $y = x^2$ has been displaced (horizontally) to the left by $\dfrac{1}{2}$ unit. The point C is where the curve cuts the y-axis, i.e. $x = 0$, so $y = \left(\dfrac{1}{2}\right)^2 = \dfrac{1}{4}$. Figure 4.4(b) shows the curve $y = (x - 1)^2$. B is the point on the x-axis where $x = 1$; this time the horizontal shift has been to the right by 1 unit. The point A is where $x = 0$, hence $y = 1^2 = 1$. In each case the curve is symmetrical about a vertical line through the point B.

What happens if we move the curve of $y = x^2$ *both* vertically *and* horizontally? By analogy with the earlier examples we might expect that the curve of the relationship $y = (x - 1)^2 + 2$ could be obtained by displacing the curve of $y = x^2$ to the right by 1 unit then displacing the resulting curve upwards by 2 units.

Figure 4.5(a), plotted from a table of values, shows that our supposition was correct. Figure 4.5(b) shows $y = \left(x - \dfrac{1}{2}\right)^2 - 2\dfrac{1}{4}$, and here the curve of $y = x^2$ has been displaced

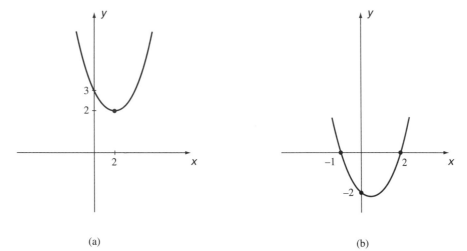

(a)   (b)

Figure 4.5   (a) $y = (x - 1)^2 + 2$ and (b) $y = \left(x - \dfrac{1}{2}\right)^2 - 2\dfrac{1}{4}$

first to the right by $\dfrac{1}{2}$ unit and then downwards by $2\dfrac{1}{4}$ units. When $x = -1$,

$y = \left(-1\dfrac{1}{2}\right)^2 - 2\dfrac{1}{4} = 2\dfrac{1}{4} - 2\dfrac{1}{4} = 0$;   when $x = 2, y = \left(1\dfrac{1}{2}\right)^2 - 2\dfrac{1}{4} = 0$;   and   when

$x = 0, y = \left(-\dfrac{1}{2}\right)^2 - 2\dfrac{1}{4} = \dfrac{1}{4} - 2\dfrac{1}{4} = -2.$

In the case of $y = (x - 1)^2 + 2$ we see that no matter what value we give to x, the expression $(x - 1)^2$ cannot be negative. The smallest value it can take is zero and this occurs when $x = 1$; this is where the value of y is also least, i.e. a minimum, and its value there is 2.

In the case of $y = \left(x - \dfrac{1}{2}\right)^2 - 2\dfrac{1}{4}$ a similar argument leads to the conclusion that y

takes its least value of $-2\dfrac{1}{4}$ when $x = \dfrac{1}{2}$. Note that $x = \dfrac{1}{2}$ is halfway between the values

$x = -1$ and $x = 2$, as we might have expected.

Using the methods of Chapter 2 we can expand the two expressions in x. In the first example

$$y = (x - 1)^2 + 2$$

becomes

$$y = x^2 - 2x + 1 + 2$$

i.e.

$$y = x^2 - 2x + 3.$$

In the second example

$$y = \left(x - \frac{1}{2}\right)^2 - 2\frac{1}{4}$$

becomes

$$y = x^2 - x + \frac{1}{4} - 2\frac{1}{4} = x^2 - x - 2 = (x + 1)(x - 2).$$

The curve crosses the x-axis at $x = -1$ and $x = 2$, as in Figure 4.5(b).

If we are given these forms of the relationships it is not easy to tell that the first curve does not cross the axis whereas the second curve crosses twice, nor is it obvious where these crossing points are. In the next section we discover ways of finding such information.

The general **quadratic expression** has the form

$$ax^2 + bx + c$$

where a, b and c are constants but $a \neq 0$.

The graph of the (quadratic) relationship $y = ax^2 + bx + c$ is called a **quadratic curve**, sometimes a **parabola**.

The curve of such a relationship can be built up from the knowledge we have gained in this section. If $a > 0$ the curve is similar in shape to $y = x^2$; if $a < 0$ it is similar to $y = -x^2$.

Exercise 4.1

1   The following tables represent relationships of the form $y = ax^2$. Write down a for each case.

x	-3	-2	-1	0	1	2	3
(a) y	45	20	5	0	5	20	45
(b) y	-10.8	-4.8	-1.2	0	-1.2	-4.8	-10.8

2   Verify that the curve of the form $y = ax^2$ that passes through the point (2, 12) must be $y = 3x^2$, i.e. $a = 3$. Find the value(s) of

(a) y when $x = -2$ (b) y when $x = 5$

(c) x when $y = 3$ (d) x when $y = \dfrac{1}{3}$.

Sketch the curve.

 3   Find the curves of the form $y = ax^2$ which pass through the following points:

(a) $(-3, 3)$ (b) $(5, 50)$

(c) $(2.3, 6.5)$ (d) $(0, 0)$.

Find the appropriate value of a in each possible case.

4   On the same axes sketch the curves

(a) $y = x^2 + 1$ (b) $y = x^2 - 3$

(c) $y = 2(x^2 + 1)$ (d) $y = -2x^2 + 1$.

5   For each curve in Question 4 write down the coordinates of the points when the axes are crossed. State clearly when an axis is not crossed.

6   Accepting that $y = -2(x + 1)^2 + 3 \equiv -2x^2 - 4x + 1$ is the curve of $y = -2x^2$ moved so that its vertex is $(-1, 3)$, identify the following in equivalent terms:

(a) $3(x - 1)^2 + 4$ (b) $-9(x + 3)^2 + 11$.

7   Prove that $x^2 + 6x + 11 \equiv (x + 3)^2 + 2$. What is the least value that $y = x^2 + 6x + 11$ can have?

8   In the previous question the coefficient of x, i.e. $+6$, was halved to give $+3$. Then the expression $x + 3$ was squared and a constant added to produce an expression equivalent to the original one. Write the following expressions as a squared term plus a constant (the constant may be negative).

(a) $x^2 + 4x + 5$ (b) $x^2 + 4x - 2$

(c) $x^2 + 10x + 10$ (d) $x^2 - 6x + 10$

(e) $-x^2 - 2x + 2$ (f) $3x^2 + 6x + 4$

(g) $x^2 + 3x + 1$ (h) $2x^2 - 8x + 5$.

9   If the curve $y = (x + 3)^2 + 2$ is shaped exactly like the curve $y = x^2$ with its vertex at $(-3, 2)$, write down the minimum value attained in Question 8 parts (a) to (d) and parts (f) to (h) and the maximum value in part (e).

10* For the quadratic curve

$$y = x^2 + px + q$$

repeat the procedures of the previous two questions and identify the criteria p and q must fulfil for the minimum value of y to lie above, on or below the x-axis.

4.2 QUADRATIC EQUATIONS AND ROOTS

The graph of $y = ax^2 + bx + c$ crosses the x-axis where $y = 0$, i.e. where

$$ax^2 + bx + c = 0.$$

This equation is a **quadratic equation** because of the x^2 term; hence a cannot be zero. In Figures 4.1(a) and 4.5 we see that there can be 0, 1 or 2 crossing points. There cannot be any more than two, as we shall discover.

The key to finding the values of x which satisfy a quadratic equation lies in the result that if the product of two numbers is zero, either one or the other or both of them must be zero, i.e.

$$\text{if } ab = 0 \text{ then either } a = 0 \text{ or } b = 0 \text{ or } a = b = 0.$$

Example   Figure 4.6(a) shows the graph of the relationship $y = (x + 1)(x - 2)$. Now $(x + 1)(x - 2) = 0$ where $x + 1 = 0$ or $x - 2 = 0$, i.e. where $x = -1$ or $x = 2$.

However, the curve in Figure 4.6(b) also crosses the x-axis at $x = -1$ and $x = 2$. Its equation can be expressed as $y = (x + 1)(2 - x)$. Note that

$$(x + 1)(2 - x) = 0 \quad \text{where} \quad x + 1 = 0 \quad \text{or} \quad 2 - x = 0$$

which gives the stated values of x.   ■

Note that the curves cross the y-axis at different points. The curve in Figure 4.6(a) crosses where $y = (0 + 1)(0 - 2) = -2$, whereas in Figure 4.6(b) it crosses where $y = (0 + 1)(2 - 0) = 2$. In general, three points are necessary to specify a quadratic curve uniquely. (Remember that only one straight line could be drawn through two given points.)

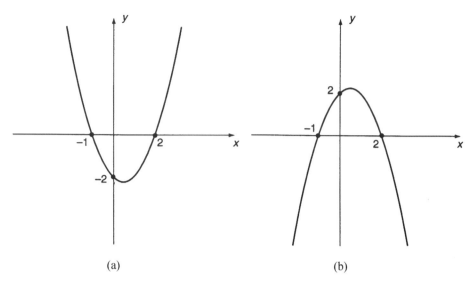

(a)   (b)

Figure 4.6   (a) $y = (x + 1)(x - 2)$ and (b) $y = (x + 1)(2 - x)$

Note also that the curve of $y = a(x + 1)(x - 2)$ always crosses the x-axis at $x = -1$ and $x = 2$, whatever the (non-zero) value of a.

Example   Expand the expressions

(i) $3(x + 1)(2 + x)$

(ii) $-3(1 - x)(2 + x)$.

Where do the graphs of (i) $y = 3(x + 1)(2 + x)$, (ii) $y = -3(1 - x)(2 + x)$ cross the x-axis?

(i) $3(x + 1)(2 + x) \equiv 3(2x + 2 + x^2 + x) \equiv 3(x^2 + 3x + 2) \equiv 3x^2 + 9x + 6$

(ii) $-3(1 - x)(2 + x) \equiv -3(2 - 2x + x - x^2) \equiv -3(-x^2 - x + 2) \equiv 3x^2 + 3x - 6$

Curve (i) crosses the x-axis at $x = -1$ and $x = -2$.
Curve (ii) crosses the x-axis at $x = 1$ and $x = -2$.   ■

If we can factorise a quadratic expression then we can determine where the graph of the corresponding relationship crosses the x-axis.

Examples   1. The expression $2x^2 + 2x - 12$ factorises into the product

$$(2x - 4)(x + 3) \equiv 2(x - 2)(x + 3).$$

Hence the relationship $y = 2x^2 + 2x - 12$ can be rewritten as

$$y = 2(x - 2)(x + 3)$$

and the curve crosses the x-axis at $x = 2$ and $x = -3$.

2. The expression $-18 + 12x - 2x^2$ factorises into the product $-2(x - 3)^2$. Hence the graph of $y = -18 + 12x - 2x^2$ touches the x-axis at $x = 3$.

3. The expression $x^2 + 6x + 10$ does not factorise and the graph of

$$y = x^2 + 6x + 10$$

does not touch or cross the x-axis. ∎

When the factors of a quadratic expression $ax^2 + bx + c$ are easy to determine, we therefore have a simple means of finding the roots of the corresponding quadratic equation

$$ax^2 + bx + c = 0.$$

If the factors are not easy to find, or if there are no factors to find, we need an alternative approach.

Completing the square

Consider the expression $x^2 + 4x + 3$. Take the coefficient of x $(+4)$ and halve it $(+2)$. Consider the expression $(x + 2)^2 \equiv x^2 + 4x + 4$; this is quite close to our given expression. Then write the given expression as $(x + 2)^2$ together with a constant added or subtracted to achieve an identity:

$$x^2 + 4x + 3 = (x + 2)^2 - 1.$$

Now $x^2 + 4x + 3 = 0$ when $(x + 2)^2 - 1 = 0$, so

$$(x + 2)^2 = 1.$$

Taking square roots, we have two options:

$$x + 2 = 1 \quad \text{and} \quad x + 2 = -1$$

giving the solutions $x = -1$ and $x = -3$.

Example

Using the method above find the solutions of the equations

(i) $x^2 - 4x - 5 = 0$

(ii) $x^2 + 4x + 5 = 0$.

Solution

(i) Consider the expression $x^2 - 4x - 5$. Half the coefficient of x is -2. We can rewrite the expression $(x - 2)^2$ as $x^2 - 4x + 4$, so

$$x^2 - 4x - 5 \equiv x^2 - 4x + 4 - 9 \equiv (x - 2)^2 - 9.$$

The equation $x^2 - 4x - 5 = 0$ becomes

$$(x - 2)^2 - 9 = 0 \quad \text{or} \quad (x - 2)^2 = 9 = 3^2.$$

This means that

$$x - 2 = 3 \quad \text{or} \quad x - 2 = -3$$

so $x = 5$ or $x = -1$.

(ii) $x^2 + 4x + 5 \equiv (x + 2)^2 + 1$.

Then $x^2 + 4x + 5 = 0$ becomes

$$(x + 2)^2 = -1.$$

Since no number has a square of -1, the quadratic equation has no solutions.   ■

Formula method

The technique of completing the square is put to more use in deriving a general formula for the solution of quadratic equations.

The equation

$$ax^2 + bx + c = 0$$

can be rewritten by dividing by a (non-zero, remember) as

$$x^2 + \frac{b}{a}x + \frac{c}{a} = 0.$$

Note first that

$$\left(\frac{b}{2a}\right)^2 = \frac{b^2}{4a^2}$$

and that

$$\left(x + \frac{b}{2a}\right)^2 = x^2 + \frac{b}{a}x + \frac{b^2}{4a^2}.$$

Adding $\dfrac{b^2}{4a^2}$ to both sides of the quadratic equation gives

$$x^2 + \frac{b}{a}x + \frac{b^2}{4a^2} + \frac{c}{a} = \frac{b^2}{4a^2}$$

so that

$$\left(x + \frac{b}{2a}\right)^2 + \frac{c}{a} = \frac{b^2}{4a^2}.$$

Hence

$$\left(x + \frac{b}{2a}\right)^2 = \frac{b^2}{4a^2} - \frac{c}{a} = \frac{b^2 - 4ac}{4a^2}.$$

If $b^2 < 4ac$ then we can proceed no further and there are no roots to be found. However, if $b^2 \geq 4ac$ then we take square roots to obtain

$$x + \frac{b}{2a} = \pm\frac{\sqrt{b^2 - 4ac}}{2a}$$

or

$$x = \frac{-b \pm \sqrt{b^2 - 4ac}}{2a}.$$

This means that there are two roots:

$$x_1 = \frac{-b + \sqrt{b^2 - 4ac}}{2a} \quad \text{and} \quad x_2 = \frac{-b - \sqrt{b^2 - 4ac}}{2a}.$$

Thus a quadratic equation can be solved, provided $b^2 \geq 4ac$, to give two possible solutions. The all-important quantity $b^2 - 4ac$ is called the **discriminant**. It is given the symbol Δ (Greek capital D) for short. But Δ must be non-negative ($\Delta \geq 0$) for the formula to yield real values of x. For the time being, refer to cases when Δ is negative as **complex**.

Examples

1. Consider the equation

$$-14x^2 + x + 25 = 0.$$

Change the signs throughout to give

$$14x^2 - x - 25 = 0.$$

Apply the formula

$$x = \frac{1 \pm \sqrt{1 + 1400}}{28},$$

i.e. $x = 1.3725$ (4 d.p.) and $x = -1.3011$ (4 d.p.).

2. We consider the equation

$$14x^2 + 65x - 25 = 0.$$

The formula gives

$$x = \frac{-65 \pm \sqrt{4225 + 1400}}{28}$$

$$= \frac{-65 \pm 75}{28} = -5 \text{ or } \frac{5}{14}.$$

In other words the sledgehammer has cracked the nut but it is worth noticing that Δ is a **perfect square** in this case. In fact this always happens whenever a quadratic factorises. ∎

Example

In the equation

$$9x^2 - 24x + 16 = 0$$

we note that $a = 9, b = -24$ and $c = 16$.
    The formula produces the result

$$x = \frac{24 \pm \sqrt{576 - 576}}{18} = \frac{24 \pm 0}{18} = \frac{24}{18} = \frac{4}{3}$$

we say that $x = \frac{4}{3}$ is a **repeated root**.

In fact the equation could be rewritten as

$$(3x - 4)^2 = 0$$

showing that $x = \frac{4}{3}$ is the only solution. Note that the curve $y = (3x - 4)^2$ touches the

x-axis at $x = \frac{4}{3}$ but does not cross it, since $y \geq 0$ for all values of x.   ■

It is useful to note that the expression $ax^2 + bx + c$ takes its least value when $x = -\dfrac{b}{2a}$

if $a > 0$; if $a < 0$ then the expression takes its greatest value when $x = -\dfrac{b}{2a}$. In either

case the curve $y = ax^2 + bx + c$ is symmetrical about the vertical line $x = -\dfrac{b}{2a}$.

The discriminant Δ provides a ready check on the nature of the roots of a given quadratic
equation.

> If $b^2 < 4ac$ there are no roots, if $b^2 = 4ac$ the root is repeated, and if $b^2 > 4ac$ there
> are two distinct roots.

Example

What is the nature of each root in the following equations?

(i) $4x^2 - 12x + 5 = 0$

(ii) $4x^2 - 12x + 9 = 0$

(iii) $4x^2 - 12x + 15 = 0$

In all cases $a = 4$ and $b = -12$.

Solution

(i) $c = 5$ and $\Delta = 144 - 80 > 0$; two roots $\left(\text{in fact they are } x_1 = \dfrac{1}{2}, x_2 = \dfrac{5}{2}\right)$

(ii) $c = 9$ and $\Delta = 144 - 144 = 0$; repeated root $\left(\text{in fact } x = \dfrac{3}{2}\right)$

(iii) $c = 15$ and $\Delta = 144 - 240 < 0$; no roots.   ■

Roots and coefficients

Suppose that we write a quadratic equation in the form

$$x^2 + \frac{b}{a}x + \frac{c}{a} = 0$$

and suppose that this equation has the solutions $x = \alpha$ and $x = \beta$. Then we know that it can be written as

$$(x - \alpha)(x - \beta) = 0.$$

If we expand the left-hand side of this equation, we obtain

$$x^2 - \alpha x - \beta x + \alpha\beta = 0$$

or

$$x^2 - (\alpha + \beta)x + \alpha\beta = 0.$$

Comparing this equation with the original equation, we see that

$$\alpha + \beta = -\frac{b}{a} \quad \text{and} \quad \alpha\beta = \frac{c}{a}.$$

Examples

1. The equation with solutions $x = 2$ and $x = 3$ can be written as

$$(x - 2)(x - 3) = 0$$
$$x^2 - 5x + 6 = 0.$$

Here $a = 1, b = -5, c = 6$.

The sum of the roots, $\alpha + \beta$, is $-\left(\dfrac{-5}{1}\right) = 5$ and the product of the roots is $\dfrac{6}{1} = 6$, as can easily be seen.

2. The equation $14x^2 + 65x - 25 = 0$ has $a = 14, b = 65, c = -25$. Hence the sum of the roots is $-\dfrac{65}{14}$ and the product of the roots is $-\dfrac{25}{14}$.

We found earlier that the roots were -5 and $\dfrac{5}{14}$; call them α and β respectively. Then

$$\alpha + \beta = -5 + \frac{5}{14} = -\frac{70}{14} + \frac{5}{14} = \frac{65}{14} = -\frac{b}{a}$$

$$\alpha\beta = -5 \times \frac{5}{14} = -\frac{25}{14} = \frac{c}{a}. \qquad \blacksquare$$

We can use the root properties above to determine equations with roots related to those of a given equation.

For example, if the roots of the equation

$$ax^2 + bx + c = 0$$

are denoted α and β, we can find the equation whose roots are 2α and 2β.

This equation can be written as $(x - 2\alpha)(x - 2\beta) = 0$, which can be expanded as

$$x^2 - 2\alpha x - 2\beta x + (2\alpha)(2\beta) = 0$$

i.e. $\qquad x^2 - 2(\alpha + \beta)x + 4\alpha\beta$

We know that $\alpha + \beta = -\dfrac{b}{a}$ and $\alpha\beta = \dfrac{c}{a}$, so the new equation can be rewritten as

$$x^2 + \frac{2b}{a} + \frac{4c}{a} = 0$$

or $ax^2 + 2bx + 4c = 0$.

You could use the formula method to verify that the roots of this equation are indeed as claimed.

Example   If the equation $ax^2 + bx + c = 0$ has roots α and β, find the equation whose roots are α^2 and β^2.

Solution

This equation can be written as

$$(x - \alpha^2)(x - \beta^2) = 0$$

i.e. $\quad x^2 - (\alpha^2 + \beta^2)x + \alpha^2\beta^2 = 0.$

Now

$$\alpha^2\beta^2 = (\alpha\beta)^2 = \left(\frac{c}{a}\right)^2 = \frac{c^2}{a^2}.$$

Also, since $(\alpha + \beta)^2 \equiv \alpha^2 + \beta^2 + 2\alpha\beta$, it follows that

$$\alpha^2 + \beta^2 \equiv (\alpha + \beta)^2 - 2\alpha\beta$$

$$= \left(-\frac{b}{a}\right)^2 - 2\frac{c}{a}$$

$$= \frac{b^2 - 2ac}{a^2}.$$

The equation we seek is therefore

$$x^2 - \frac{(b^2 - 2ac)}{a^2}x + \frac{c^2}{a^2} = 0$$

or

$$a^2x^2 - (b^2 - 2ac)x + c^2 = 0. \qquad \blacksquare$$

You could use this approach to find the equation whose roots are 2 and 3, predict the equation whose roots are 4 and 9 and verify directly that this is so.

Exercise 4.2

1 Expand the following:

(a) $(x - 5)(x - 15)$ (b) $(2x + 9)(3x - 5)$

(c) $-4(x - 3)(2x + 5)$ (d) $(x - 5)(2x + 3)(3x - 7).$

2 Factorise the following quadratics or conclude that there are no simple factors.

(a) $\quad 2x^2 + 5x - 42$ (b) $\quad 3x^2 + x + 9$

(c) $4x^2 - 15x - 4$ (d) $6x^2 - 53x - 70$

(e) $x^2 - x + 16$ (f) $x^2 - 11x + 30$

(g) $x^2 - 11x + 31$ (h) $3x^2 + 18x + 30$

(i) $3x^2 + 18x + 27$ (j) $3x^2 + 18x + 24$

3 Consider $y = x^2 + 8x + q$ for $q = 12$ (1)17, i.e. q in steps of 1 from 12 to 17.

(a) Sketch the curves for the cases $q = 12, 15, 16$, where factors can be easily found. Where is the x-axis crossed or touched?

(b) For the non-factorising cases $q = 13, 14$ verify by sketching that the x-axis is still crossed.

(c) Establish that the graph cannot touch or cross the x-axis when $q > 16$.

4 By halving the coefficient of x and completing the square, solve the following equations:

(a) $x^2 + 5x + 6 - 0$ (b) $x^2 - x - 30 = 0$

(c) $x^2 + 17x + 72 = 0$ (d) $x^2 - 42x + 440 = 0.$

With the following quadratic equations, divide the first by the coefficient of x^2 then proceed as before.

(e) $6x^2 + x - 2$ (f) $2x^2 - 3x - 20$

(g) $11x^2 + 20x + 9$ (h) $10x^2 + 3x - 18.$

 5 Use the formula to determine the roots of the following quadratic equations, if they exist. If no roots exist, say so and give the value of the discriminant.

(a) $x^2 - 13x - 21 = 0$ (b) $x^2 - 13x + 21 = 0$

(c) $x^2 + 5x + 60 = 0$ (d) $3x^2 - 11x + 17 = 0$

(e) $16 - 19y - 5y^2 = 0$ (f) $17x^2 + 6xy - 9y^2 = 0$
(solve for the ratio x/y)

In the following case find z^2 then take the square root as appropriate:

(g) $z^4 - 7z^2 + 1 = 0$

Now do the same for w^3 and take the cube root:

(h) $w^6 - 6w^3 + 3 = 0$

 6 Factorise the following quadratic expressions, noting that in the corresponding quadratic equation Δ is a perfect square.

(a) $187x^2 + 141x - 76$ (b) $345x^2 - 412x - 29$

(c) $10\ 807x^2 - 113x - 11\ 124$ (d) $189x^2 - 386x + 65$

(e) $-71x^2 + 2632xy - 185y^2$

7   When the quadratic equation $ax^2 + bx + c = 0$ has real roots, α and β, we can write $ax^2 + bx + c \equiv a(x - \alpha)(x - \beta) = 0$. When α and β are surds, instead of simple fractions, the expression $a(x - \alpha)(x - \beta)$ is usually called the **synthetic factorisation** of $ax^2 + bx + c$. Find synthetic factorisations of the following:

(a) $x^2 - 91x + 20$ (b) $2x^2 - 240xy + 3y^2$

(c) $z^3 - 5z^2 + 2z + 2$, noting that $(z - 1)$ is a factor.

8   For the following equations, without solving, write down the sum of the roots, the product of the roots and the sum of the squares of the roots.

(a) $x^2 - 12x - 4 = 0$ (b) $2x^2 + x - 19 = 0$

(c) $23 - 5x - 4x^2 = 0$ (d)* $\pi - (\pi^2 + 1)x - x^2/\pi = 0.$

9   Write down the quadratic equations whose roots are respectively double those in each part of Question 8.

10* If the quadratic equation $ax^2 + bx + c = 0$ has roots α and β, show that the equation whose roots are 3α and 3β is

$$ax^2 + 3bx + 9c = 0.$$

What is the form of the quadratic equation whose roots are $k\alpha$ and $k\beta$, $k \neq 0$?

11* The quadratic equation $ax^2 + bx + c = 0$ has roots α and β, neither of which is zero. Determine the quadratic equation whose roots are $1/\alpha$ and $1/\beta$. Verify the result by comparing the quadratic whose roots are 3 and 4 with the quadratic whose roots are $\frac{1}{3}$ and $\frac{1}{4}$.

12* Following from Questions 10 and 11, determine the quadratic equations whose roots are

(a) α/β and β/α (b) $1/\alpha^2$ and $1/\beta^2$.

4.3 COMMON PROBLEMS INVOLVING QUADRATICS

Quadratic equations with one meaningful solution

Even when a quadratic equation has two real roots, only one of them may make sense in the context of the problem which gave rise to the equation.

Example The length of a rectangle is 2 m greater than its width. If the area of the rectangle is $8\,\text{m}^2$ find the lengths of the sides.

Solution

If we let the width of the rectangle be x m then the length is $(x+2)$ m. Hence the area is $x(x+2)\,\text{m}^2$ and therefore

$$x(x+2) = 8$$

i.e. $\qquad x^2 + 2x - 8 = 0$

or $\qquad (x+4)(x-2) = 0.$

The roots are $x = -4$ and $x = 2$. However, $x = -4$ makes no sense in the context of a length, so the only meaningful solution is $x = 2$. Hence the width of the rectangle is 2 m and its length 4 m. ∎

Simultaneous equations

If we have two equations in two unknowns, x and y, we may eliminate one of the variables by substitution. In the next example the resulting equation in the other unknown is a quadratic.

Example Solve the equations

$$x + y = 9 \tag{1}$$
$$xy = 18. \tag{2}$$

Solution

Rearranging (1) we obtain

$$y = 9 - x$$

and substituting this into (2) gives

$$x(9 - x) = 18$$

so that

$$9x - x^2 = 18$$
or $$x^2 - 9x + 18 = 0.$$

Factorising the left-hand side gives

$$(x - 3)(x - 6) = 0.$$

The solutions are $x = 3$ and $x = 6$.
When $x = 3$, $y = 6$ and when $x = 6$, $y = 3$, as can be found using (1) or (2).
Whichever equation we use to find y, we use the other equation to verify the values. ∎

Note that if the problem had been to find two numbers of sum 9 and product 18, there would have been only one answer namely, the numbers 3 and 6.

A geometric problem which gives rise to the same equations is to find the dimensions of a rectangle whose area is $18\,\text{m}^2$ and whose perimeter is 18 m. If adjacent sides have lengths x m and y m, the perimeter is $(2x + 2y)\,\text{m}$ and $x + y = 9$; the area is $xy\,\text{m}^2$ therefore $xy = 18$. Refer to Figure 4.7.

Figure 4.7   Finding the side lengths of a rectangle

Examples   1. To find the dimensions of a rectangle whose area is $36\,\text{m}^2$ and whose perimeter is 24 m, we proceed as above and obtain the equations

$$xy = 36 \qquad (1)$$
$$x + y = 12. \qquad (2)$$

Rearranging (2) to $y = 12 - x$ and substituting in (1) gives

$$x(12 - x) = 36$$

i.e. $\qquad x^2 - 12x + 36 = 0$

or $\qquad (x - 6)^2 = 0.$

The only solution is $x = 6$, and we can readily see that $y = 6$ also. Hence the rectangle is in fact a square of side 6 m.

2. To find the dimensions of a rectangle whose area is 21 m^2 and whose perimeter is 18 m, we follow the now familiar steps and obtain the equations

$$xy = 21 \qquad\qquad\qquad (1)$$
$$x + y = 9. \qquad\qquad\qquad (2)$$

Hence $\qquad y = 9 - x$

and $\qquad x(9 - x) = 21$

or $\quad x^2 - 9x + 21 = 0.$

This equation has no real solutions; the discriminant

$$\Delta = 9^2 - 4 \times 21 = 81 - 84 < 0.$$

No such rectangle exists. ∎

Finally, we use the result that a quadratic expression $ax^2 + bx + c$ achieves its least value if $a > 0$, or its greatest value if $a < 0$, when $x = -\dfrac{b}{2a}$. This result follows because

$$ax^2 + bx + c \equiv a\left\{x^2 + \frac{b}{a}x + \frac{c}{a}\right\}$$
$$\equiv a\left\{\left(x + \frac{b}{2a}\right)^2 + \frac{c}{a} - \frac{b^2}{4a^2}\right\}.$$

The least value the expression $\left(x + \dfrac{b}{2a}\right)^2$ can take is zero and this occurs when $x = -\dfrac{b}{2a}$. Then the least possible value for the expression within braces {} is $\left\{\dfrac{c}{a} - \dfrac{b^2}{4a^2}\right\}$. For any other value of x we would add something on and make it larger. At this special value of x, we have

$$ax^2 + bx + c \equiv a\left\{\frac{c}{a} - \frac{b^2}{4a^2}\right\} \equiv c - \frac{b^2}{4a} = \frac{4ac - b^2}{4a}.$$

If $a > 0$ this value represents the least value of the original quadratic expression, since the corresponding quadratic curve is $\cup$-shaped.

If $a < 0$ then this value is the greatest value of the original quadratic expression, since the curve is $\cap$-shaped.

If we return to the quadratic formula

$$x = \frac{-b \pm \sqrt{b^2 - 4ac}}{2a}$$

we see that $x = -\dfrac{b}{2a}$ when $b^2 = 4ac$, which is the condition for the equation to have equal roots.

Example

What is the maximum area that can be enclosed by a rectangle whose perimeter is 18 m?

Solution

If, as usual, we let the lengths of the sides by x m and y m then

$$x + y = 9.$$

Suppose that the area is A m^2 then

$$xy = A$$

and the elimination of y from these two equations leads to the quadratic equation

$$x(9 - x) = A.$$

Now the expression

$$x(9 - x) \equiv 9x - x^2$$

is a quadratic for which the coefficient of x^2 (denoted a) is negative. Here $a = -1, b = 9$ and the greatest value of the expression occurs when $x = -\dfrac{b}{2a} = \dfrac{9}{2} = 4.5$. Hence $y = 9 - 4.5 = 4.5$.

The maximum value of $xy = (4.5)(4.5) = 20.25$. Therefore the rectangle of maximum area is a square of side 4.5 m and the maximum area is 20.25 m^2. ∎

In general, two positive numbers whose sum is fixed have a maximum product when they are equal. Notice that the two expressions $x + y$ and xy do not really change their values if x and y are interchanged because

$$x + y = y + x \quad \text{and} \quad xy = yx.$$

They are said to be **symmetrical** in x and y. If one of these expressions has a fixed value it follows that the other has a maximum or a minimum value when $x = y$.

Intersections of a line with a parabola

When a straight line intersects a parabola the coordinates of the points of intersection can be found by solving a quadratic equation.

Example
Find the points of intersection of the straight line $y = x + 1$ and the parabola $y = x^2 - x - 2$.

Solution

Where the curves meet, the value of x gives the same value of y from both formulae. Hence

$$x^2 - x - 2 = x + 1$$
i.e. $\quad x^2 - 2x - 3 = 0$
or $\quad (x + 1)(x - 3) = 0$

hence $x = -1$ and $x = 3$ are solutions.
When $x = -1, y = 0$ and when $x - 3, y = 4$; see Figure 4.8(a). ∎

Note that if the resulting quadratic equation has equal roots when the line **touches** the parabola and is said to be a **tangent** to the parabola at the point of intersection. In Figure 4.8(b) the line $y = x - 4$ meets the parabola $y = x^2 - 3x$ at $x = 2$. If the quadratic equation has no real roots then the line does not meet the parabola at all. Sometimes the equation found by solving simultaneously the equations of two parabolas reduces to a linear equation (no x^2 terms).

Example
Where do the parabolas $y = x^2 - 2x$ and $y = x^2 + 4$ meet?

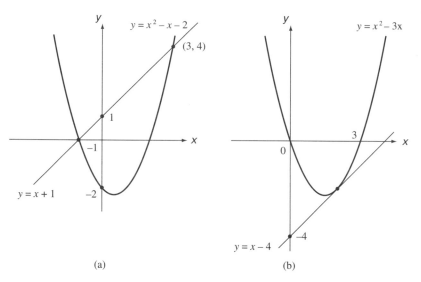

Figure 4.8   Intersections of a line with a parabola

Solution

Solving the equations simultaneously we obtain

$$x^2 + 4 = x^2 - 2x$$

i.e. $2x = -4$

or $x = -2.$

When $x = -2, y = 8$. Hence the parabolas meet at the point $(-2, 8)$.   ■

Quadratic inequalities

Consider the inequality $y > x^2 + 4$. We know that $y = x^2 + 4$ is the equation of a parabola. The points for which $y > x^2 + 4$ lie in the shaded region of Figure 4.9(a). The region below the curve is where $y < x^2 + 4$.

To unravel a quadratic inequality it is often useful to solve a related quadratic equation.

Example   For which values of x is $x^2 - x - 2 < 0$?

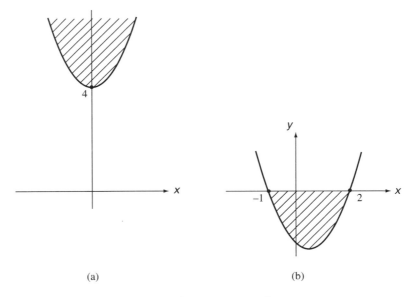

(a)   (b)

Figure 4.9   (a) $y > x^2 + 4$ and (b) $x^2 - x - 2 < 0$

Solution

First we solve the quadratic equation

$$x^2 - x - 2 = 0.$$

The left-hand side will factorise, so we may write

$$(x - 2)(x + 1) = 0.$$

Hence $x = 2$ or $x = -1$.

From our knowledge of the shape of the curve $y = x^2 - x - 2$ we can state that the required interval is $-1 < x < 2$. Refer to Figure 4.9(b). ∎

Sometimes we can use the diagram to lead us towards the solution. As with straight line inequalities, the curve separates the x–y plane into two regions of opposite inequality.

Example   For which values of x is $x + 1 < x^2 - x - 2$?

Solution

Consider first the equation

$$x + 1 = x^2 - x - 2.$$

We have seen in the example on page 157 that the solutions are $x = -1$ and $x = 3$. Reference to Figure 4.8(a) shows that the two regions are $x < -1$ and $x > 3$. ■

The technique of completing the square can be used to verify some inequalities.

Example   Show that for all values of x

$$x^2 + 2x + 3 > 0.$$

Solution

Since

$$x^2 + 2x + 3 \equiv (x + 1)^2 + 2$$

then the least value that the quadratic expression can take is 2, when $x = -1$. Since $2 > 0$ it is clear that the expression is always greater than zero. ■

Exercise 4.3

1 A rectangle of area $154 \, \text{m}^2$ has one side 3 m longer than the other. What are the lengths of the sides?

2 A picture frame is designed so the ratio of length l to breadth b is the same as the ratio of length plus breadth to length, i.e.

$$\frac{l}{b} = \frac{l + b}{l}.$$

Let $x = l/b$ and find the value of this ratio; it is called the *golden ratio*.

3 Solve the simultaneous equations

$$x + y = 20, \qquad xy = 91.$$

Why are the solutions for x and y interchangeable?

4 A positive number x and a negative number y add together to produce the number 6. When multiplied their product is -160. What are x and y?

5 A medieval German riddle was to find two numbers whose sum is 12 and whose product is 40. By obtaining a quadratic equation prove that no real numbers can satisfy the requirement.

6 Solve the simultaneous equations

$$x + 2y = 5, \quad x^2 + 2xy - y^2 = 1.$$

7

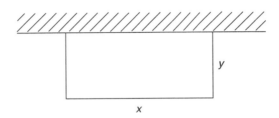

A rectangular enclosure of sides x m and y m, as shown in the diagram, is being constructed from 20 m of fencing adjacent to an existing garden wall which forms one side. If the area enclosed is to be 48 m^2 what are x and y?

8 In an optics problem to determine the focal length f of a lens, the following equation arises:

$$f(f - 2) = f^2 + 10.$$

(a)   Show that $f = -5$.

However, in the original calculation the parameter $x = 1/f$ was used, having emerged from the lens equation.

(b)   Divide the original equation by f^2, since $f \neq 0$, and obtain the two solutions for x. What do you conclude for f?

9 When heated, a metal rod expands according to the excess temperature T, which for values of $T > 40$ satisfies a quadratic equation of the form

$$10^{-5}T^2 + 10^{-2}T = K.$$

Determine T when $K = 1$.

10   For the general quadratic $ax^2 + bx + c$ establish the identity

$$ax^2 + bx + c \equiv a\left(\left(x + \frac{b}{2a}\right)^2 + \frac{(4ac - b^2)}{4a^2}\right).$$

11   What is the minimum value of the expression

$$\left(x + \frac{b}{2a}\right)^2 + \frac{(4ac - b^2)}{4a^2}$$

and what conditions must be fulfilled for it to be positive, zero or negative? For what value of x is it a minimum?

12   For the following quadratics identify whether their graphs point up or down and find the coordinates of their maximum or minimum. In each case draw a rough sketch but do not plot the graphs.

(a) $-x^2 + 2x + 7$ (b) $3x^2 - 5x + 4$

(c) $-19x^2 + 21x + 45$ (d) $5.612x^2 - 23.125x + 79.322.$

13*   Given the following choices A to F for the maximum or minimum of the general quadratic $ax^2 + bx + c$, decide whether the equation $ax^2 + bx + c = 0$ has real distinct roots, real equal roots, or no real roots.

	Positive	Zero	Negative
Maximum	A	B	C
Minimum	D	E	F

14   The graph of $y = 6 - x$ and $y = x^2 + x - 2$ is drawn below.

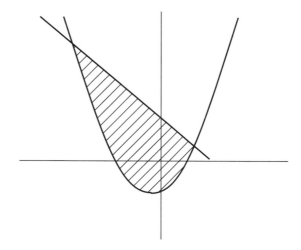

(a)   Determine the coordinates of their intersection points.

(b)   What inequality holds in the shaded region?

(c)   Where does the curve have its minimum value?

15 The straight line $y = x + \dfrac{3}{4}$ is a tangent to the parabola $y = x^2 + 1$. Via a sketch determine the point of contact.

16 Where do the parabolas $y = 3x^2 - 5x - 2$ and $y = -x^2 - 4x + 12$ meet? Draw a sketch and shade the region of points (x, y) which satisfy the inequality

$$-x^2 - 4x + 12 < y < 3x^2 - 5x - 2.$$

17 The parabola whose equation is $y = ax^2 + bx + c$ passes through the points $(0, 4)$, $(1, 0)$ and $(6, 10)$. Substitute for x and y and obtain three linear equations for a, b and c. Solve them and determine the equation of the parabola. Draw a sketch.

4.4 FEATURES OF CUBICS

An expression of the form

$$ax^3 + bx^2 + cx + d$$

where a, b, c and d are constants is a **cubic** expression. The simplest examples are x^3 and $-x^3$. Figure 4.10 shows the graphs of $y = x^3$ and $y = -x^3$. They could be plotted from a table of values.

The curves here are related to each other in that reflecting the curve of Figure 4.10(a) in *either* the x-axis or the y-axis gives the curve of Figure 4.10(b). Note also that $(-x)^3 \equiv -x^3$. Notice that there is no minimum or maximum. The values of y steadily increase left to right along the curve $y = x^3$ whereas the values of y steadily decrease left to right along the curve $y = -x^3$. Anyone driving along a road shaped like $y = x^3$ from the south-west would experience a change in the direction of bending at the origin; the car would have to be changed from turning to the right to turning to the left as it went through the origin; the reverse would happen for the case $y = -x^3$. The feature at the origin is called a **point of inflection**.

Unfortunately, these curves are *not* typical of cubics in general in the way that $y = x^2$ and $y = -x^2$ *were* typical quadratics. However, the following rule applies:

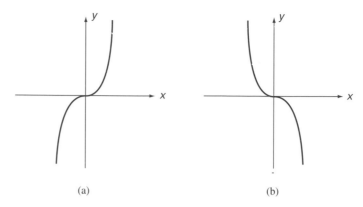

Figure 4.10 (a) $y = x^3$ and (b) $y = -x^3$

All cubic curves with $a > 0$ start in the south-west quadrant and end in the north-east quadrant, whereas those with $a < 0$ start in the north-west quadrant and end in the south-east quadrant.

Before we look at graphs of more general cubics there are some features of the curve $y = x^3$ that we should point out. First, there is only *one* value of x which corresponds to each value of y, i.e. cube roots are unique. For example, if $y = -8$ then $x^3 = -8$ and the only solution is $x = \sqrt[3]{-8}$, i.e. $x = -2$.

First we make a comparison between the curves of $y = x^2$ and $y = x^3$, as in Figure 4.11. Notice that whereas $y = x^2$ is symmetrical about the y-axis, $y = x^3$ is not. At the

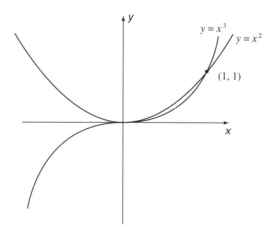

Figure 4.11 Comparing $y = x^2$ and $y = x^3$

origin the two curves meet and as x increases the curve $y = x^2$ increases more rapidly in the early stages, then the curve $y = x^3$ begins to 'catch up' and the two curves meet again at $x = 1$ ($1^2 = 1$ and $1^3 = 1$). For $x > 1$ the cubic curve begins to 'accelerate away' and the gap between the two curves widens. For negative values of x, x^3 is negative whereas x^2 is positive.

Certain cubic curves can be obtained from the curve $y = x^3$ by horizontal or vertical translation. Consider the curves shown in Figure 4.12. In diagram (a) the graph is $y = x^3 - 1$; the curve of $y = x^3$ has been displaced downwards by 1 unit. When $x = 0, y = -1$ and when $x = 1, y = 1 - 1 = 0$. In diagram (b) the graph is $y = (x - 1)^3$; the curve $y = x^3$ has been displaced to the right by 1 unit. When $x = 0, y = (-1)^3 = -1$ and when $x = 1, y = 0^3 = 0$.

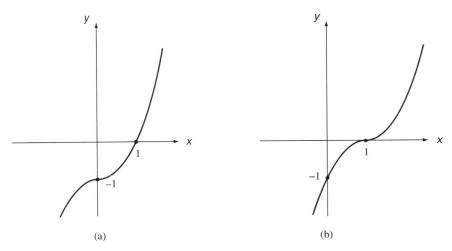

(a)   (b)

Figure 4.12   (a) $y = x^3 - 1$ and (b) $y = (x - 1)^3$

Although both curves cut the axes at the same points, they are quite different from each other. In general it is necessary to specify *four* points through which the cubic curve $y = ax^3 + bx^2 + cx + d$ passes in order to determine the four coefficients a, b, c, d.

Example

We sketch the curve $y = (x - 1)^3 + 2$.

Note that when $x = 0, y = (-1)^3 + 2 = -1 + 2 = 1$ and that when $x = 1, y = 0^3 + 2 = 2$.

Using the ideas above we can obtain the curve from $y = x^3$ by moving it 1 unit to the right then 2 units upwards; see Figure 4.13.

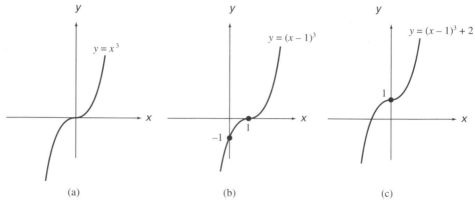

Figure 4.13   (a) $y = x^3$, (b) $y = (x - 1)^3$ and (c) $y = (x - 1)^3 + 2$

Note that $y = 0$ when $(x - 1)^3 = -2$, so

$$(x - 1) = \sqrt[3]{(-2)} \equiv -1.2599 \text{ (4 d.p.)}.$$

Hence $x = 1 + \sqrt[3]{(-2)} \equiv -0.2599$ (4 d.p.).   ∎

Cubic curves in general

Remember that the general cubic curve is $y = ax^3 + bx^2 + cx + d$. Figure 4.14(a) shows the curve

$$y = (x + 1)(x - 1)(x - 2).$$

When $x = -1, x = 1$ or $x = 2$ then $y = 0$ and these are the only values of x for which the value of y does equal zero. When $x = 0, y = (1) \times (-1) \times (-2) = 2$.

When travelling along the curve from left to right we notice first a **local maximum** just before $x = 0$ then a **point of inflection** $\left(\text{in fact at } x = +\dfrac{2}{3} \right)$ than a **local minimum** between $x = 1$ and $x = 2$. These features, in that order, are typical for a cubic curve with $a > 0$.

Figure 4.14(b) shows the curve

$$y = (x + 2)(x - 2)(1 - x).$$

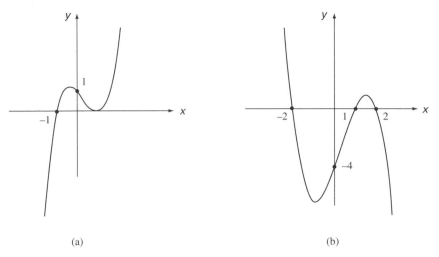

(a)   (b)

Figure 4.14   (a) $y = (x + 1)(x - 1)(x - 2)$ and (b) $y = (x + 2)(x - 2)(1 - x)$

When $x = -2$, $x = 2$ and $x = 1$, then and only then does $y = 0$; when $x = 0, y = (2) \times (-2) \times (1) = -4$.

When travelling along the curve from left to right we encounter first a local minimum then a point of inflection, then a local maximum. This order is typical for a cubic curve with $a < 0$. How to locate precisely for which values of x these features occur is dealt with in Chapter 11.

The expanded formulae for the curves are respectively $y = x^3 - 2x^2 - x + 2$ and $y = -x^3 + x^2 + 4x - 4$. This form makes it difficult to determine where the curves cross the x-axis. But to factorise the cubic expressions into their original forms would not be simple either (if we did not already know the answers). In general, cubic curves do not readily yield their features without the use of the calculus-based methods of Chapter 11.

How many times does a cubic curve cross the x-axis? Figure 4.15(a) shows the curves $y = x^3 - 2x^2 - x + 3$ and $y = x^3 - 2x^2 - x - 1$. Each curve crosses the x-axis once only; they have been obtained by displacing the curve of Figure 4.14(a) vertically upwards by 1 unit and vertically downwards by 3 units respectively.

The curves do not appear to be a constant distance apart, bunching together as x becomes more negative and as x becomes more positive. However, the vertical gap between them remains constant as x varies. (Plot some points and see for yourself.)

Figure 4.15(b) shows another possibility; on this occasion there is one crossing point and one point at which the curve touches the x-axis. In fact this curve is

$$y = (x - 1)^2(x + 1)$$

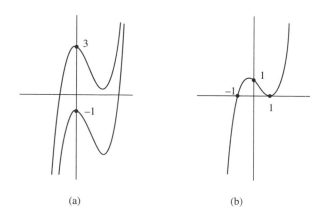

(a)   (b)

Figure 4.15 (a) $y = x^3 - 2x^2 - x + 3$ (b) $y = (x-1)^2(x+1)$

In summary a cubic curve can either (i) cross the x-axis three times, (ii) cross the x-axis only once, or (iii) cross the x-axis once and touch the x-axis once.

Exercise 4.4

1 Sketch the following curves on the same set of axes:

(a) $y = -x^3$ (b) $y = 1 + (2 - x)^3$
(c) $y = -(x+3)^3 - \sqrt{2}$.

What can you say about the curves?

2 Determine the unique solutions of the following equations:

(a) $3(x-2)^3 + 2 - 0$ (b) $9(5+3x)^3 - 11 = 0$

and find values of x to satisfy

(c) $(1+x^2)^3 = \dfrac{27}{8}$ (d) $25 - 10x + x^2 = \dfrac{64}{5-x}$.

3* Which of the following are true or false about cubic curves of the form $y = ax^3 + bx^2 + cx + d$?

(a) A cubic must always have at least one real root, i.e. the x-axis crossed at least once.
(b) A cubic has three roots.

(c)   A cubic has two roots only when one of them coincides with a tangent point on the x-axis.

(d)   Cubic curves always have individual maximum and minimum points.

4   Sketch the curves

(a) $y = (x + 1)(x + 2)(x - 1)$ (b) $y = -x(x - 3)^2$.

5   On the same axes plot the graphs of $y = x^2, x^3$ and x^4 for the values $-2 \le x \le 2$.

6   A cubic expression is best computed using **nested multiplication**. This reduces to a minimum the number of arithmetic operations required. As an example, $3x^3 - 2x^2 + 11x + 50 \equiv [(3x - 2)x + 11]x + 50$, as you can check. Use nested multiplication to complete the table below.

x	-2	-1	0	1	2
y					

7   When x is large (positive or negative) the highest power term in a cubic expression predominates. Illustrate this by finding the ratio $\dfrac{y}{3x^3}$ in Question 6 for the following values of x:

$$x = \pm 10, \pm 10^2, \pm 10^3.$$

8   Given that $y - x^3 - 2x^2 + ax + b = 0$ when $x = 1$ and $x = 2$, what are the values of a and b?

9   Given the identity

$$(x - 3)^3 - (x - 4)^3 \equiv 3x^2 + cx + d$$

select values of x to eliminate one or other of the terms on the left-hand side and hence find c and d.

10*   On the same axes, sketch the cubics

$$y = x(x + 1)(x - 2) \quad \text{and} \quad y = (x + 2)(x - 1)(x - 4).$$

Determine their two intersection points and express the region enclosed between the curves in terms of an inequality.

SUMMARY

- **Quadratic expression**: an expression $ax^2 + bx + c$; a, b, c constant. Its graph can be obtained from that of x^2 by a combination of scaling, translation and (if $a < 0$) reflection in the x-axis.

- **Quadratic curve**: has its vertex at $x = -\dfrac{b}{2a}$; this is the lowest point if $a > 0$, the highest point if $a < 0$.

- **Quadratic equation**: $ax^2 + bx + c = 0$ has no roots if $b^2 - 4ac < 0$, a repeated root if $b^2 - 4ac = 0$ and two roots if $b^2 - 4ac > 0$.

- **Solution of a quadratic equation**: the main methods are as follows.

Factorisation

$$x^2 + bx + c \equiv (x - \alpha)(x - \beta), \quad \text{so the roots are } x = \alpha, x = \beta.$$

Completing the square

$$x^2 + bx + c + \left(x + \frac{b}{2}\right)^2 + \frac{b^2}{4} + c.$$

Formula

$$x = \frac{-b \pm \sqrt{b^2 - 4ac}}{2a}.$$

- **Sum and product of roots**: for the quadratic equation $ax^2 + bx + c = 0$ we have:

Sum of roots $\qquad \alpha + \beta = \dfrac{-b}{a}$

Product of roots $\qquad \alpha\beta = \dfrac{c}{a}$.

- **Cubic expression**: an expression of the form $ax^3 + bx^2 + cx + d$.

- **Cubic graphs**: generally unlike the graph of x^3.

> If $a > 0$ the graph starts in the south-west quadrant and ends in the north-east quadrant.

> If $a < 0$ the graph starts in the north-west quadrant and ends in the south-east quadrant.

- **General cubic graph**: travelling from left to right we meet the following sequences of points.

$a > 0$	$a < 0$
local maximum	local minimum
point of inflection	point of inflection
local minimum	local maximum

Answers

Exercise 4.1

1 (a) 5   (b) -1.2

2 (a) 12 (b) 75 (c) 1 or -1 (d) $\frac{1}{3}$ or $-\frac{1}{3}$

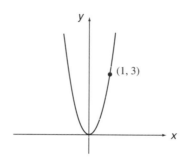

3 (a) $\frac{1}{3}$ (b) 2 (c) 1.2287 (to 4 d.p.) (d) Any a

4

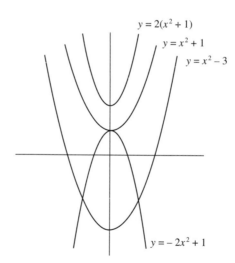

$y = 2(x^2 + 1)$

$y = x^2 + 1$

$y = x^2 - 3$

$y = -2x^2 + 1$

5 (a) (0, 1); x-axis not crossed   (b) $(0, -3)$, $(\pm\sqrt{3}, 0)$

   (c) (0, 2) only; x-axis not crossed   (d) $(0, 1)$, $(\pm 1/\sqrt{2}, 0)$

6 (a) $y = 3x^2$, centre (1, 4)   (b) $y = -9x^2$, centre $(-3, 11)$

7 2.

8 (a) $(x+2)^2 + 1$ (b) $(x+2)^2 - 6$

(c) $(x+5)^2 - 15$ (d) $(x-3)^2 + 1$

(e) $-(x+1)^2 + 3$ (f) $3(x+1)^2 + 1$

(g) $(x+3/2)^2 - \dfrac{5}{4}$ (h) $2(x-2)^2 - 3$

9 (a) 1   (b) -6 (c) -15 (d) 1

(e) 3   (f) 1   (g) $-\dfrac{5}{4}$ (h) -3

10 Centre $\left(-\dfrac{p}{2}, q - \dfrac{p^2}{4}\right)$

The minimum is above, on or below the axis according to $\Delta > 0, = 0, < 0$ where $\Delta = q - p^2/4$.

Exercise 4.2

1 (a) $x^2 - 20x + 75$ (b) $6x^2 + 17x - 45$

(c) $-8x^2 + 4x + 60$ (d) $6x^3 - 35x^2 + 4x + 105$

2 (a) $(x+6)(2x-7)$ (b) no simple factors

(c) $(x-4)(4x+1)$ (d) $(x-10)(6x+7)$

(e) no simple factors   (f) $(x-6)(x-5)$

(g) no simple factors   (h) $3(x^2+6x+10)$ i.e. no simple factors

(i) $3(x+3)^2$ (j) $3(x+4)(x+2)$

3

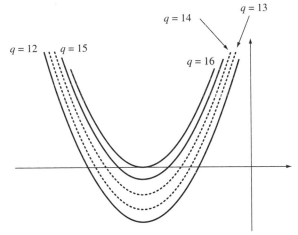

Graphs of $y = x^2 + 8x + q$

$q = 12$ crosses x-axis at $x = -2, -6$

$q = 15$ cross x-axis at $x = -3, -5$

$q = 16$ touches x-axis at $x = -4$

$x^2 + 8x + q \equiv (x + 4)^2 + (q - 16) > 0$ when $q > 16$ for all x

4 (a) $-2, -3$ (b) $-5, 6$

(c) $-8, -9$ (d) $20, 22$

(e) $\dfrac{1}{2}, -\dfrac{2}{3}$ (f) $4, -\dfrac{5}{2}$

(g) $-\dfrac{9}{11}, -1$ (h) $\dfrac{6}{5}, -\dfrac{3}{2}$

5 (a) $\dfrac{1}{2}(13 \pm \sqrt{253})$ (b) $\dfrac{1}{2}(13 \pm \sqrt{85})$

(c) No roots, $\Delta = -215$ (d) No roots, $\Delta = -83$.

(e) $\dfrac{1}{10}(-19 \pm \sqrt{681})$ (f) $\dfrac{x}{y} = \dfrac{1}{17}(-3 \pm 9\sqrt{2})$

(g) $\pm 0.38197, \pm 2.61803$ (h) $(3 \pm \sqrt{6})^{1/3} = 1.7598, 0.8196$

6 (a) $(11x - 4)(17x + 19)$ (b) $(15x + 1)(23x - 29)$

(c) $(101x - 103)(107x + 108)$ (d) $(7x - 13)(27x - 5)$

(e) $(-71x + 5y)(x - 37y)$

7 (a) $\left(x - \dfrac{91}{2} + \dfrac{\sqrt{8201}}{2}\right)\left(x - \dfrac{91}{2} - \dfrac{\sqrt{8201}}{2}\right)$

(b) $2\left(x - 60 + \dfrac{\sqrt{14394}}{2}\right)\left(x - 60 - \dfrac{\sqrt{14394}}{2}\right)$

(c) $(z - 1)(z - 2 + \sqrt{6})(z - 2 - \sqrt{6})$

8 (a) $12, -4, 152$ (b) $-\dfrac{1}{2}, -\dfrac{19}{2}, \dfrac{77}{4}$

(c) $\dfrac{5}{4}, -\dfrac{23}{4}, \dfrac{209}{16}$ (d) $-\pi(\pi^2 + 1), -\pi^2, \pi^2(\pi^4 + 2\pi^2 + 3)$

9 (a) $x^2 - 24x - 16 = 0$ (b) $x^2 + x - 38 = 0$

(c) $46 - 5x - 2x^2 = 0$ (d) $4\pi - 2(\pi^2 + 1)x - x^2/\pi = 0$

10 $ax^2 + kbx + k^2c = 0$

11    $cx^2 + bx + a = 0$

$$(x-3)(x-4) = x^2 - 7x + 12 = 0, \text{ so } x = 3, 4$$

$$12\left(x - \frac{1}{3}\right)\left(x - \frac{1}{4}\right) = (3x - 1)(4x - 1)$$

$$= 12x^2 - 7x + 1 = 0, \text{ so } x = \frac{1}{3} \text{ or } \frac{1}{4}$$

12   (a) $x^2 - \left(\dfrac{b^2 - 2ac}{ac}\right)x + 1 = 0$ (b) $c^2x^2 - (b^2 - 2acx) + a^2 = 0$

Exercise 4.3

1   11 m and 14 m

2    $\dfrac{1}{2}(1 + \sqrt{5}) \simeq 1.618$

3   7 and 13; the solutions for x and y are interchangeable because each can play either role

4   -10 and 16 or vice versa

5    $\Delta = -16$, no solution possible

6    $x = 1, y = 2$ or $x = 29, y = -12$

7    $x = 8, y = 6$ or $x = 12, y = 4$

8    $x = -\dfrac{1}{5}, 0$ or $f = -5$, infinity

9    $T = 91.61$ (only solution)

11

$$\text{Minimum} = \frac{4ac - b^2}{4a^2}$$

Positive $4ac > b^2$

Zero $4ac = b^2$

Negative $b^2 > 4ac$

$$x = -\frac{b}{2a}$$

12 (a)　　(1, 8)

(b)　　$\left(\dfrac{5}{6}, \dfrac{23}{12}\right)$

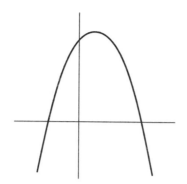

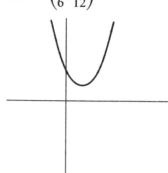

(c)　　(0.553, 50.803)

(d)　　(2.060, 55.500)

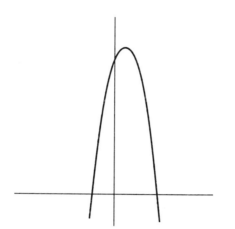

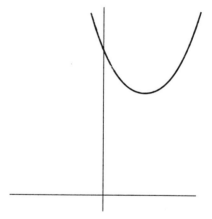

13 (a)　　Distinct real roots (A)

(b)　　Equal real roots (B)

(c)　　No real roots (C)

(d)　　No real roots (D)

(e)　　Equal real roots (E)

(f)　　Distinct real roots (F)

14 (a)　　(2, 4), (−4, 10)　　　(b)　　(x, y) satisfy the inequality $x^2 + x - 2 < y < 6 - x$

(c)　　$\left(-\dfrac{1}{2}, -2\dfrac{1}{4}\right)$

15

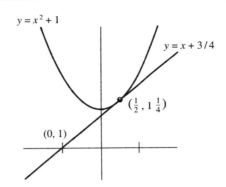

$y = x^2 + 1$

$y = x + 3/4$

$\left(\frac{1}{2}, 1\frac{1}{4}\right)$

$(0, 1)$

16

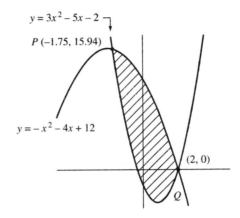

$y = 3x^2 - 5x - 2$

$P\,(-1.75, 15.94)$

$y = -x^2 - 4x + 12$

$(2, 0)$

Q

Coordinates of P are $\left(-1\frac{3}{4}, 15\frac{15}{16}\right)$.

17 $a = 1, b = -5, c = 4$
$y = x^2 - 5x + 4$

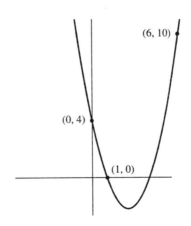

$(6, 10)$

$(0, 4)$

$(1, 0)$

Exercise 4.4

1

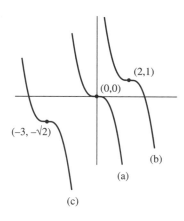

All are *quasi-parallel* in that they are replications of $y = -x^3$ centred on the points indicated.

2 (a) $x = \left(-\dfrac{2}{3}\right)^{1/3} + 2$ (b) $x = \dfrac{1}{3}\left\{\left(\dfrac{11}{9}\right)^{1/3} - 5\right\}$

(c) $x = \pm\dfrac{1}{\sqrt{2}}$ (d) $x = 1$

3 (a) True. if $a > 0$, the graph is in the NE quadrant if $x \gg 0$ (large, positive) and in the SW quadrant if $x \ll 0$ (large, negative). The x-axis must be crossed. Likewise for $a < 0$ but with the SE and NW quadrants for $x \gg 0$ and $x \ll 0$.

(b) False, e.g.

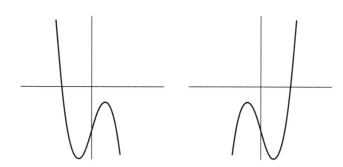

(c)   True, e.g.

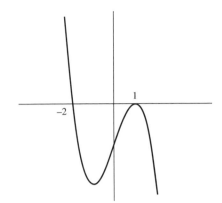

(d)   False, e.g. x^3

4   (a)

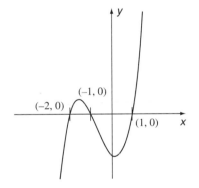

(b)

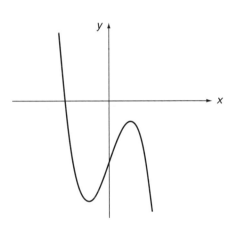

5

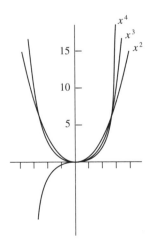

6

x	-2	-1	0	1	2
y	-4	34	50	62	88

7　0.9867　1.0867　for $x = \pm 10$ respectively

　　0.9937　1.0070　for $x = \pm 100$ respectively

　　0.9993　1.0007　for $x = \pm 1000$ respectively

8　$a = -1, b = 2$

9　$c = -21, d = 37$

10

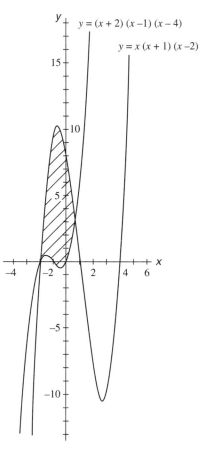

$x = \pm\sqrt{5} - 1$

(x, y) satisfy $x(x + 1)(x - 2) < y < (x + 2)(x - 1)(x - 4)$

5 GEOMETRY

INTRODUCTION

The classification and properties of basic shapes in two and three dimensions has always been an essential ingredient of our knowledge of the world and our ability to solve real-life problems. Today, with an increasing role for computers in design, geometry is again coming to the fore as the need for the more economical use of scarce materials becomes more acute. Increasingly impressive computer simulations rely on the properties of plane shapes and solids.

OBJECTIVES

After working through this chapter you should be able to

- recognise the different types of angle
- identify the equal angles produced by a transversal cutting parallel lines
- identify the different types of triangle
- state and use the formula for the sum of the interior angles of a polygon
- calculate the area of a triangle
- use the rules for identifying congruent triangles
- know when two triangles are similar
- understand radian measure and convert from degrees to radians and vice versa
- state and use the formulae for the area and circumference of a circle
- calculate the volumes of the regular solids
- calculate the surface areas of the regular solids

5.1 SHAPES IN TWO DIMENSIONS

The first step is to introduce some basic geometric concepts. The fundamental concept is that of a point.

A **point** occupies a position but has neither dimension nor size. It is represented by a dot for clarity but the dot is not the point itself. It is usually designated by a capital letter.

A **line** has length as its only dimension: it has no thickness. It may be represented by the path of a pen on paper. In theory a straight line is unlimited in extent.

A **ray** is that part of a straight line extending from a given point in one direction whereas a **straight line segment** is that part of the straight line between two specified points on it. A line segment is designated by the letters of its endpoints; thus the line segment with endpoints A and B is designated AB.

A **surface** has two dimensions but no thickness: if any two points on a surface can be connected by a straight line, all of whose points are on the surface, then surface is called a **plane**. In everyday language the plane is a *flat* surface. In the first three sections we confine our attention to geometry in the plane.

When two lines intersect they have a point in common—the **point of intersection**; if they never intersect they are said to be **parallel**.

When two rays have a common endpoint they form an **angle**. The rays form the sides of the angle. The common point is called the **vertex** of the angle. The angle can be named by its vertex capital letter with a circumflex overhead, e.g. $\hat{A}$, or by a smaller letter inside the angle, e.g. α, or by three capital letters, e.g. $B\hat{A}C$; see Figure 5.1.

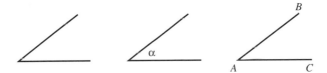

Figure 5.1   Notation for angles

Generally speaking, $B\hat{A}C$ and $C\hat{A}B$ are alternative notations for the same angle. The magnitude of an angle is measured by the amount of rotation from one ray to the other. A full rotation takes a ray back to its starting point. A **degree** is $\frac{1}{360}$ of the full rotation about a point. In this notation thirty degrees is written $30°$.

There are several classification of angles which are important. An angle of $90°$ (or one-quarter of a full turn) is called a **right angle**; it is marked ∟. An angle between $0°$ and $90°$ is called an **acute angle**; an angle between $90°$ and $180°$ is called an **obtuse angle**. An angle of $180°$ (one-half of a full turn) is a **straight angle** and one between $180°$ and $360°$ is a **reflex angle**; see Figure 5.2.

When two angles are **complementary** their sum is $90°$ and when two angles are **supplementary** their sum is $180°$.

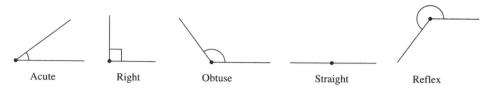

Figure 5.2 Types of angle

Four angles are formed when two straight lines intersect; those which have a common endpoint and share a common ray are called **adjacent** whereas those which have only a common vertex are called **vertically opposite**, or just **opposite**.

In Figure 5.3(a) angles a and b are adjacent, as are angles b and c, c and d, and d and a. Angles a and c are opposite as are angles b and d. Notice that opposite angles are equal. In this case the adjacent angles are supplementary.

The adjacent angles in Figure 5.3(b) are not supplementary. The angles marked in Figure 5.3(c) are not equal; they are called vertical angles.

When two lines intersect to form four right angles they are said to be **perpendicular**; see Figure 5.4. We write $AB \perp CD$.

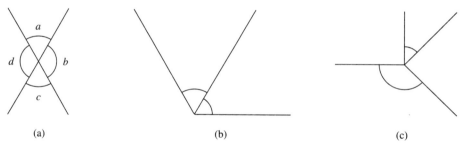

(a)   (b)   (c)

Figure 5.3 Pairs of angles

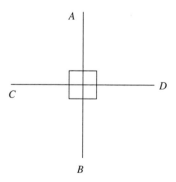

Figure 5.4 Perpendicular lines

Parallel Lines

Parallel lines are indicated by arrows, as shown in Figures 5.5, 5.6 and 5.7.

A line which intersects a set of two or more lines is called a **transversal**. Two sets of equal angles are produced when two parallel lines are intersected by a third line.

In Figure 5.5(a) the marked angles are equal; they are **corresponding angles**. They are on the same side of the transversal and on the same side of the parallel lines. In Figure 5.5(b) the angles marked with the same letters are equal; each pair consists of corresponding angles.

In Figure 5.6(a) the marked angles are equal; they are **alternate angles**. They are non-adjacent angles between the parallel lines and on opposite sides of the transversal. In Figure 5.6(b) the angles marked with the same letters are equal; each pair consists of alternate angles.

Two lines are parallel if any pair of corresponding angles are equal or if any pair of alternate angles are equal. Two lines are parallel if they are parallel to the same line. Two lines are parallel if they are both perpendicular to the same line.

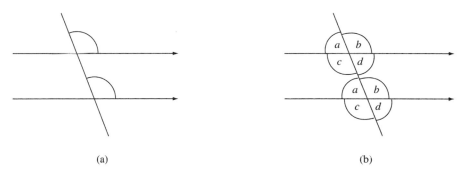

(a)   (b)

Figure 5.5   Corresponding angles

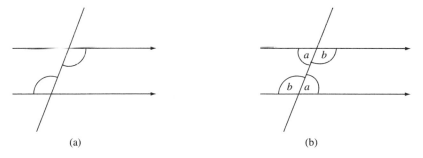

(a)   (b)

Figure 5.6   Alternate angles

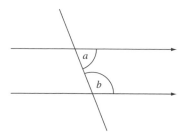

Figure 5.7 Supplementary angles

In Figure 5.7, the angles *a* and *b* are supplementary. This fact can also be used to test whether two lines are parallel.

Triangles

A **triangle** is a closed figure with three sides which are straight line segments. It is labelled by a capital letter at each vertex. If one of the angles at the vertices is 90° the figure is a **right-angled triangle**; if all angles are acute we have an **acute-angled triangle**; if one of the angles is obtuse the triangle is **obtuse-angled**; see Figure 5.8.

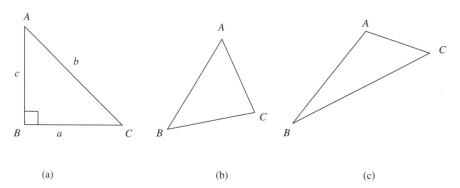

(a)   (b)   (c)

Figure 5.8   Types of triangle: (a) right-angled, (b) acute and (c) obtuse

Here is an important result concerning the sum of the angles:

The sum of the angles of any triangle is 180° (two right angles).

In a right-angled triangle, the side opposite the right angle is the **hypotenuse**. An important result concerning right-angled triangles is **Pythagoras' theorem**. This states that the square of the length of the hypotenuse is equal to the sum of the squares of the lengths of the other two sides. With reference to Figure 5.8(a) the theorem can be stated as:

$$(AC)^2 = (AB)^2 + (BC)^2$$

The theorem will be proved in Chapter 6.

Sometimes the lengths of the sides of a triangle are labelled by lower case letters opposite the corresponding angle; see Figure 5.8(a). In this case the theorem can be stated as:

$$b^2 = c^2 + a^2$$

If all three sides are equal the triangle is called **equilateral**. In Figure 5.9(a) the sides AB, BC and AC are equal; the angles $B\hat{A}C$, $A\hat{B}C$ and $B\hat{C}A$ are equal. Note that if all three angles of a triangle are equal then it is equilateral and its sides are of equal length. Each angle is 60°.

A triangle with two sides equal is called **isosceles**. The unequal side is the **base** and the opposite angle is called the **vertex angle**. The two other angles are equal. In Figure 5.9(b) the sides AB and BC are equal; the angles $B\hat{A}C$ and $B\hat{C}A$ are equal. If two angles of a triangle are equal then it is isosceles and the sides opposite these angles are equal. A triangle with three unequal sides is a **scalene** triangle; see Figure 5.9(c).

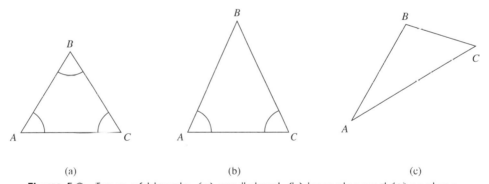

Figure 5.9 Types of triangle: (a) equilateral, (b) isosceles and (c) scalene

A useful result can be found by introducing the idea of an **exterior angle** of a triangle. It is formed when one of the sides is extended. Hence, in Figure 5.10(a) the angles $P\hat{A}C$ and $A\hat{C}Q$ are examples of exterior angles. The angles we have discussed up to now are **interior angles**, of which $B\hat{A}C$ is an example.

An exterior angle is equal to the sum of the two opposite interior angles. Figure 5.10(b) shows an example: angle $A\hat{C}Q$ is equal to the sum of angles $B\hat{A}C$ and $A\hat{B}C$.

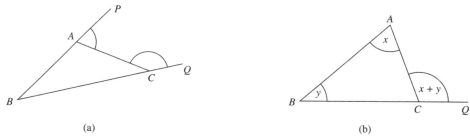

Figure 5.10 Exterior angles

Special lines in a triangle

A line segment that bisects an interior angle of a triangle is called an **angle bisector**; see Figure 5.11(a) where the line segment BD bisects the angle $A\hat{B}C$.

A line segment that joins a vertex to the midpoint of the opposite side is called a **median** of the triangle. In Figure 5.11(b) BD is a median; the lengths of the line segments AD and DC are equal and we could write $AD = DC$. Note how we mark the equality of these lengths. (If more than one pair of lengths on a diagram are equal, we may also use a single short line | or a treble |||.)

If a line bisects a side at right angles it is called a **perpendicular bisector** of that side; in Figure 5.11(c) PQ is a perpendicular bisector of AC.

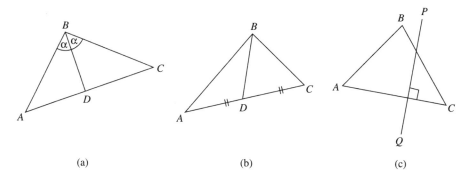

Figure 5.11 Bisectors

An **altitude** of a triangle is a line segment from a vertex which is perpendicular to the opposite side. In Figure 5.12(a) BD is an altitude. In Figure 5.12(b) the triangle is obtuse and the altitude lies outside the triangle; it meets the opposite side produced.

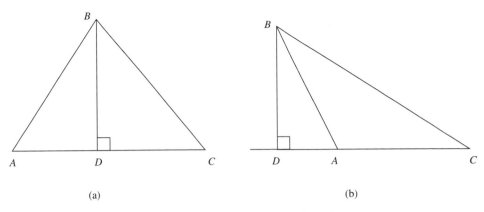

(a)   (b)

Figure 5.12   Altitude of a triangle

Polygons

Any closed geometric figure whose sides are straight line segments is a **polygon**. As we have seen, a three-sided polygon is a triangle. If all the sides of a polygon are equal it is called **equilateral**; and if all the angles are equal it is a **regular** polygon. The **perimeter** of a polygon is the sum of the lengths of its sides. Figure 5.13 shows some examples of polygons.

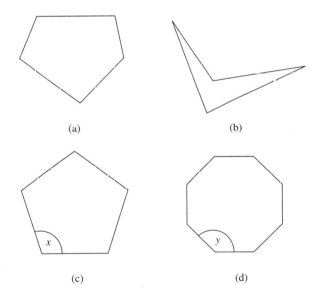

(a)   (b)

(c)   (d)

Figure 5.13   Examples of polygons: (a) pentagon, (b) quadrilateral with reflex angle, (c) regular pentagon ($x = 108°$) and (d) regular octagon ($y = 135°$)

Diagram (b) is an example of a polygon with a re-entrant vertex; each of the other diagrams shows a convex polygon.

We sometimes quote the angle sum in the following way:

> The sum of the interior angles of a polygon is $(2n - 4) \times 90°$ where n is the number of sides.

Quadrilaterals

A **quadrilateral** is a polygon with four sides; see Figure 5.14. **Rectangles** have all four interior angles equal to $90°$. **Squares** are rectangles with all sides equal. A **rhombus** has all sides equal.

If a quadrilateral has opposite sides parallel it is a **parallelogram**; it is a consequence that the opposite sides are of equal length. Furthermore, one pair of opposite sides being parallel and of equal length is sufficient to determine a parellelogram. Note that the three previous cases are all parallelograms. To indicate that two lines are parallel we use the symbol II; for example, in Figure 5.14(a) *AD* II and *BA* II *CD*. When only one pair of opposite sides are parallel, the figure is called a **trapezoid**.

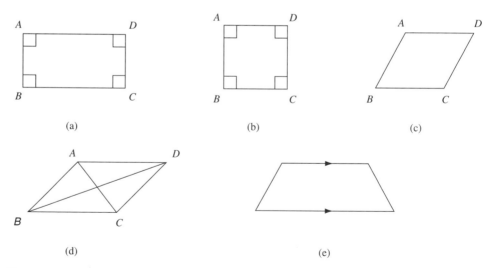

Figure 5.14 Some quadrilaterals: (a) rectangle, (b) square, (c) rhombus, (d) parallelogram and (e) trapezoid

A line segment joining opposite vertices is a **diagonal**; in Figure 5.14(d) the diagonals are *AC* and *BD*. In a parallelogram the diagonals bisect each other; in a square they are also perpendicular.

Areas

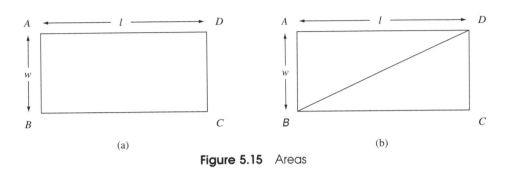

Figure 5.15   Areas

The **area** of a rectangle is the product of its length and its breadth. In Figure 5.15(a) the area *A* of the rectangle is given by $A = lw$.

In Figure 5.15(b) the rectangle has been divided into two equal triangles. The area of each triangle is clearly $\frac{1}{2}lw$. In triangle *BCD* the side *BC* is the base and the side *CD* is the altitude or **height**. The area of any triangle can be found by a similar principle. In Figure 5.16(a) and (b) the area of the triangles *ABC* is $\frac{1}{2}ah$:

> The area of a triangle is $\frac{1}{2} \times$ base $\times$ height.

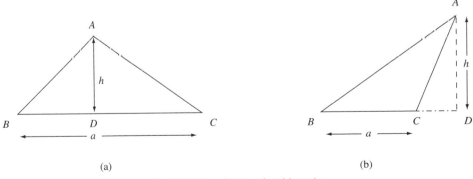

Figure 5.16   Area of a triangle

In the first case the area is the sum of the areas of triangles ABD and ADC, i.e. $\frac{1}{2}BD \times h + \frac{1}{2}DC \times h = \frac{1}{2}BC \times h$. In the second case the area is the difference between the areas of the triangles ABD and ACD, i.e. $\frac{1}{2}BD \times h - \frac{1}{2}CD \times h = \frac{1}{2}BC \times h$.

Since a parallelogram can be divided into two equal triangles its area is bh; see Figure 5.17(a). In the case of the trapezoid shown in Figure 5.17(b) the area is $\frac{1}{2}(a + b)h$.

The area of a general polygon can be found by dividing it into triangles and adding the areas of the triangles, as shown in Figure 5.17(c).

Take any interior point O, then draw straight lines from that point to each of the vertices as shown. The angles at O add up to 360°, i.e. four right angles. We now add together the angles in all the triangles such as $\triangle AOB$. In all there are n triangles, one for each edge, so the total angle sum is $2n$ right angles.

The sum of the interior angles = the sum of the triangle angles *minus* four right angles, i.e. $(2n - 4)$ right angles, as we stated earlier.

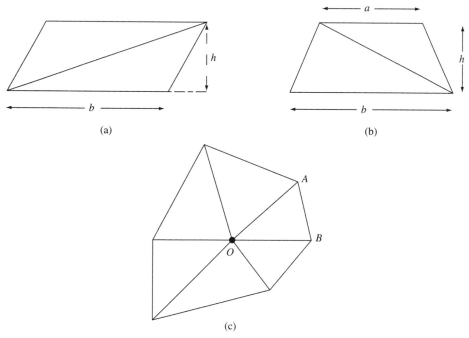

Figure 5.17 Further areas

If the polygon is regular then each interior angle is

$$\frac{2n-4}{n} = 2 - \frac{4}{n} \text{ right angles}$$

e.g. a regular pentagon has five angles of 108°.

Sometimes regular polygons can form patterns known as **tessellations**, as in Figure 5.18. Only certain kinds of regular polygon can actually form neat joints like those shown.

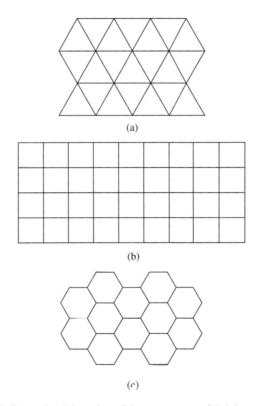

(a)

(b)

(c)

Figure 5.18   Tessellations: (a) triangles, (b) squares and (c) hexagons (honeycomb)

Exercise 5.1

1   Determine the magnitudes of the marked angles in each of the following cases:

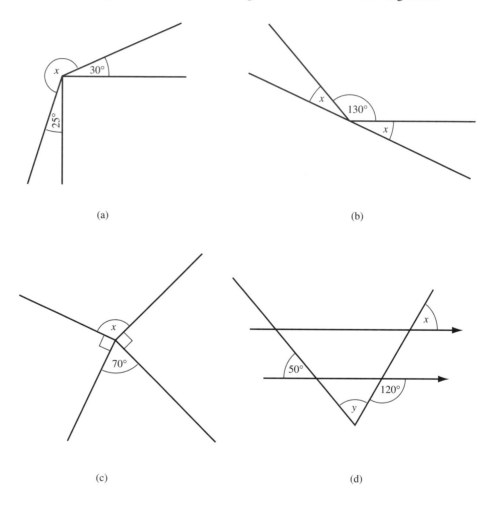

(a)

(b)

(c)

(d)

(Remember that the sum of angles in a triangle is 180°.)

2   *AB*, *CD* and *EF* are straight lines passing through the axis *O′O* at the angles shown. The straight line *PQ* has variable angle α. For what values of α is *PQ* unable to meet the other lines?

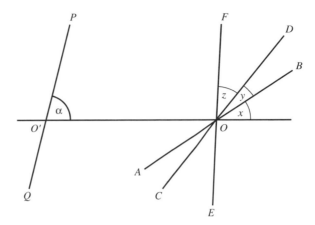

3

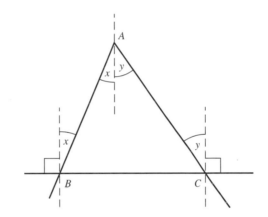

The dashed lines in the diagram are all perpendicular to BC. Find $A\hat{B}C$ and check the sum of the angles of $\triangle ABC$.

4* (a) The line segment AB contains an interior point P where $AP:PB = 3:4$. If $AB = 28$ cm, what are AP, PB?

(b) In this triangle PQ is parallel to BC. How long is PQ?

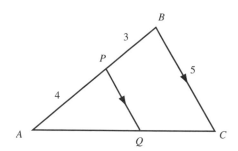

(c)   The triangles ABC and CDE are isosceles with $CA=CB$ and $CD=CE$. $AC=3$ cm, $CE=5$ cm and $DE=6$ cm. How long are the other sides?

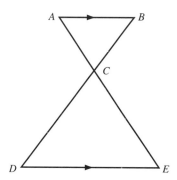

(d)   The triangle ABC has sides $AB=11$ cm, $AC=5$ cm and $BC=9$ cm. On AB produced (i.e. drawn from A to B and extended beyond) is a point P, which is 5.5 cm from B. A further straight line PQ is drawn parallel to BC and meets AC produced at Q. Draw the diagram and calculate the length of AQ.

5   The following triples represent side lengths of triangles. Determine which of them are right-angled.

(a)   (3, 4, 5)   (b)   (4, 5, 6)   (c) $(\sqrt{2}, \sqrt{3}, \sqrt{5})$

(d)   (4, 7, 11)   (e)   (5, 12, 13)   (f) $(6, 7, \sqrt{85})$

(g)   (24, 7, 25)   (h)   (9, 12, 15)

6   What conclusion can you draw from parts (a) and (h) of Question 5?

7 ABC is an equilateral triangle of side 10 cm. AB is produced to D so that $BD=AB$. Prove first that ACD is a right-angled triangle then calculate the distance CD. How far is the point P from CD?

8 Write down the sum of all interior angles and hence determine all the angles in the following polygons:

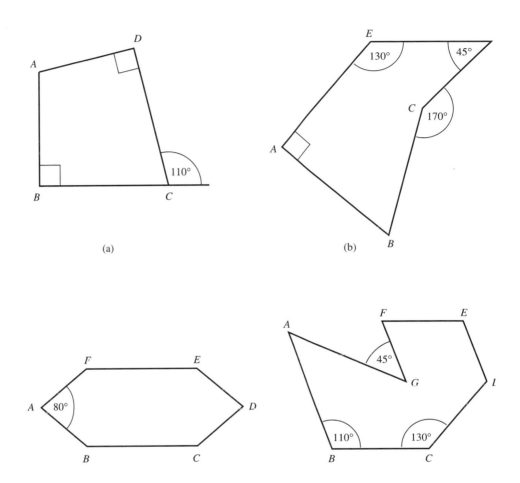

(a)

(b)

9   What is the area of each of the following figures?

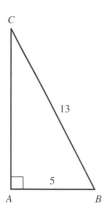

(a)

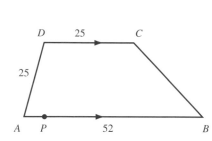

P is vertically below *D* and *AP* = 7

(b)

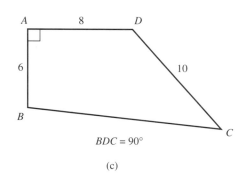

BDC = 90°

(c)

ABCD is a parallelogram

(d)

 10

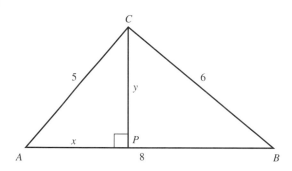

(a)   Find $x^2 + y^2$ from $\triangle APC$.

(b)   Apply Pythagoras' theorem to $\triangle PBC$ and determine x.

(c)   Hence find y.

(d)   What is the area of $\triangle ABC$?

11   Construct a triangle whose sides are in the proportion $8:6:5$ and show that its measured area agrees with the result in Question 10.

12   An isosceles triangle has two equal sides of x units and a base of length y units. What is its area in terms of x and y?

13   (a)   What is the area of an equilateral triangle of side a?

(b)   Extend this result to find the area of a rectangular hexagon (six sides) of side a.

14   *ABCDEFGH* is a regular octagon of side 1 unit. By removing the areas of the shaded triangles from the square *XYZT*, find its area.

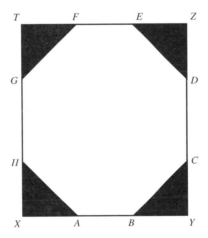

15   How many sides does a regular polygon need to have before the vertex angles exceed $150°$?

16   Construct a regular hexagon by first drawing a circle, centre O. Move the compass point to A Then mark off B along the circumference, an amount equal to the radius OA. Move the

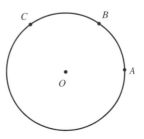

compass point to B then mark off C, etc. Proceed until you return to A then draw the straight lines AB, BC, etc.

17* Three regular polygons meet at a common vertex. If polygon 1 has a sides then its vertex angles equal $\left(2 - \dfrac{4}{a}\right)$ right angles. Assuming that polygon 2 has b sides and polygon 3 has c sides, show that:

$$\frac{1}{a} + \frac{1}{b} + \frac{1}{c} = \frac{1}{2}$$

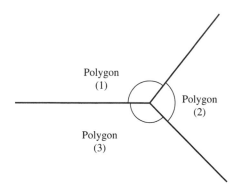

Polygon (1)

Polygon (2)

Polygon (3)

18* Using the result of the previous exercise, explain what happens when:

(a) a, b, c are all equal   (b) $a = 12$, $b = 6$

(c) $a = 3$, $b = c$ (d) $a = 6$, $b = 7$

5.2 CONGRUENCE AND SIMILARITY

Two shapes are **congruent** if one is an *exact* duplicate of the other, in that they have the same size and shape. One could be superimposed exactly on the other.

If two circles have the same radius then they are congruent. Things are more complicated for triangles.

Congruent triangles

When two triangles are congruent, corresponding sides in the triangle are equal and corresponding angles are equal. Figure 5.19 shows two triangles which are congruent. You can verify this by tracing a copy of either triangle and confirming that it is possible to superimpose this tracing *exactly* on the other triangle.

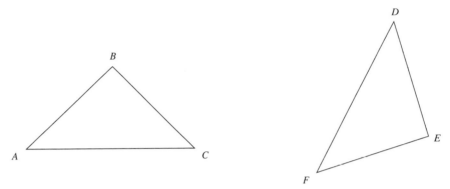

Figure 5.19 Congruent triangles

Sides AB and DE are equal in length, as are sides BC and EF and sides AC and DF. Angles $A\hat{B}C$ and $D\hat{E}F$ are equal in magnitude as are angles $B\hat{C}A$ and $E\hat{F}D$, and angles $C\hat{A}B$ and $F\hat{D}E$.

To emphasise which sides and which angles correspond, we say that triangles ABC and DEF are congruent; then the first two letters in each case, AB and DE, represent sides of equal length, the third letter in each case shows that the angle at vertex C is equal in magnitude to the angle at vertex F.

Figure 5.20(a) shows two congruent triangles marked to demonstrate which sides are equal in length and Figure 5.20(b) shows the marking to indicate which angles are equal in magnitude.

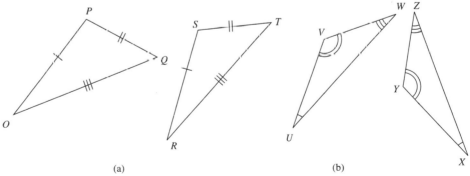

(a)   (b)

Figure 5.20 Indicating congruence

Tests for congruent triangles

How do we test whether two triangles are congruent without tracing over one of them? The tests presented here are based on the idea that the information given is sufficient to construct one triangle, and only one triangle, apart from its position and orientation.

Four such tests are presented then each is examined. A triangle of unique size and shape is specified by stating the magnitudes of one of the following sets:

- Three sides   (SSS)
- Two sides and the included angle   (SAS)
- Two angles and the side joining them   (AAS)
- Right angle, hypotenuse and another side   (RHS)

Three sides (SSS)

In Figure 5.21(a) triangle *ABC* is constructed by drawing the side *AB* to a specified length; its position is not important. With centre at *B* an arc of a circle is drawn having radius equal to the second specified length. With centre at *A* an arc of a circle is drawn having radius equal to the third specified length. The two arcs meet at *C*.

Note that we cannot merely specify any three values and expect to draw a triangle. In Figure 5.21(b) the arcs drawn from *B* and *A* do not meet, so no triangle can be drawn. Note also that the order in which we use the specified lengths is unimportant; it is allowed to turn over the tracing of one triangle to make it coincide with the other triangle.

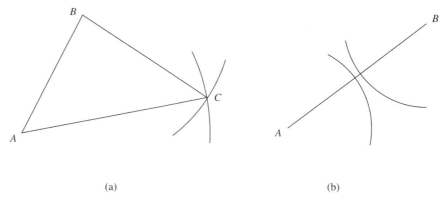

(a)   (b)

Figure 5.21   Constructing a triangle: SAS

Two sides and an included angle (SAS)

Refer to Figure 5.22(a). Side *AB* is drawn to one of the specified lengths; side *AC* is drawn to the second specified length so that it makes the required angle with *AB*. The 'free' ends *B* and *C* are joined to form the triangle. Again, there is the possibility of a 'mirror image', Figure 5.22(b).

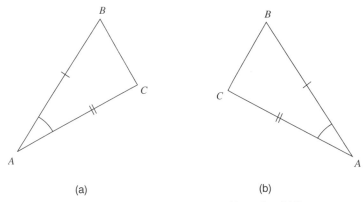

Figure 5.22   Constructing a triangle: SAS

Two angles and the joining side (AAS)

The side AB is drawn to a specified length; this immediately gives us two vertices, A and B. A line AE is drawn from A at one of the given angles and a second line AD is drawn from B at the other given angle. As shown in Figure 5.23(a), these lines meet at C, the third vertex.

Right angle, hypotenuse and side (RHS)

The side AB is drawn to the smaller of the given lengths. The line BD is drawn at right angles to AB. An arc is drawn of a circle centre A having radius equal to the larger of the given lengths. As shown in Figure 5.23(b), the arc meets the line BD at C, the third vertex.

We can use the idea of congruent triangles to demonstrate some useful results.

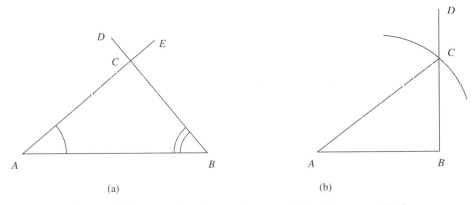

Figure 5.23   Constructing a triangle: (a) AAS and (b) RHS

Example

If two sides of a triangle are of equal length then the angles opposite them are of equal magnitude. In Figure 5.24 the isosceles triangle ABC is such that $AB = CB$.

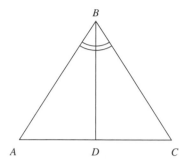

Figure 5.24 Angles opposite equal sides are equal

From vertex B the line which bisects angle $A\hat{B}C$ is drawn to meet the opposite side AC at the point D.

Consider the triangles ABD and CBD. We are told that $AB = CB$ and the side BD is common to both triangles. Since $A\hat{B}D = C\hat{B}D$, by construction the triangles are congruent according to SAS. Hence $B\hat{A}D = B\hat{C}D$, as required.

Note also that $A\hat{D}B = C\hat{D}B$, so each must be equal to 90°. Hence BD is perpendicular to AC and the angle bisector BD is also an **altitude** of triangle ABC. And $AD = DC$, so BD is the **perpendicular bisector** of the side AC; it is also a **median** of the triangle. ∎

The converses of these last two results are also true: (a) if a triangle has two angles of equal magnitude then the sides opposite them are of equal length, and (b) if a triangle has all its angles of equal magnitude (and therefore equal to 60° because their sum is 180°) then its sides are of equal length.

Similarity

Two shapes are **similar** if one is an enlargement of the other; they have the same shape and their corresponding angles are equal.

An Ordnance Survey map is a scaled-down representation of a geographical area. A popular scale is $1:50\,000$, which means that two points a distance of 1 cm apart on the map represent two points on the land a distance $50\,000$ cm (i.e. 0.5 km) apart.

If three points P, Q, R on the land form a triangle which we may assume lies on a plane surface then their corresponding points on the map form a triangle with the same angles. (If the points P, Q, R are a great distance apart then the curvature of the earth will modify this result.)

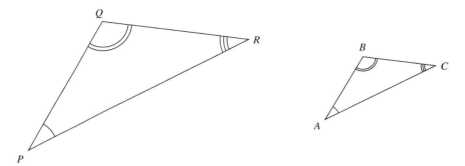

Figure 5.25 Similar triangles

Figure 5.25 shows the triangles PQR and ABC, which is our representation of P, Q, R on a scale $1:2$ (we use a simple scale for ease of understanding).

Since corresponding angles are equal, i.e. $P\hat{Q}R = A\hat{B}C$, $Q\hat{R}P = B\hat{C}A$, $R\hat{P}Q = C\hat{A}B$, the triangles PQR and ABC are similar. The lengths of corresponding sides are in the same ratio, i.e.

$$\frac{AB}{PQ} = \frac{BC}{QR} = \frac{AC}{PR} = \frac{1}{2}$$

To emphasise which sides correspond, we may write that triangles $\frac{ABC}{PQR}$ are similar.

For example, $\frac{\textbf{ABC}}{\textbf{PQR}}$ tells us that $\frac{AB}{PQ}$ is the given ratio and $\frac{\textbf{ABC}}{\textbf{PQR}}$ tells us that the angle at vertex C ($B\hat{C}A$) is equal in magnitude to the angle at vertex R ($Q\hat{R}P$).

We may summarize by saying that if two triangles ABC and PQR are such that $\frac{AB}{PQ} = \frac{BC}{QR} = \frac{CA}{RP}$ then the triangles are similar and $A\hat{B}C = P\hat{Q}R$, $B\hat{C}A = Q\hat{R}P$ and $C\hat{A}B = R\hat{P}Q$.

Conversely, if two triangles ABC and PQR are such that $A\hat{B}C = P\hat{Q}R$, $B\hat{C}A = Q\hat{R}P$ and $C\hat{A}B = R\hat{P}Q$ then the triangles are similar and $\frac{AB}{PQ} = \frac{BC}{QR} = \frac{CA}{RP}$.

Sometimes it is not immediately obvious that two triangles are similar.

Example

In Figure 5.26 ABC is a right-angled triangle and BD is perpendicular to AC. Consider triangle ABD. It has a right angle at D and $D\hat{A}B = C\hat{A}B$. Since two of its angles have the same magnitude as two of the angles of triangle ABC and since the angle sum of any triangle is $180°$, the third angles must be equal in magnitude, i.e. $A\hat{B}D = A\hat{C}B$. Hence triangles ADB and ABC are similar.

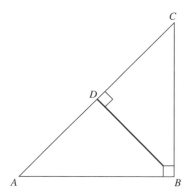

Figure 5.26 Triangles *ADB, BDC* annd *ABC* are similar

Now consider triangle *BCD*. It too has a right angle at *D* and $D\hat{C}B = A\hat{C}B$. Since two of its angles have the same magnitude as two of the angles of triangle *ABC* then $C\hat{B}D = C\hat{A}B$, hence triangles *BDC* and *ABC* are similar.

The idea of similar triangles can be used to estimate heights.

Example

Figure 5.27 shows a pole *DE* placed vertically in the shadow of a vertical tree *BC*. *AC* is the shadow cast by the tree *BC* and *AE* is the shadow cast by the pole *DE*. It should be obvious that triangles *ADE* and *ABC* are similar. Hence $\dfrac{BC}{DE} = \dfrac{AC}{AE}$ and therefore $BC = DE \times \dfrac{AC}{AE}$.

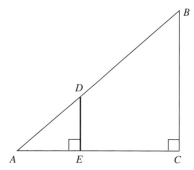

Figure 5.27 Estimating heights

Suppose we use a pole DE of length 2 m and we measure the lengths AE as 4 m and AC as 40 m. Then $BC = 2 \times \dfrac{40}{4} = 20$ m. We therefore estimate the height of the tree to be 20 m.

Note that the area of the triangle $ABC = \dfrac{1}{2}(AC \times BC) = \dfrac{1}{2} \times 40 \times 20 = 400$ m^2 and the area of the triangle $ADE = \dfrac{1}{2}(AE \times DE) = \dfrac{1}{2} \times 4 \times 2 = 4$ m^2. The ratio of the areas $(100 : 1)$ is the *square* of the ratio of the lengths $(10 : 1)$. ■

In general, if two geometric figures are similar then the ratio of their areas is the square of the ratio of two corresponding lengths.

Exercise 5.2

1

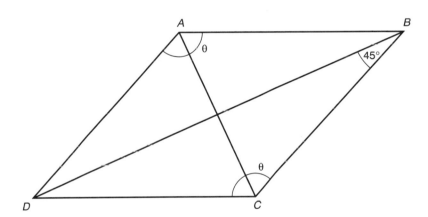

Show that in the parallelogram $ABCD$

(a)   triangles ABD and DBC are congruent, which we write as $\triangle ABD \equiv \triangle DBC$.

(b) $\triangle ADC$ is isosceles.

Show also that if $D\hat{B}C = 45°$, then $ABCD$ is a square.

2   Consider the following diagram. Show that $\triangle ABC$ and $\triangle DEC$ are congruent.

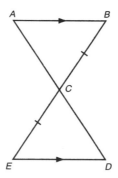

3 In the diagram from Question 2 draw straight lines from B to the midpoint of AC, say F, and from D to the midpoint of EC, say G.

Show that

$$\triangle EGD \equiv \triangle AFB.$$

To what triangle is $\triangle CGE$ congruent?

4 A parallelogram can be described as a four-sided polygon with two sides equal and parallel. Consider the four points $ABDE$ in Question 2, noting that AB and ED are equal and parallel.

(a)   Show that

$$\triangle ACE \equiv \triangle BCD.$$

(b)   Compare the sides ED and BD.

5

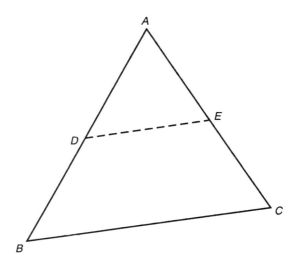

Given that D and E are the midpoints of AB and AC, show that:

(a)   triangles ADE and ABC are similar

(b) $DE \parallel BC$ and $DE = \dfrac{1}{2}BC$.

6 $ABCD$ is an irregular quadrilateral and E, F, G and H are the midpoints of the sides as shown.

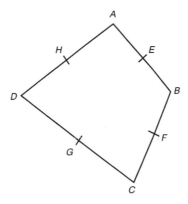

Draw in DB, HE and GF.

(a)   Using the results of Question 5, what do you conclude about the relationship between HE and GF?

(b)   What sort of quadrilateral is $EFGH$?

(c)   Had we started with AC, EF and HG, what could we have concluded?

7 ABC is a right-angled triangle with $B\hat{A}C = B\hat{C}A = 45°$. Identify a triangle in the diagram which is congruent to $\triangle DEC$.

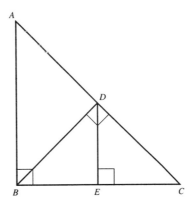

8 What is the ratio of the areas of $\triangle DEC$ to $\triangle ABC$ in Question 7?

9 $ABCD$ is a parallelogram and the lines ED and FC are perpendicular to CD.

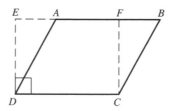

Show that:

(a)   the triangles DEA and CFB are congruent

(b) $EFCD$ is a rectangle

(c) $ABCD$ and $EFCD$ have the same area

This proves the result that the area of a parallelogram is equal to its base (DC) multiplied by its height (ED or FC).

10 $ABCD$ is a parallelogram. Show that triangles AEB and DFC are congruent.

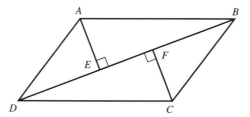

11 A rhombus is a quadrilateral with all sides equal, or if one prefers, a parallelogram with two adjacent sides equal. $ABCD$, shown below is a rhombus. Show that all four triangles AHB, BHC, CHD, DHA are congruent. What is the value of angle $A\hat{H}B$?

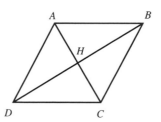

12 Draw a straight line *AB* and mark its midpoint *E*. Draw another straight line *CD*, not parallel to *AB*, whose midpoint is also *E*, as in the following diagram.

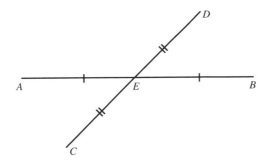

Show that:

(a) △*AED* ≡ △*CEB* and △*AEC* ≡ △*DEB*

(b) *ADBC* is a parallelogram

Now show that the vertically opposite angles in a parallelogram are equal, e.g. $A\hat{C}B = A\hat{D}B$. Deduce that adjacent angles in a parallelogram, e.g. $A\hat{C}B$ and $C\hat{B}D$, are supplementary (i.e. add up to 180°).

5.3 CIRCLES

A **circle** is the set of all points in a plane which are the same distance from a fixed point in the plane. The fixed point is the **centre** of the circle. The length of the perimeter of a circle is called its **circumference** (but we often use 'circumference' to mean the boundary) and a straight line segment from the centre to any point on the circumference is the **radius** of the circle.

A straight line segment from a point on the circumference through the centre to another point on the circumference is called a **diameter**. The length of a diameter is twice the length of a radius. In Figure 5.28(a) *O* is the centre, *AOB* is a diameter and *OP* is a radius. Often we speak of a circle of radius 4 cm when we mean that the length of a radius is 4 cm. In the same way we speak of a circle of diameter 8 cm.

Note that two circles which have the same radius are congruent. Two circles which have different radii are, nonetheless, similar.

A diameter divides the circle into two equal parts called **semicircles**; one half of a semicircle is a **quadrant**. The interior of the circle together with the circle itself is called a **disc**.

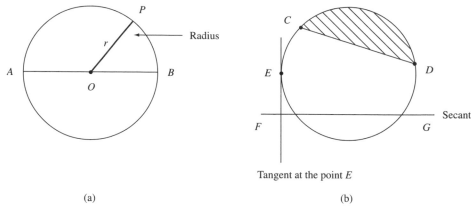

(a)   (b)

Figure 5.28 The circle

A straight line segment connecting two points on the circumference is a **chord**; *CD* in Figure 5.28(b) is a chord. A chord that is not a diameter divides the circle into two unequal parts; the smaller is called the **minor arc** and the larger is the **major arc**. A line which intersects a circle twice is called a **secant**; *FG* is a secant. Where a line touches the circle (i.e. has only one point in common) it is a **tangent**. Figure 5.28(b) shows the tangent which has *E* as its **point of contact**; we call it the tangent at *E*. A tangent is perpendicular to the radius from the centre of the circle to the point of contact.

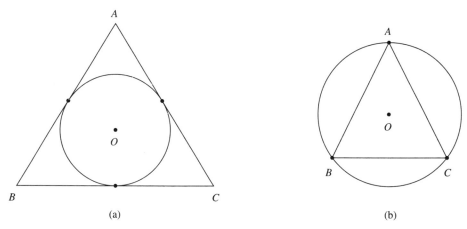

(a)   (b)

Figure 5.29 Circles and triangles: (a) incircle, the triangle circumscribes the circle; (b) circumcircle, the triangle is inscribed within the circle

The region inside the circle is also divided into two parts by a chord. The smaller part is called the **minor segment**, shaded in Figure 5.28(b), and the larger part is called the **major segment**. The major segment always contains the centre of the circle.

If a polygon is drawn around a circle in such a way that all its sides are tangents to the circle, we call it a circumscribed polygon. In Figure 5.29(a) *ABC* is a **circumscribed triangle** to the circle. We also refer to the circle as being **inscribed** relative to the triangle. In Figure 5.29(b) the situation is reversed, the circle circumscribes the triangle and the triangle is inscribed in the circle.

Two circles with the same centre and different radii are **concentric**; see Figure 5.30(a).

The area inside a circle cut off by two radii is called a **sector**; see Figure 5.30(b). In fact, the art produces two sectors; the smaller sector is shown shaded. Unless otherwise stated, we usually specify the sector which contains an acute angle at the centre, in this case *POQ*.

A very important result is that the ratio of the circumference of a circle to its diameter is constant. This ratio is written π and has an approximate value of 3.14159; in some situations we make use of the cruder approximation of $\frac{22}{7}$. In fact the exact value of π cannot be expressed as a fraction and therefore not as a finite decimal. The circumference of a circle is equal to $2\pi r$ where r is the radius. It can also be shown that the area of a circle A is equal to πr^2.

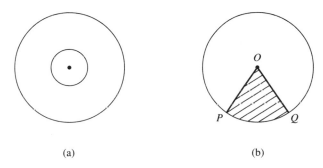

(a)   (b)

Figure 5.30 Centres and radii: (a) concentric circles have the same centre but different radii: (b) two radii divide a circle into minor sector (shaded) and major sector (unshaded)

The circumference of a circle $= 2\pi \times \text{radius}$

Area of a circle $= \pi \times (\text{radius})^2$

Examples

1. A circle of diameter 20 cm has radius 10 cm and circumference $2\pi \times 10$ cm $= 20$ π cm $= 62.8$ cm (3 s.f.). Its area is $\pi(100)$ cm^2, i.e. 314 cm^2 to the nearest integer.

2. Find the area of the region between two concentric circles whose radii are 5 cm and 10 cm. This region is called an **annulus** or a ring.

$$\text{The area of the larger circle} = \pi \times 10^2 = 100\pi \text{ cm}^2$$

$$\text{The area of the smaller circle} = \pi \times 5^2 = 25\pi \text{ cm}^2$$

Therefore the area of the annulus is 75π cm^2.

In general, if the larger radius is r_2 and the smaller radius is r_1 then the annular area is

$$\pi r_2^2 - \pi r_1^2 \equiv \pi(r_2^2 - r_1^2) \qquad \blacksquare$$

A **tangent** to a circle is a straight line drawn from an exterior point T to the circle and which touches the circle at a point on the circumference P. PT must be perpendicular to the radius OP; see Figure 5.31(a).

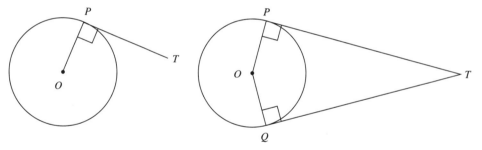

Figure 5.31 Circles and lines: (a) one tangent and (b) two equal tangents

In Figure 5.31(b) OP and OQ are equal—they are both radii—and angle $O\hat{P}T = O\hat{Q}T = 90°$; OT is a common line. The triangles OPT and OQT are therefore congruent, so the tangents TP and TQ are equal.

Circular Measure

In Figure 5.32(a) AB is the arc of a circle whose centre is O and whose radius is r. If the length of the arc $AB = r$, then $O\hat{A}B$ is defined to be an angle of magnitude one **radian**. Therefore if a sector of a circle contains an angle of one radian, its arc length is equal to the radius of the circle.

In general, if the arc length of a sector of a circle of radius r is s and the arc subtends an angle of θ radians at the centre of the circle, then $\theta = \dfrac{s}{r}$; see Figure 5.32(b).

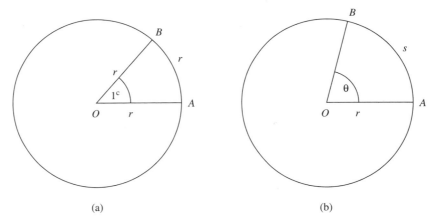

Figure 5.32 Radian measure

Since the circumference of the circle is $2\pi r$, the angle subtended at the centre (i.e. 1 revolution) is $\dfrac{s}{r} = \dfrac{2\pi r}{r} = 2\pi$ radians. Thus 1 revolution $= 2\pi$ radians $= 360°$, so

$$\pi \text{ radians} = 180°$$

$$1 \text{ radian} = \frac{180°}{\pi} \simeq 57.3°$$

The notation for one radian is 1^c. To convert an angle in degrees to radian measure we multiply by $\left(\dfrac{\pi}{180}\right)$. To convert an angle in radians to degree measure we multiply by $\left(\dfrac{180}{\pi}\right)$.

Examples

1. Convert the following to radians:

(a) 180° (b) 45° (c) 60° (d) 30° (e) 120° (f) 900°

Solution

(a) $\quad 180° = \pi^c$

(b) $45° = \left(\dfrac{\pi}{180} \times 45\right)^c = \left(\dfrac{\pi}{4}\right)^c$

(c) $\quad 60° = \left(\dfrac{\pi}{180} \times 60\right)^c = \left(\dfrac{\pi}{3}\right)^c$

(d) $30° = \left(\dfrac{\pi}{180} \times 30\right)^c = \left(\dfrac{\pi}{6}\right)^c$

(e) $\quad 120° = \left(\dfrac{\pi}{180} \times 120\right)^c = \left(\dfrac{2\pi}{3}\right)^c$ (f) $\;900° = \left(\dfrac{\pi}{180} \times 900\right)^c = (5\pi)^c$

2. Convert the following to degrees:

(a) $\left(\dfrac{\pi}{2}\right)^c$

(b) $\left(\dfrac{3\pi}{4}\right)^c$

(c) $\left(\dfrac{5\pi}{6}\right)^c$

Solution

(a) $\left(\dfrac{\pi}{2}\right)^c = 90°$

(b) $\left(\dfrac{3\pi}{4}\right)^c = \dfrac{3}{4} \times 180° = 135°$

(c) $\left(\dfrac{5\pi}{6}\right)^c = \left(\dfrac{5}{6} \times 180\right)° = 150°$ ∎

The **length of arc of a sector** of a circle of radius r which subtends an angle θ at the centre is given by:

$$s = r\theta$$

The area of the sector will be a fraction of the area of the circle, i.e. a fraction of πr^2. If the angle at the centre is θ then this fraction is $\dfrac{\theta}{2\pi}$, so the **area of a sector** is given by:

$$A = \dfrac{1}{2} r^2 \theta$$

Example

Find the length of arc of a sector of a circle of radius 3 cm which contains an angle of 65°. What is the arc of the sector?

First we convert the angle to radians:

$$65° = 65\left(\dfrac{\pi}{180}\right)^c$$

Then the arc length is

$$s = 3 \times 65\left(\dfrac{\pi}{180}\right) = 3.403 \text{ cm}$$

and the area is

$$A = \frac{1}{2} \times 3^2 \times 65\left(\frac{\pi}{180}\right) = 5.105 \text{ cm}^2 \blacksquare$$

Exercise 5.3

1   Find the shaded areas in each diagram.

(a)

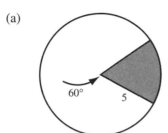

(b)

Inner radius = 6 cm
Outer radius = 8 cm

(c)

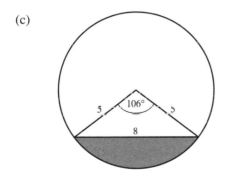

(d)

$AP = 25$ cm
$PQ = 14$ cm
$PB = 11\frac{2}{3}$ cm
$P\hat{A}Q = 32.5°$

2 What is the area of this figure? The curved arcs are semicircles.

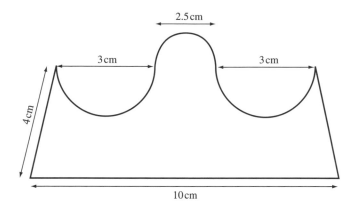

3 This circle has radius 12 cm and $PT = 35$ cm. How far is T from the centre of the circle?

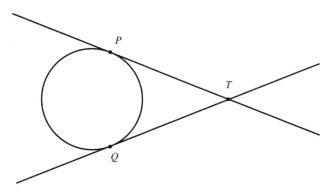

4*

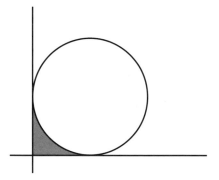

(a)   The circle depicted has radius r. What is the magnitude of the shaded area in terms of r?

(b)   The largest circle that can be drawn inside the shaded region touches the original circle and has common tangents. What do you notice about its tangents and in particular the relative distances of centre and tangent points from the origin? Use your observations to find the ratio between the radius of this new circle and the radius of the original circle.

5   A circle of radius 3 cm is inscribed on an isosceles triangle ABC. D, E, F are the points where the circle touches the sides of triangle ABC.

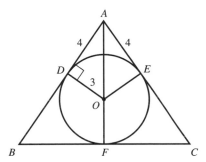

Use Pythagoras' Theorem to find the length of the sides of triangle ABC by

(a)   determining AO

(b)   setting $BF = BD = x$ units and considering triangle ABF.

6   Determine the lengths requested in the following examples:

(a) BC, given $y = 7$ (b) AB, CD

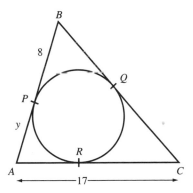

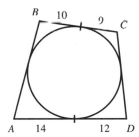

(c)   *x*, given *r* = 5,
      r, given *x* = 10

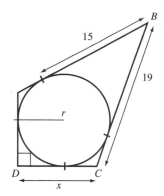

7* The equilateral triangle *ABC* has perpendiculars drawn from each vertex to the midpoints of the other sides. These meet at a common point *I*; note the related symmetries.

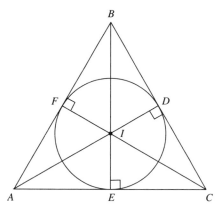

(a)   Show that the smaller triangles, *IFB*, etc., are similar to △*BEA*.

(b)   What is the ratio of their respective areas?

(c)   Show that *I* is two-thirds of the way down *BE* using the similar triangles in (a).

(d)   Assuming that a circle can be drawn, centre *I*, which touches the sides of *ABC* at *D, E, F*, i.e. inscribed in △*ABC*, determine the ratio of the circle's area to the area of △*ABC*.

8* Now consider the circle which circumscribes an equilateral triangle as well. Determine the ratio area of inscribed circle : area of triangle : area of circumscribed circle.

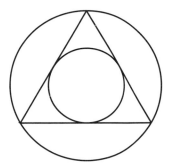

9* Determine the equivalent ratio for a square, i.e. area of incircle : area of square : area of circumcircle.

10* Incircles and circumcircles also exist for any regular polygon. Determine the equivalent ratio for a hexagon, i.e. area of incircle : area of hexagon : area of circumcircle.

11 Determine the unknowns in the following cases:

(a)

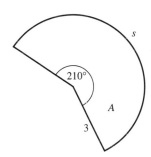

Area A, arc length s

(b)

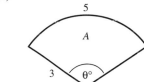

Angle θ, area A

(c)

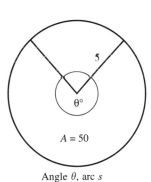

Angle θ, arc s

(d)

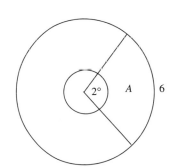

Radius r, area A

5.4 SHAPES IN THREE DIMENSIONS

A **solid** is a portion of three-dimensional space bounded by plane (i.e. flat) and/or curved surfaces. Practical examples of solids are a box, a ball and a wedge.

A **polyhedron** (plural polyhedra) is a solid bounded by plane surfaces only. A box and a wedge are two examples.

Polyhedra

If every face is a rectangle then the polyhedron is known as a **cuboid** or **box**; see Figure 5.33(a). If every face is a square then the solid is known as a **cube**. The plane surfaces are called **faces**, the lines where two faces meet are called **edges**, and the points where the edges meet are called **vertices**. A line which joins two vertices that are not in the same face is a **diagonal**; see Figure 5.33(b). If every face is a parallelogram then the solid is known as a **parallelepiped**, Figure 5.33(c). (The cuboid is sometimes called a rectangular parallelepiped.)

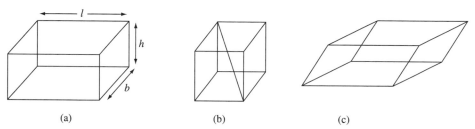

Figure 5.33   Four-sided prisms: (a) cuboid, (b) cube and (c) parallelepiped

In Figure 5.33(a) the length of the box is l, the breadth (or width) b and the height h; these are the **three dimensions** of the box. In the case of the cube, $l = b = h$. Each of the solids in Figure 5.33 has six faces, twelve edges and eight vertices.

In Figure 5.34 the box $ABCDEFGH$ has faces $ABCD$, $EFGH$, $ABFE$, $DCGH$, $AEHD$ and $BFGC$; edges AB, BC, CD, DA, EF, FG, GH, HE, BF, AE, CG, DH; vertices A, B, C, D, E, F, G, H; and diagonals AG, BH, CE, DF.

A **prism** is a polyhedron with two 'opposite' faces which are polygons of the same size and shape and parallel to each other; these faces are called **bases**. The other (lateral) faces are parallelograms. If the lateral faces are rectangles then the solid is a **right prism**; the lateral faces will be at right angles to the bases.

Figure 5.35(a) shows a right triangular prism and Figure 5.35(b) shows a right hexagonal prism. Each of these solids stands on a base whereas the prism in Figure 5.35(c) rests on one of its lateral faces. If a prism is cut by a plane parallel to the base, the cross-section of either exposed face is congruent to that base; see Figure 5.36(a).

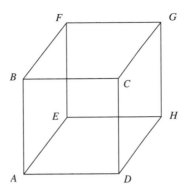

Figure 5.34   A box

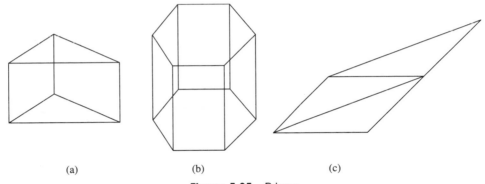

(a)   (b)   (c)

Figure 5.35   Prisms

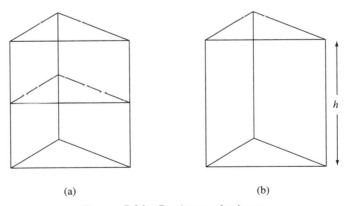

(a)   (b)

Figure 5.36   Features of prisms

The area of a base is also the **cross-sectional area** of the prism, and the length of an edge joining corresponding points on the base and the upper face is the **height** of the prism, h in Figure 5.36(b).

A **pyramid** has a polygon as its base; the other faces must be triangles which meet at a point called the vertex of the pyramid. Figure 5.37(a) shows a pyramid on a rectangular base *ABCD*. The **sloping faces** *ABE*, *ADE*, *BCE* and *DCE* are triangles meeting at the vertex *E*.

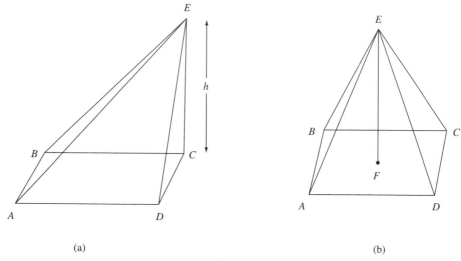

(a) (b)

Figure 5.37 Pyramids

The **height** of the pyramid, h, is the distance of the vertex from the base. In Figure 5.37(b) the pyramid has a square base. The line segment *EF* is perpendicular to the base; it is the **altitude** of the pyramid. Its magnitude is the height h. In this case, the base is a square and *F* is the centre of the base.

The **tetrahedron** has four triangular faces; see Figure 5.38. It can be regarded as a pyramid on a triangular base.

A **regular polyhedron** is a solid all of whose faces are regular polygons with the same number of edges meeting at each vertex. One example is a cube in which all faces are squares and three squares meet at each vertex. A regular tetrahedron has equilateral triangles for all its faces.

Cylinders

We shall confine our attention to a **right circular cylinder**, commonly called a cylinder. A cylinder is a solid with its bases as parallel circles and the line joining the centres of the

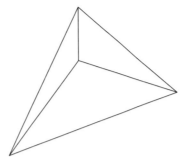

Figure 5.38 Tetrahedron

circles being perpendicular to the bases. The length of this line is the height of the cylinder and the radius of either base is the radius of the cylinder. For the cylinder shown in Figure 5.39 the height is h and the radius r. Any cut parallel to a base exposes circular surfaces. The cross-sectional area is the area of a base.

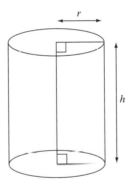

Figure 5.39 Cylinder

A cylinder may be thought of as the limiting case of a prism when the number of sides increases indefinitely. The limiting case of a pyramid with a regular polygon as a base is a cone.

Cones

A **right circular cone**, commonly called a cone, has a circular base. A fixed point called the **vertex** lies vertically above the centre of the base and the conical surface is swept out by a straight line segment with one end at the vertex and the other end tracing out the circumference of the base.

Alternatively, the cone can be formed by rotating a right-angled triangle about one of its shorter sides. This side becomes the altitude of the cone and the other short side forms a radius of the base. The length of the altitude is the height of the cone and the length of the hypotenuse of the triangle is the **slant height** of the cone; see Figure 5.40.

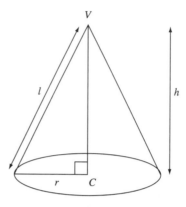

Figure 5.40   Cone

Spheres

A sphere is the three-dimensional analogue of the circle. It is a solid such that every point on its surface is the same distance from a fixed point called the **centre**. This fixed distance is the radius of the sphere. It can be formed by rotating a semicircle through a complete turn about its diameter.

A segment of a sphere is formed when the sphere is cut by one plane or two parallel planes; see Figure 5.41.

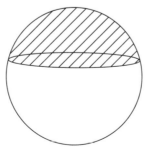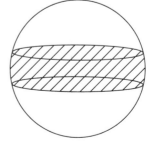

Figure 5.41   Spheres

Surface areas and volumes

We quote standard formulae for surface areas S and volumes V of some of the solids we have mentioned.

- **Box**: the surface area of a box is the sum of the areas of its six faces which form three pairs; see Figure 5.33(a). It is given by

$$S = 2(lb + bh + lh)$$

The volume of a box is given by

$$V = lbh$$

- **Cube**: in this case $l = b = h$ and the formulae simplify to

$$S = 6l^2, \qquad V = l^3$$

- **Prism**: if the area of the base is A and the height h, then

$$V = Ah$$

- **Pyramid**: if the area of the base is A and the height h, then

$$V = \frac{1}{3}Ah$$

- **Cylinder**: refer to the cylinder shown in Figure 5.42(a). The base has area πr^2 where r is the radius of the base. Thus

$$V = \pi r^2 h$$

The total surface area is

$$A = 2\pi r^2 + 2\pi rh$$

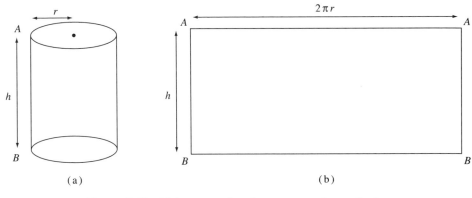

Figure 5.42 Volume and surface area of a cylinder

To find the total surface area we note there are two identical circular faces at the top and bottom along with a vertical side. Imagine a soup can and suppose that we cut the label along a vertical line AB. The label can be spread out into a rectangle, Figure 5.42(b); it has a length equal to the circumference of the base, namely $2\pi r$, and a breadth equal to the height of the cylinder, namely h. Its area is then $2\pi rh$. Hence the total surface area is given by

$$A = 2\pi r^2 + 2\pi rh$$

Cone

The volume of a cone of radius r and height h is given by

$$V = \frac{1}{3}\pi r^2 h$$

Note that, like a pyramid, the volume is given by $\frac{1}{3} \times$ base area $\times$ height.

To find the curved surface area of the cone in Figure 5.43(a) we cut the conical surface along the edge AB. The surface is then opened out to the circular sector in Figure 5.43(b). The sector is a part of the circle of radius l. The full area of the circle is πl^2. The arc length BCB' is $2\pi r$, since it was the circumference of the base of the cone. However, the circumference of the circle in Figure 5.41(b) is $2\pi l$. The area of the sector shown is therefore the fraction $\left(\dfrac{2\pi r}{2\pi l}\right)$ of the area πr^2, i.e.

$$\left(\frac{2\pi r}{2\pi l}\right) \times \pi l^2 = \pi rl$$

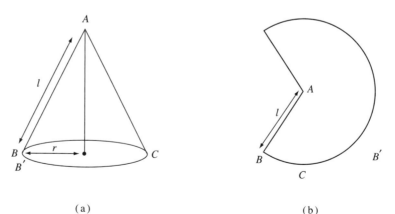

(a)   (b)

Figure 5.43   Volume and surface area of a cone $AB = CD = DE = FA$, $BC = EF$

The total surface area of the cone is therefore

$$A = \pi r^2 + \pi r l$$

Sphere

The surface area of a sphere of radius r is given by

$$S = 4\pi r^2$$

Its volume is given by

$$V = \frac{4}{3}\pi r^3$$

Exercise 5.4

1 Here are a tetrahedron and a parallelepiped with the same linear dimensions.

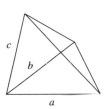 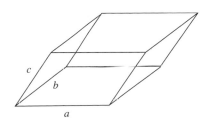

(a)   How many tetrahedra (plural of tetrahedron) are needed to build the parallelepiped?

(b)   Are the linear dimensions of these tetrahedrons exactly the same?

2 This solid consists of a cuboid 10 cm long of square cross-section, 4 cm × 4 cm, with two prismatic endpieces, each 6 cm long.

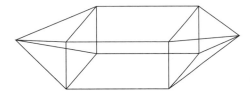

For the composite solid, determine:

(a)   the volume   (b)   the surface area

3 A metal cuboid with linear dimensions l, b and h cm is heated to a higher temperature and the linear dimensions all increase by 1%. To the nearest percent find the increase in:

(a)   the volume   (b)   the surface area

4 A cone of height $3r$ and radius r rests on top of a hemisphere of radius r. In terms of r find the volume of the composite solid.

5 A girder 8 m long has a symmetric cross-section as shown. If the girder is made from reinforced steel of density 8 kg cm^{-3}, determine its weight in kilograms.

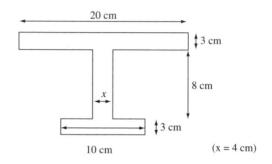

6 A spherical meteorological balloon holds 70 litres of helium. How many more litres need to be pumped in to increase the radius by 3.1%? (1 litre = 100 cm^3)

7 A hexagonal prism of igneous rock crystal of density 3.7 g cm^{-3} is 30 cm long and each hexagonal side is 4.6 cm in length. What is its weight?

8* This figure represents a globe, a three-dimensional map of the Earth with some longitude lines drawn in, the 45° N parallel and the equator. The equator is usually a **great circle**, i.e. one whose centre is the centre of the globe; any parallel of latitude is a **small circle**.

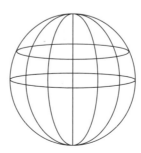

If the radius of the Earth is r, show that:

(a)   the radius of the 45° N small circle is $r/\sqrt{2}$

(b)   the distance from longitude 0° to longitude 180° following the 45° N parallel is $\pi r/\sqrt{2}$

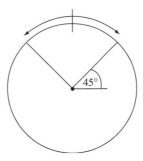

(c)   the distance between the two points in (b) is only $\pi r/2$ when travelling over the North Pole (about 30% less)

This demonstrates the principle of great circle travel that ships and aircraft follow. Try it by stretching a string over a globe, e.g. from London to New York or Sydney.

SUMMARY

- **Angles:**

	Acute	$<90°$
	Acute	$<90°$
$90°<$	Obtuse angle	$<180°$
$180°<$	Reflex	$<360°$

- **Parallel lines:** three sets of equal angles are generated when a transversal cuts parallel lines. They are vertically opposite, corresponding and alternate

- **Triangles:** the sum of the interior angles is 180°. Triangles may be classified by the number of equal, sides:

Scalene:	no sides equal
Isosceles:	two sides equal
Equilateral:	all sides equal

- **Congruent triangles:** look for equality in one of the following sets.

  Two sides and the included angle
  Two angles and the side joining them
  Three sides
  Right angle, hypotenuse and another side

- **Similar triangles:** corresponding angles are equal; the lengths of corresponding sides are in the same ratio.

- **Area of a triangle:** $\frac{1}{2} \times$ base $\times$ height

- **Interior angles of a polygon:** sum to $(2n-4) \times 90°$

- **Sectors and arcs:**

$$\pi^c = 180°$$

$$\text{Arc length} = r\theta$$

$$\text{Sector area} = \frac{1}{2}r^2\theta$$

- **Properties of a circle:** If the radius is r then

$$\text{Area} = \pi r^2$$

$$\text{Circumference} = 2\pi r$$

- **Volumes of solids:**

Cuboid:	base area × height
Tetrahedron:	$\frac{1}{3}$ × base area × height
Pyramid:	$\frac{1}{3}$ × base area × height
Sphere:	$\frac{4}{3}\pi r^3$
Cylinder:	$\pi r^2 h$
Cone:	$\frac{1}{3}\pi r^2 h$

- **Surface areas of solids:**

Sphere:	$4\pi r^2$
Closed cylinder:	$2\pi r^2 + \pi r^2 h$
Closed cone:	$\pi r^2 + \pi r l$

Answers

Exercise 5.1

1 (a) 215° (b) 25° (c) 110° (d) $x = 60°$, $y = 70°$

2 $\alpha = x$, PQ will not meet AB; $\alpha = x + y$; PQ will not meet CD; $\alpha = x + y + z$, PQ will not meet EF

3 $A\hat{B}C = 90° - x$, $A\hat{C}B = 90° - y$, sum $= 180°$

4 (a) $AP = 12$ cm, $PB = 16$ cm

(b) 20/7 units

(c) $AB = 3.6$ cm, $BC = 3$ cm, $CD = 5$ cm

(d)

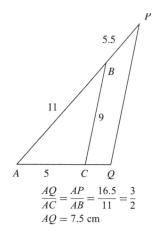

$$\frac{AQ}{AC} = \frac{AP}{AB} = \frac{16.5}{11} = \frac{3}{2}$$

$$AQ = 7.5 \text{ cm}$$

5 (a) Yes (b) No (c) Yes (d) No

(e) Yes (f) Yes (g) Yes (h) Yes

6 A triangle whose sides are $(3x, 4x, 5x)$ is right-angled.

7 $CD = 17.32$ cm, P is 5 cm from CD

8 (a) $\hat{A} = 110°$, $\hat{C} = 70°$

(b) $\hat{B} = 85°$, $\hat{C} = 190°$

(c) $\hat{B} = \hat{C} = \hat{E} = \hat{F} = 140°$, $\hat{D} = 80°$

(d) $\hat{A} = 45°$, $\hat{B} = 110°$, $\hat{C} = 130°$, $\hat{D} = 120°$, $\hat{E} = 110°$, $\hat{F} = 70°$, $G = 315°$

9 (a)   30   (b)   924 $(DP = 24)$

   (c)   74   (d)   8.186 (3 d.p.)

10 (a)   25   (b) $53/16 = 3.313$ (3 d.p.)

   (c)   3.745 (3 d.p.)   (d)   14.98 (2 d.p.)

12 $\dfrac{y}{2}\left(x^2 - \dfrac{y^2}{4}\right)^{1/2}$

13 (a) $\dfrac{\sqrt{3}a^2}{4}$ (b) $\dfrac{3\sqrt{3}a^2}{2}$

14 $2(1 + \sqrt{2})$

15 (a)   12

18 (a)   all equal 6; hexagons meeting (i.e. honeycomb)

   (b) $c = 4$; dodecagon (12), hexagon (6) and square (4) meeting

   (c)   triangle and two dodecagons

   (d)   hexagon, heptagon (7) and regular 42-sided figure meeting

Exercise 5.2

2 $\triangle CGE = \triangle CFB$ (and $FB \parallel EG$)

3 $\triangle CFB$

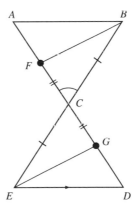

4 $AE = BD$ and $AE \parallel BD$

6 (a) equal and parallel

(b) parallelogram

(c) $AC = 2EF = 2HG$ and $AC \parallel EF \parallel HG$

7 $\triangle DEB$

8 $1:4$

11 $A\hat{H}B = 90°$

12 Total of all angles is 360°, so result follows.

Exercise 5.3

1 (a) $25\pi/6$ (b) 28π

(c) 11.126 (3 d.p.)   (d) 31.62 cm^2 (2 d.p.)

2 31.73 cm^2 (2 d.p.)

3 37 cm

4 (a) $r^2(1 - \pi/4)$ (b) $\dfrac{\sqrt{2} - 1}{\sqrt{2} + 1} = 0.1616$ (4 d.p.)

5 (a) 5 cm   (b) $AB = AC = 10$ cm, $BC = 12$ cm

6 (a) 18 (b) $24, 21$

(c) $r = 5, x = 9$; $r = 6, x = 10$

7 (b) $1:3$ (c) $IE:IB = IE:IA = AE:AB = 1:2$

(d) $\pi : 3\sqrt{3}$

8 $1:4$

9 $\pi : 4 : 2\pi$

10 $3\pi/4 : 3\sqrt{3}/2 : \pi$

11 (a) $s = 10.996, A = 16.493$ (b) $\theta = 95.49°, A = 7.5$

(c) $\theta = 4^c = 229.2°, s = 20$ (d) $r = 3, A = 9$

Exercise 5.4

1   (a)   6   (b)   Yes

2   (a)   224 cm^3 (b)   261.19 cm^2

3   (a)   3%   (b)   2%

4   $\frac{5}{3}\pi r^3$

5   780.8 kg

6   6.71 litres

7   6.102 kg

6 PROOF

INTRODUCTION

The development of mathematics relies on the framework provided by proof. Many of the mathematical procedures we carry out can take place with the knowledge that underlying what we do is a bedrock of proof. When we obtain a solution to a particular equation we are sure it is *the* unique solution because the method has been proved to yield a unique solution. Yet it is easy to take for granted that the proof is there. It might be an exaggeration to say that the whole of trigonometry rests on Pythagoras' theorem, but in reality it is not much of an exaggeration. An appreciation of the main methods of proof helps us to make our own mathematical arguments rigorous and it gives us confidence in drawing conclusions from the application of mathematics in our area of specialisation.

OBJECTIVES

After working through this chapter you should be able to

- distinguish between an axiom and a theorem
- understand how a theorem can be derived from a set of axioms
- appreciate the development of a corollary from a theorem
- follow the proof of Pythagoras' theorem
- understand the proofs of the theorems on concurrency of sets of lines related to triangles
- understand the proofs of theorems relating to equality of angles related to circles
- recognise the methods of *reductio ad absurdum*, induction and negation
- understand the logical development of a proof
- appreciate the economy of effort in a proof
- appreciate the absolute nature of a proof
- understand the distinction between a necessary condition and a sufficient condition

6.1 DEDUCTIVE REASONING AND PYTHAGORAS' THEOREM

The philosophy of mathematics rests upon **proof**, and the aim of this chapter is to set in place some of the building blocks of proof. We will keep the arguments simple, although some proofs are often difficult.

We will accept reasonably-held definitions and properties as well as seemingly fundamental truths or results. For example, we accept that every second integer, positive or negative, is exactly divisible by 2. Similarly we accept that every third integer is exactly divisible by 3.

By doing this we can assemble building blocks into bigger mathematical structures called **theorems**. The theorem in the first example has important consequences for mathematical analysis and computer science. We prove it by **deductive reasoning**. At a more basic level, certain fundamental truths are called **axioms**; see the second example.

Examples

1. The product of three successive **positive integers** is always exactly divisible by 6, i.e. the result of dividing the product by 6 is an integer. For example,

$$\frac{2 \times 3 \times 4}{6} = 4, \qquad \frac{7 \times 8 \times 9}{6} = 84$$

Proof: Take the first integer in the sequence to be n. Then consider the product $n(n + 1)(n + 2) = K$.

We can draw two conclusions:

(a) Either both n and $n + 2$ or, exclusively, $n + 1$ must be even; therefore 2 divides K.
(b) The number 3 divides every third integer, so it divides one, and only one, of n, $n + 1$ and $n + 2$; therefore, 3 divides K.

Any number divisible exactly by 2 and 3 is also divisible by 6, so the result follows.

2. **Archimedes' axiom** states that, given any real number x, there exists an integer whose value exceeds x.

This is easily proved. Let N be the largest integer not greater than x, so $N \leq x$ and $x - N \leq 1$. Now consider $N + 1$; obviously $N + 1 \geq x$, but $N + 2 > x$ and the result is proved. ∎

In other words, Archimedes' axiom claims there is no largest integer. One way of thinking of the result is to imagine the integers as equally spaced benchmarks on an infinitely long ruler which represents the set of all real numbers, called $\mathbb{R}$. Refer to Figure 6.1.

As x is any point upon this line, benchmarks must exist to the left and the right, i.e. there are integer values both above *and* below. The axiom is concerned only with integers above x.

Figure 6.1 Integers

The first example demonstrated that the results of a theorem may not always be obvious; almost invariably they lead to further results, and this is how mathematics is built up.

In geometry we have already used some deductive reasoning with congruence and similarity, and we have used the axiom which states that the interior angles in a triangle must add up to 180°.

Perhaps the most important theorem in the whole of mathematics is Pythagoras' theorem. In its familiar form with reference to the right-angled triangle *PQR* shown in Figure 6.2, then Pythagoras' theorem states that the square on the hypotenuse *PR* is equal to the sum of the squares of the other two sides, *PQ* and *QR*. In other words,

$$PQ^2 + QR^2 = PR^2$$

This can be interpreted in Figure 6.2 as the sum of the areas of the two smaller squares is equal to the area of the largest square.

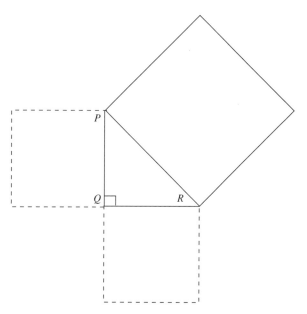

Figure 6.2 Pythagoras' theorem

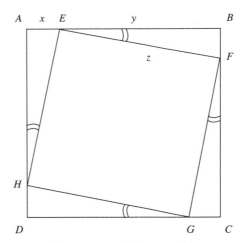

Figure 6.3 Proving Pythagoras' theorem

Example

To prove Pythagoras' theorem we use the construction shown in Figure 6.3, where $ABCD$ is a square. The points E, F, G, H are marked out at equal distances along the sides, i.e. $AE = BF = CG = DH = x$ and $EB = FC = GD = HA = y$. $ABCD$ therefore has side $x + y$.

Now we examine the interior of $ABCD$. First of all, $\hat{A} = \hat{B} = \hat{C} = \hat{D} = 90°$, and the triangles EBF, FCG, GDH and HAE must be congruent, having two sides and the included angle equal. It follows therefore that the third sides must be equal, i.e. $EF = FG = GH = HE = z$, say.

So $EFGH$ is a rhombus. It looks as though it could be a square. We establish this to be true.

Note that the straight angle $A\hat{E}B = A\hat{E}H + H\hat{E}F + F\hat{E}B = 180°$. Now $F\hat{E}B = A\hat{H}E$ and $A\hat{H}E + A\hat{E}H = 90°$ so that $A\hat{E}B = H\hat{E}F + 90°$. Therefore $H\hat{E}F = 90°$.

The same reasoning shows that $E\hat{F}G = F\hat{G}H = G\hat{H}E = 90°$. $EFGH$ must therefore be a square of size z.

We now compare areas:

$$\text{The area of } ABCD = (x + y)^2 \equiv x^2 + 2xy + y^2$$
$$\text{But the area of } ABCD = 4 \times \text{area of } \triangle EFB + \text{area of } EFGH$$
$$= 4 \times \frac{1}{2} xy + z^2 = 2xy + z^2$$
$$\text{Therefore} \quad x^2 + 2xy + y^2 = 2xy + z^2$$

Cancelling the term $2xy$ we obtain the result

$$x^2 + y^2 = z^2$$

and the theorem is proved.

■

The entire theory of trigonometry is based upon Pythagoras' theorem as is the theory of coordinate geometry and the measurement of distance in two and three dimensions. We will use the theorem many times.

It is worth noting that when the numbers x and y are integers, z is not usually an integer. If x, y and z are all integers they form a **Pythagorean set**.

Examples

1. **Pythagorean sets:** four examples of Pythagorean sets are:

	x	y	z	
(a)	3	4	5	$(3^2 + 4^2 = 5^2)$
(b)	5	12	13	
(c)	7	24	25	
(d)	12	35	37	

Integer or fractional multiples of these sets also form Pythagorean sets; for example, 6, 8, 10 and 6, 17.5, 18.5.

2. **Using Pythagoras' theorem:** in Figure 6.4 H lies vertically below G and both the triangles are right-angled. Use Pythagoras' theorem to find the distance AG.

First, $(EC)^2 = (ED)^2 + (DC)^2$ so that $(ED)^2 = (EC)^2 - (DC)^2 = 20^2 - 12^2 = 256$.

Therefore $ED = 16$ and $FD = 34$. Also $FG^2 = \left(22\frac{1}{2}\right)^2 - 18^2 = 182\frac{1}{4} = \left(13\frac{1}{2}\right)^2$.

Then $FG = 13\frac{1}{2}$, so $GH = FH - FG = 32 - 13\frac{1}{2} = 18\frac{1}{2}$.

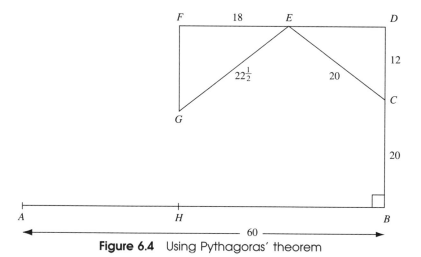

Figure 6.4 Using Pythagoras' theorem

But $AH = AB - HB = AB - FD = 26$, therefore

$$AG = \sqrt{AH^2 + GH^2} = \left((26)^2 + \left(18\frac{1}{2} \right)^2 \right)^{1/2} = 31.91$$ ∎

Exercise 6.1

1 The product of an integer one less than a given integer and an integer one more than the given integer appears to be one less than the square of the given integer, e.g.

$$7 \times 5 = 6^2 - 1 = 35, \quad 8 \times 10 = 9^2 - 1 = 80$$

Prove that this result is always true by labelling the given integer N and forming the product described.

2 Show that $N - 1$ is a factor of $N^3 - 1$ where N is a positive integer. Hence prove that an integer one less than a perfect cube, e.g. 63 or 124, is **composite**, i.e. not prime, with only one exception.

3 (a) Prove that the product of four consecutive integers multiplied together is always divisible by 24.
(b) What is the greatest common divisor of five consecutive integers multiplied together?

4 Identify which of the following are Pythagorean sets:

(a) 14, 48, 50 (b) $3, 8\frac{3}{4}, 9\frac{1}{4}$

(c) 0.9, 1.2, 1.5 (d) 9, 17, 20

(e) 8, 15, 17 (f) 21, 28, 35

5 A ladder 5.2 m long is put up against the wall of a building with its foot set 2 m out from the base. How far up the building does it stretch?

6 Determine the unknowns in the following diagrams. A is vertically above B in part (b).

(a)

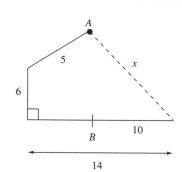

(b)

7   Right-angled triangle ABC is isosceles, $AB = AC$.

(a)   What are $A\hat{B}C$ and $A\hat{C}B$?
(b)   What is the ratio BC/AB?

8   Determine the unknown lengths in this figure.

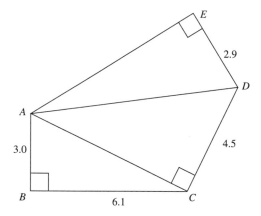

9   This circle has radius r and area πr^2. Determine the areas of both the inner square and the outer square as drawn and deduce that π must lie between 2 and 4.

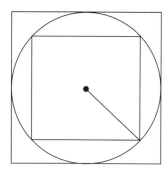

10* The cut-off points in this figure are all in the same ratio, $\dfrac{AE}{EN} = \dfrac{BF}{FC} = \dfrac{FI}{IE}$, etc., and the inner shaded square has exactly half the area of the outer square. What is the cut-off ratio?

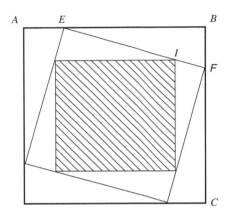

6.2 THEOREMS IN CLASSICAL PLANE GEOMETRY

Deductive reasoning, in which one result leads to another and so on until an end result emerges, was well understood in classical times. Classical plane geometry is based upon this kind of reasoning, as we saw in proving Pythagoras' theorem. We shall reinforce the deductive reasoning process by proving other theorems which will be useful later when the geometry is put into practice.

Starting with the triangle and relying upon no more than the concepts of similarity and congruence, we can prove several important theorems. Some construction, or drawing, may be necessary.

Examples

1. **The bisectors of the angles of a triangle meet at a point, the incentre, which is equidistant from each side.** In triangle ABC we bisect $\hat{CAB}$ and $\hat{ABC}$, as shown in Figure 6.5. The bisectors of the angles can be assumed to meet at a point I. Draw the lines ID and IE perpendicular to AB and BC respectively.

   Look now at the triangles IBD and IBE. Since $D\hat{B}I = E\hat{B}I$ and the angles at D and E are 90°, the triangles are similar.

   Notice too that IB is common to both of them. Hence $\triangle IBD \equiv \triangle IBE$, i.e. the triangles are congruent, so $ID = IE$.

   The same reasoning applies to the triangles IAD and IAF, where F is the foot of the perpendicular from I to AC, so that $\triangle IAD \equiv \triangle IAF$ and $ID = IF$.

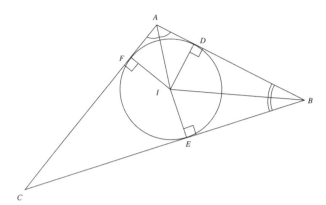

Figure 6.5 Angle bisectors are coincident

Hence $ID = IE = IF$ and I is a point equidistant from each side; I is therefore the centre of a circle which touches each side with ID, IE and IF as radii.

Draw IC and observe that for the triangles IEC and IFC

(i) $IE = IF$
(ii) IC is common
(iii) $I\hat{E}C = I\hat{F}C = 90°$

Hence $\triangle IEC \equiv \triangle IFC$ so that $I\hat{C}E = I\hat{C}F$ and $B\hat{C}A$ is bisected.

You can see that the argument must go the other way too. Namely, if a circle which touches each side is drawn inside the triangle and has centre I, then the lines IA, IB, IC bisect the angles at the vertices A, B and C. By drawing the picture as shown in Figure 6.5 the triangles IBD and IBE have two sides and a corresponding right angle in common. Therefore $I\hat{B}D = I\hat{B}E$ so that $A\hat{B}C$ is bisected, and so on.

We conclude from the two-way argument that the bisectors of the angles lead to an interior touching circle and that any interior touching circle leads to angle bisection. Because angles are bisected uniquely, the interior circle must also be unique.

2. **The perpendicular bisectors of the sides of a triangle meet at a point, the circumcentre.** Refer to Figure 6.6(a). First bisect the sides AB and BC perpendicularly at the points D and E, and let these perpendicular bisectors meet at O. Now draw the lines AO and BO, two of the dashed lines in the figure. For the triangles AOD and BOD notice that

(i) $AD = DB$
(ii) $A\hat{D}O = B\hat{D}O = 90°$
(iii) OD is common

With two sides and a common angle equal $\triangle AOD \equiv \triangle BOD$, it follows that $OA = OB$. Now draw the line segment OC, and consider the triangles BOE and COE. Then

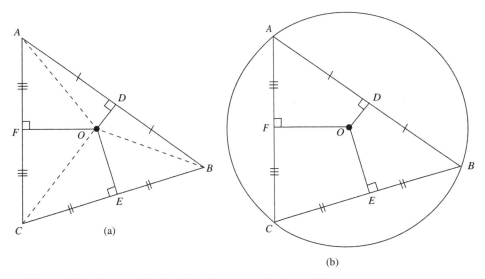

Figure 6.6 Perpendicular bisectors meet at a point

(i) $CE = EB$
(ii) $C\hat{E}O = B\hat{E}O = 90°$
(iii) OE is common

Hence $OC = OB$ and therefore $OC = OB = OA$. Therefore OC, OB annd OA can be regarded as radii of a circle centre O passing through A, B and C; see Figure 6.6(b). We draw the perpendicular OF to C, where F is not yet known to be the midpoint of AC. However, we do not know that for triangles AOF and COF

(i) OF is common
(ii) $AO = CO$
(iii) $A\hat{F}O = C\hat{F}O = 90°$

Hence $\triangle AOF \equiv \triangle COF$ and therefore $AF = FC$, so F is the midpoint of AC. The perpendicular bisectors OD, OE and OF therefore meet at a point, namely O.   ■

We have now established the result that there is a circle which passes through the vertices of a triangle. Putting it more generally, there is a circle passing through any three non-collinear points (i.e. not on the same straight line) in a plane. Is is unique?

Start with three non-collinear points which form a triangle ABC. Take O to be the centre of a circle passing through A, B and C. Clearly, $OA = OB = OC$. Now draw bisectors from O to meet the sides AB, BC, AC, at D, E, F. You can then prove that $\triangle AOD \equiv \triangle AOD$ and that $A\hat{D}O = B\hat{D}O = 90°$, etc.

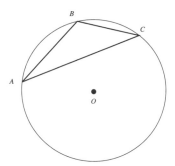

Figure 6.7   Perpendicular bisectors meeting outside the triangle

Note that if ABC is an obtuse triangle then the point O lies outside the triangle, as in Figure 6.7.

The following two theorems are stated without proof. They can be proved by methods similar to those used in the previous examples.

- The altitudes of a triangle are concurrent (i.e. meet at a point). The point where they meet is called the **orthocentre** (Figure 6.8).

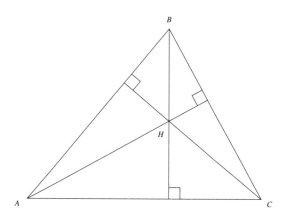

Figure 6.8   Orthocentre of a triangle

- The lines joining the vertices of a triangle to the midpoints of the opposite sides (i.e. the medians of the triangle) are concurrent (Figure 6.9).

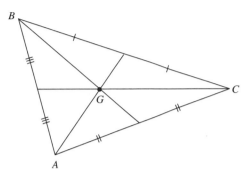

Figure 6.9 Medians are concurrent

G is called the **centroid** of the triangle, i.e. the point about which a cardboard cut-out of the triangle ABC would balance. In the eighteenth century Euler proved that the points O, H and G all lie on a straight line, called the **Euler line**.

Circle Theorems

The classical theorems involving the circle are proved using deductive reasoning together with certain other triangle properties.

Example

'Angle in a segment' theorem: in Figure 6.10 A and B are two points on the circumference of a circle centre O. The point C on the **major arc** AOB is chosen so the radius OC is interior to $A\hat{C}B$.

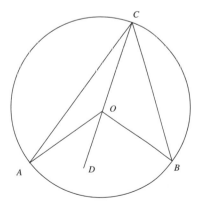

Figure 6.10 'Angle in a segment' theorem

Observe that the triangles ACO and BCO are each isosceles because $AO = CO = BO$, all radii of the circle. Hence

$$O\hat{A}C = O\hat{C}A \quad \text{and} \quad O\hat{C}B = O\hat{B}C$$

Now we produce CO to D. Via the exterior angles of the triangles and the isosceles properties of the triangles, we see that

$$A\hat{O}D = O\hat{A}C + O\hat{C}A = 2O\hat{C}A$$
$$D\hat{O}B = O\hat{C}B + O\hat{B}C = 2O\hat{C}B$$

Adding the equations, we obtain

$$A\hat{O}B = 2(O\hat{C}A + O\hat{C}B) = 2A\hat{C}B$$

In other words the angle at the centre of circle (subtended by the points A and B) is twice the angle subtended on the circumference (at the point C). ∎

The only condition we put on C was that the radius OC had to be inside $A\hat{C}B$. This meant that at any point C on the circumference the angle at the circumference is half that at the centre no matter where C placed, provided OC was inside $A\hat{C}B$.

It can also be proved that there is no need for OC to be interior to $A\hat{C}B$ (see Figure 6.11). In other words

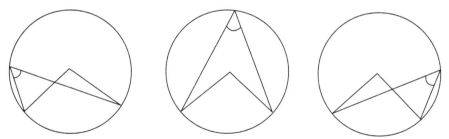

Figure 6.11 Angle at the centre is twice the angle at the circumference

$$A\hat{C}B = \frac{1}{2}A\hat{O}B$$

for any point C on the major arc AB.

The theorem also holds on the **minor arc** AB (Figure 6.12). In this case $A\hat{C}B = \frac{1}{2}A\hat{O}B$ where $A\hat{C}B$ is obtuse and $A\hat{O}B$ reflex.

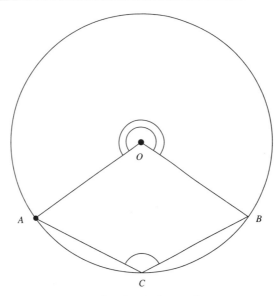

Figure 6.12   Angle in a minor arc

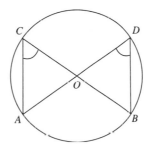

Figure 6.13   Angles in the same segment

Corollary (follow-on theorem). Two angles in the same segment of a circle are equal. For the proof we refer to Figure 6.13, where C and D are two points on the major arc AB. We see that

$$A\hat{O}B = 2A\hat{C}B \quad \text{and} \quad A\hat{O}B = 2A\hat{D}B$$

Therefore $A\hat{C}D = A\hat{D}B$. The corollary is proved. ∎

Look now at Figure 6.14. From the 'angle in a segment' theorem

$$A\hat{C}B = \frac{1}{2}A\hat{O}B \text{ (obtuse) and } A\hat{D}B = \frac{1}{2}A\hat{O}B \text{ (reflex)}.$$

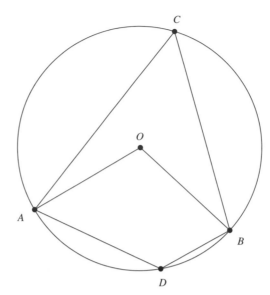

Figure 6.14   Cyclic quadrilateral

Adding these equations

$$A\hat{C}B + A\hat{D}B = \frac{1}{2}360° = 180°$$

The points A, B, C, D form a quadrilateral, the vertices of which lie on a circle. We say that they are **concyclic** or that $ABCD$ is a **cyclic quadrilateral**.

Because the angles in a quadrilateral add up to 360° and $A\hat{C}B + A\hat{D}B = 180°$, it is also true that $D\hat{A}C + D\hat{B}C = 180°$.

In other words, the opposite angles in a cyclic quadrilateral add up to 180° and are therefore supplementary. Conversely, if a quadrilateral has the property that opposite angles are supplementary then it is cyclic.

Examples

1. A further consequence of the 'angle in a segment' theorem is found by reference to Figure 6.15. Observe that the triangles PXQ and RXS are similar because $S\hat{P}Q = S\hat{R}Q$ (being subtended on the chord SQ) and $P\hat{Q}R = P\hat{S}R$ (vertically opposite angles are equal).

   The similarity means that $RX/PX = XS/XQ$, so

   $$RX \times XQ = PX \times XS$$

   which is the **intersecting chords theorem**.

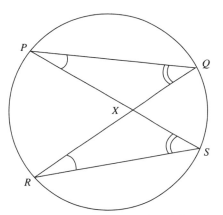

Figure 6.15 Intersecting chord theorem

2. In Figure 6.16(a) *GH* is a tangent to the circle at *A* and *AOC* is a diameter of the circle (whose centre is *O*). We shall prove that the angle between the chord *AB* and the tangent is equal to the angle in the alternate segment, i.e.

$$B\hat{A}H = A\hat{C}B$$

Remember that $A\hat{C}B$ is equal to any other angle in the segment *ACB*, so we can make our task easier by placing *C* at the opposite end of the diameter from *A*, which is, of course, the case in Figure 6.16(b).

First we note that

$$B\hat{A}H + B\hat{A}C = 90°$$

and that

$$A\hat{C}B + B\hat{A}C = 90°.$$

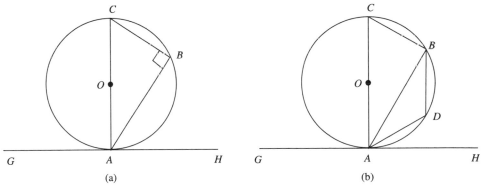

(a)   (b)

Figure 6.16 Alternate segment theorem

Therefore

$$B\hat{A}H = A\hat{C}.$$

If D is in the minor segment (Figure 6.16(b)) and the chord is AB then we need to prove that $G\hat{A}B = B\hat{D}A$.

Now $G\hat{A}B + B\hat{A}H = 180°$ and $B\hat{D}A + A\hat{C}B = 180°$, since $ABCD$ is a cyclic quadrilateral. And since $B\hat{A}H = A\hat{C}B$, it follows that $G\hat{A}B = B\hat{D}A$.

This is the **alternate segment theorem**. ∎

We conclude with a statement of two results involving tangents to a circle.

Tangent Theorem

In Figure 6.17 it can be shown that $CT = CX$, i.e. the lengths of the tangents from a given point to a given circle are equal. The theorem is left for you to prove.

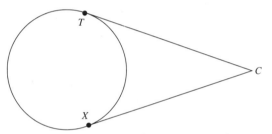

Figure 6.17 Tangent theorem

Tangent–chord Theorem

With the notation of Figure 6.18 the theorem states that $CP \times CQ = CT^2$. This is also left for you to prove by verifying that triangles CTP and CQT are similar using the alternate segment theorem. The tangent–chord theorem is a consequence of an earlier theorem, and is therefore a **corollary** of that theorem.

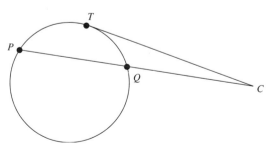

Figure 6.18 Tangent chord–theorem

Exercise 6.2

1   Complete the proof on page 249 to establish the perpendicularity of the bisectors of sides *AB*, *BC* and *AC* of triangle *ABC*; the bisectors are drawn from *O*, where *OA = OB = OC*.

2*   Prove that the altitudes of a triangle are concurrent via the following approach. The dotted triangle *PQR*, drawn exterior to *ABC*, has the property that *QP* is parallel to *AB*, *RP* is parallel to *AC* and *RQ* is parallel to *BC*.

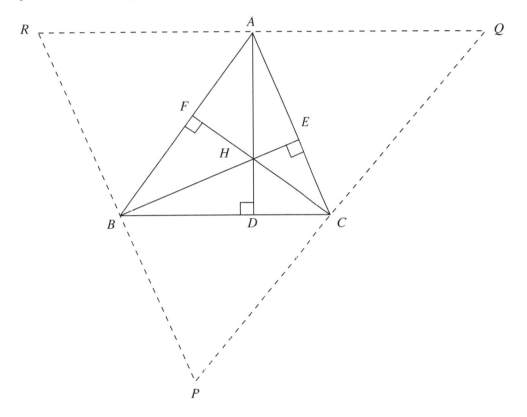

(a)   Establish that *RACB* and *AQCB* are parallelograms and that *DA* bisects *RQ* perpendicularly.

(b)   Establish similar properties for *BE* and *CF*.

(c)   What is the circumcentre of triangle *PQR*?

3 A' and B' are the midpoints of sides BC and AC respectively. To prove the medians of a triangle are concurrent, use similar triangles to establish that:

(a) $A'B' = \dfrac{1}{2}BA$ and $A'B'$ is parallel to BA

(b) G is at the point of trisection of AA' and BB', i.e.

$$GA' = \frac{1}{3}AA' \quad \text{and} \quad GA' = \frac{1}{3}BB'$$

Draw CC' where C' is the midpoint of BA and show that G lies on CC' with $GC' = \dfrac{1}{3}CC'$.

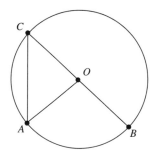

4 (a) Prove the 'angle in a segment' theorem, i.e. $A\hat{C}B = \dfrac{1}{2}A\hat{O}B$ for the case where BC is a diameter of the circle.

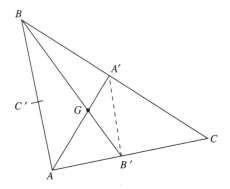 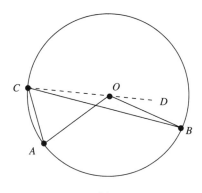

                (a)   (b)

(b) Note that $A\hat{C}D = A\hat{C}B + B\hat{C}D$ and $A\hat{O}D = A\hat{O}B + B\hat{O}D$. Hence prove that $A\hat{C}B = \dfrac{1}{2}A\hat{O}B$.

(c)

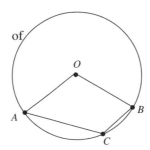

(d) If AB is a diameter, what is the value of $A\hat{C}B$?

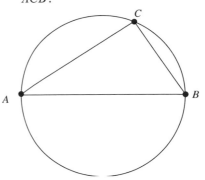

Prove that $A\hat{C}B = \dfrac{1}{2}A\hat{O}B$, reflex.

5* Determine the length YQ in this diagram, where $RX=4$, $XS=3$, $PX=5$ and XQ is the diameter of the small circle upon which lies Y. The triangle XYS is isosceles.

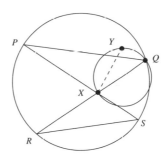

6 Prove the tangent–chord theorem.

7* In the figure shown here the line DE is parallel to the base BC of the triangle ABC. AC is a tangent to the circle at E.

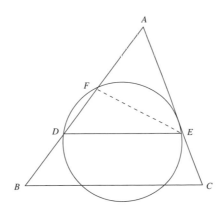

Prove that:

(a) $C\hat{E}D = E\hat{F}D$

(b) $B\hat{C}A + E\hat{F}D = 180°$

(c)   B, C, E, F are concyclic

8 In the following diagrams determine the unknown angles or lengths:

(a)

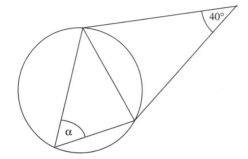

(b)

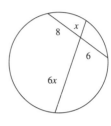

(c)

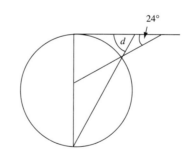

(d)

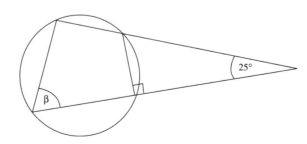

(e)

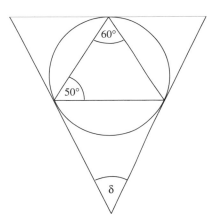

6.3 ARGUMENTS IN ARITHMETIC AND ALGEBRA

In arithmetic and algebra you have used the following symbols:

$=$ equal to (or equals)

$>$ greater than (strictly)

$<$ less than (strictly)

$\geq$ greater than or equal to (but not less than)

$\leq$ less than or equal to (but not greater than)

$\approx$ approximately equal to

These symbols qualify arithmetic or algebraic expressions, but on their own they do not prove anything unless a logical structure is imposed upon them. We can, however, use them to explore simple ideas and properties that lead to useful results. Again we will assume intuitive results to underpin our arguments.

We know that if the number a is less than the number b we write $a < b$. The inequality is unaffected by adding or subtracting a third number c to or from both sides, i.e.

$$\text{if } \quad a < b \quad \text{then} \quad a + c < b + c \quad \text{and} \quad a - c < b - c$$

Also we can multiply or divide such an inequality by any positive number $c > 0$, i.e.

$$\text{if } \quad a < b \quad \text{then} \quad ac < bc \quad \text{and} \quad \frac{a}{c} < \frac{b}{c}$$

Such reasoning may seem obvious or even trivial, but it can be developed to prove results that are much less obvious. Consider the following examples.

Example

If a, b, c, $d > 0$ are such that

$$\frac{a}{c} < \frac{b}{d} \quad \text{then} \quad \frac{a}{c} < \frac{a+b}{c+d} < \frac{b}{d}.$$

We start by cross-multiplying the given inequality, which is perfectly acceptable since $c > 0$ and $d > 0$. Hence

$$ad < bc$$

Add ac to each side to obtain

$$ad + ac < bc + ac$$
$$a(c + d) < c(a + b)$$

Now divide first by c and then by $c + d$ to obtain

$$\frac{a}{c} < \frac{a+b}{c+d}$$

We have assumed intuitively that $c + d > 0$ because $c > 0$ and $d > 0$. Similarly, by adding bd to both sides of the inequality $ad < bc$ we can prove that

$$\frac{a+b}{c+d} < \frac{b}{d}$$

Hence we have proved the result required. ∎

To illustrate the result consider the particular example

$$\frac{1}{2} < \frac{13}{24}$$

Then the result claims that

$$\frac{1}{2} < \frac{1+13}{2+24} < \frac{13}{24}$$

i.e. $\dfrac{1}{2} < \dfrac{7}{13} < \dfrac{13}{24}$, which is true.

Example

We can use the same ideas to justify the belief that if $x > 1$ then $x^n > 1$, where n is a positive integer power, and that $x^n > x^m$, if m and n are positive integers and $n > m$.

We know that if $a > 1$ and $b > 1$, then $ab > 1^2 = 1$, hence if $x > 1$, then $x^2 > x$, (multiplying by x) so that $x^2 > 1$.

Every successive power is a factor of x larger than its predecessors so the result is clear. Also $x^n = x^{n-m} \times x^m$, therefore, because $n - m > 0$, $x^{n-m} > 1$ hence $x^n > x^m$.

You should note that we are close to a proof but have not actually established that $x^n > 1$, given that $x > 1$, where n is an integer > 1. ■

Inequalities are useful in limiting the search for the factors of a large integer N.

Example

Prove that a composite integer N must possess a factor which does not exceed $\sqrt{N}$.

Solution

Assuming that $N = ab$, where a and b are integers, then one of the following is true:

(a) $a > \sqrt{N}$, $b > \sqrt{N}$ (b) $a \leq \sqrt{N}$, $b > \sqrt{N}$

(c) $a > \sqrt{N}$, $b \leq \sqrt{N}$ (d) $a \leq \sqrt{N}$, $b \leq \sqrt{N}$

In the case of (a) it follows that $ab > N$, which is not true. Similarly in case (d) $ab \leq N$ which is only possible if $a = b = \sqrt{N}$ and N is a perfect square.

Apart from the special case in (d) either (b) or (c) must apply, so one of the factors cannot exceed $\sqrt{N}$. ■

In a search for factors of integers $N \leq 1000$ you need go no further than 31 since $(31)^2 = 961 < 1000$ while $(32)^2 = 1024 > 1000$.

Example

The geometric mean of two numbers is less than or equal to their arithmetic mean.

Solution

If x and y are two numbers then $\frac{1}{2}(x + y)$ is the arithmetic mean and $\sqrt{xy}$ is the geometric mean.

We know that the square of any number must be non-negative, so consider the quantity $(a - b)^2$, where a and b are any two numbers.

Since $(a - b)^2 \geq 0$ then $a^2 - 2ab + b^2 \geq 0$ so that $a^2 + b^2 \geq 2ab$, whatever the values of a and b. If we take $a^2 = x$, $b^2 = y$ so that $x, y \geq 0$, and $\sqrt{x} \geq 0$, $\sqrt{y} \geq 0$ then the result can be written as $\frac{1}{2}(x + y) \geq \sqrt{xy}$ or $\sqrt{xy} \leq \frac{1}{2}(x + y)$, and this proves the result. ■

The result can be demonstrated by inserting some numbers:

$$\sqrt{3 \times 2} = \sqrt{6} \le \frac{1}{2}(3 + 2) \quad \text{or} \quad 2.449 < 2.500$$

Exercise 6.3

1 (a)   Establish that

(i) $\dfrac{4}{7} < \dfrac{7}{11} < \dfrac{3}{4}$

(ii) $\dfrac{2}{5} < \dfrac{7}{17} < \dfrac{5}{12} < \dfrac{8}{19} < \dfrac{3}{7}$

(b)   Rank the following fractions in increasing order of size:

$$\frac{2}{3}, \quad \frac{3}{5}, \quad \frac{8}{11}, \quad \frac{4}{7}$$

(c)   Noting that $71 = 47 + 24$ and that $145 = 97 + 48$, prove that

$$\frac{4}{9} < \frac{71}{145} < \frac{1}{2}$$

(d)   Verify all the above results on your calculator.

2   Given that $\sqrt{xy} \le \dfrac{1}{2}(x + y)$, and that $x > 0$, $y > 0$, $x + y \le 1$, prove that

$$\frac{1}{x} + \frac{1}{y} \ge 4$$

3   If $x > 1$ prove that

(a) $x + \dfrac{1}{x} \ge 2$

(b) $x(x^2 + 3) \ge 3x^2 + 1$

If $x > 3$ determine the smallest value of $x + \dfrac{9}{x}$.

4 (a)   Prove that

$$x^n \ge 1 + n(x - 1)$$

when $n = 2, 3, 4$ and $x > 1$.

(b)   If we can assume that the result in (a) holds for a chosen positive integer n, prove that

$$x^{n+1} \geq 1 + (n+1)(x-1)$$

(c)   If $0 < y < 1$, by setting $\dfrac{1}{y} = x$ and assuming the result in (b) to be true for all positive integers, prove that $0 < y^n < 1$.

5   (a)   Show that $\dfrac{a}{b} < \dfrac{a+c}{b+d} < \dfrac{c}{d}$ provided only that b, $d > 0$, with no conditions on a and c.

(b)   If we relax any of the conditions on b and d as well, e.g. with $a = -1$, $b = 1$, $c = -3$, $d = -2$, show that $\dfrac{a+c}{b+d}$ is no longer necessarily sandwiched between $\dfrac{a}{b}$ and $\dfrac{c}{d}$.

6*   Referring to the result in Question 5 with $a = 2$, $b = 3$, $c = x$, $d = x$, where $x > 0$, draw some conclusions about the fraction

$$y = \frac{x+2}{x+3}$$

Sketch a graph of y to demonstrate the result.

6.4 THE METHODOLOGY OF PROOF

Deductive reasoning based upon the principles so far examined can be organised into systematic arguments. These can be applied, repeatedly if necessary, to prove mathematical theorems and establish results. We shall clarify the reasoning process by considering necessity and sufficiency before looking at two methodologies of proof, namely *reductio ad absurdum* and *proof by induction*.

First of all we will need to formalise into a tighter framework some ideas that we have already been using. This will make subsequent reasoning simpler and reducible to symbolic treatment.

A **proposition** is a mathematical statement or property with a precise meaning which can be demonstrated true or false, e.g. *the number 8 is greater than 7*. Suppose that two propositions are being considered and suppose that it is the case that if one of these were true it would inevitably follow that the other one were true too; then we say that the first proposition **implies** the second. For example, if the first proposition is *the number 8 is* *greater than* 7, and the second proposition is *the number* 8 *is greater than* 6, then the first implies the second. In the essence the first proposition 'swallows' the second, i.e. it totally includes all its provisions with spare capacity, e.g. the number $6\frac{1}{2}$. We will see that the existence of spare capacity will form the basis of counter-examples which can disprove propositions.

Necessity and sufficiency

If a proposition p implies a proposition q, then we write $p \Rightarrow q$ or $q \Leftarrow p$. The converse (or opposite) to this would be $p \nRightarrow q$ or p *does not imply q*.

Example

Let $p : x = 4$, i.e. the proposition p states that a given number (x) is equal to 4. Also, let $q : x^2 = 16$, i.e. the proposition q states that the square of a given number (x) is equal to 16. Evidently $p \Rightarrow q$ because $4^2 = 16$.

But $(-4)^2 = 16$ as well so $q \nRightarrow p$. Putting the argument verbally, for $x^2 = 16$ it is **sufficient** that $x = 4$, but **not necessary** as -4 does equally well.   ∎

Sometimes two propositions can imply one another as the following example shows.

Example

$$p : |x| = 4 \qquad q : x^2 = 16$$

We see that $p \Rightarrow q$ and $q \Rightarrow p$ because both propositions mean precisely the same. When this is true, we write

$$p \Leftrightarrow q$$

The $\Leftrightarrow$ sign, *implies and is implied by*, has wide usage in mathematics. We say that for p to hold it is both **necessary and sufficient** that q holds, and vice versa. More briefly, p holds **if and only if** q holds.   ∎

When more than one proposition is involved we can extend the principles of implication as follows. If $p \Rightarrow q$ and $q \Rightarrow r$, then $p \Rightarrow r$.

Examples

1. If a positive integer is divisible by 4, this is *sufficient* to guarantee that it is even, but it is *not necessary* since, as an example, 6 is even but it is not divisible by 4.

2. $p : x > 8 \qquad q : x > 7 \qquad r : x > 6$

Here $p \Rightarrow q$ and $q \Rightarrow r$, so $p \Rightarrow r$.   ∎

Mutual implication is governed by similar guiding principles, i.e.

$$\text{if } \quad p \Leftrightarrow q, \quad q \Leftrightarrow r \quad \text{then} \quad p \Leftrightarrow r$$

Closely related to arguments of necessity and sufficiency is the use of the words *some*, *any* and *all*. In the first example of this section we found *some x*, namely -4, with $(-4)^2 = 16$. This was quite sufficient to disallow an implication.

It is true in Figure 6.19 that $A\hat{C}B = \dfrac{1}{2}A\hat{O}B$. Now $A\hat{O}B$ is fixed, but C is *any* point on the major arc AB. The result is therefore true for *all* points C, i.e. C can be anywhere on the major arc AB.

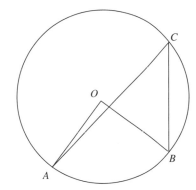

Figure 6.19 Angle at the circumference

Reductio ad absurdum

Known by its Latin name, which means *reduction to the absurd*, the method is based upon assuming false a result that we wish to prove true. The reasoning is then followed through until a contradiction is obtained.

Example

Prove there is no largest prime number.

Solution

We start by accepting the counterproposition P: a largest prime number exists. Call this number p. There is now a finite number of prime numbers, namely $2, 3, 5, \ldots, p$. Multiply them all together and add 1 to the total to obtain

$$q = (2 \times 3 \times 5 \times \cdots \times p) + 1$$

q is an absolutely enormous number, but it must exist by virtue of Archimedes' axiom. Two possibilities arise for q, namely

(a) q is prime

(b) q is composite

For case (a) $q > p$. This would invalidate P in that p is the not the largest prime number. For case (b) q must have factors. But q is odd and not divisible by $3, 5, 7, \ldots, p$ because it is one more than an exact divisor of each. If q does have factors they must exceed p. If the smallest factor is r, then r is prime, otherwise r would factorise further.

Either way we have

(a) q prime, $q > p$ and P is invalid

(b) q composite, $r > p$, r prime and P is invalid

We must therefore reject the proposition P and conclude that there is no largest prime number. ∎

Proof by induction

We know that if $p \Rightarrow q$ and $q \Rightarrow r$ then $p \Rightarrow r$. If we make the reasonable assumption that we can extend this argument to a chain or sequence of propositions $p_1, p_2, \ldots, p_n$ where $p_1 \Rightarrow p_2, p_2 \Rightarrow p_3, \ldots, p_n$, then we can conclude that $p_1 \Rightarrow p_n$. The **method of induction** operates as follows:

(i) p_n is assumed to be true

(ii) p_{n+1} is proved to be true when p_n is true, i.e. $p_n \Rightarrow p_{n+1}$, for any $n \geq 1$

(iii) p_1, separately, is proved to be true directly

It follows then that

$$p_1 \Rightarrow p_2, \quad p_2 \Rightarrow p_3, \ldots, p_{n-1} \Rightarrow p_n, \quad p_n \Rightarrow p_{n+1}, \ldots$$

Stepping forward through the integers, the result holds for all n.

Sometimes the method may be modified if p_1 is not true. For example, $n^n > n!$ for $n = 2, 3, 4, \ldots$; 1^1 is not greater than $1!$ so we must state that p_n is true for $n \geq 2$.

Example   Prove that

$$1 + 2 + 3 + \cdots + n = \frac{1}{2}n(n+1)$$

i.e. the sum of the first n positive integers is equal to $\frac{1}{2}n(n+1)$.

Before proceeding, note that we can easily *verify* the result for a particular value of n, e.g. $n = 3$; $1 + 2 + 3 = \frac{1}{2} \times 3 \times 4 = 6$. You should be fully aware that verification works for specially chosen cases only and it could not possibly be carried out for *all* values of n; *verification is not proof*.

Proof

By the method of induction

(i) $p_n: 1 + 2 + 3 + \cdots + n = \dfrac{1}{2}n(n+1)$ is assumed to be true for some value of n.

(ii) We must show that $p_n \Rightarrow p_{n+1}$, but what is p_{n+1}? Clearly it is a faithful replication of p_n with $n+1$ replacing n, $n+2$ replacing $n+1$, etc. Therefore we have the result

$$p_{n+1}: 1 + 2 + 3 + \cdots + n + (n+1) = \frac{1}{2}(n+1)(n+2) \qquad (*)$$

If we add $(n+1)$ to equation (*) then the left-hand side becomes

$$1 + 2 + 3 + \cdots + n + (n+1)$$

and the right-hand side becomes

$$\frac{1}{2}n(n+1) + (n+1) = (n+1)\left(\frac{n}{2}+1\right) = \frac{1}{2}(n+1)(n+2)$$

We have thus produced the requirements of p_{n+1}, so we may write $p_n \Rightarrow p_{n+1}$.

(iii) Now we show that p_1 is true. When $n=1$, the equation becomes $1 = \dfrac{1}{2} \times 1 \times 2 = 1$, which is clearly true.

The result is therefore proved.   ■

Less formally, proof by induction consists of the following reasoning process:

(i)   Establish or recognise a pattern that appears to be true with n (assume p_n).

(ii) Step forward by whatever means to the next stage to obtain the same pattern but with n being replaced by $n+1$ ($p_n \Rightarrow p_{n+1}$).

(iii) Verify the pattern for a chosen n, usually $n=1$; i.e. prove p_1.

Negation

Associated with the notion of a proposition p being true is the property that it is not true. We write $\bar{p}$, called p bar, as the proposition that p is not true. In every sense it has to be the complete opposite of p. For instance

if $p: x > 6$,   i.e. x is a number > 6

then $\bar{p}: x \le 6$,   i.e. x is a number ≤ 6.

In other words $\bar{p}$ is the exact opposite to p in such a way that p and $\bar{p}$ together exhaust all possibilities, which in this case is the set of all real numbers $\mathbb{R}$.

The next example demonstrates the principles; x will be used to denote a real number, $x \in \mathbb{R}$, i.e. x belongs to the set of real numbers and n will denote an integer.

Example

p	$\bar{p}$
$x > 7$	$x \leq 7$
n even	n odd
n not divisible by 5	n divisible by 5
$-2 \leq x < 3$	$x < -2$ or $x \geq 3$

Note that the method of *reductio ad absurdum* aims to establish the truth of p by assuming $\bar{p}$ to be true. The implications of $\bar{p}$ being true are followed forward until a contradiction occurs. When this happens we know that $\bar{p}$ must be false and therefore p must be true.

Our aim in this book is to use the principles of proof and propositional logic only as far as we need them in foundation level mathematics. If you are intending to study discrete and decision mathematics, possibly as a preparation for computer science, you will need a deeper treatment than we offer here.

The following example shows how to use negation and implication in mathematical reasoning.

Example

For the two propositions p and q determine whether $p \Rightarrow q, p \Leftarrow q, \bar{p} \Rightarrow q, p \Leftrightarrow q$, etc., or whether none of them are appropriate. Here x, y denote real numbers, i.e. $x, y \in \mathbb{R}$, and n is a positive integer, i.e. $n \in \mathbb{N}$.

p	q	p	q
(a) $x = 1$	$x^2 = 1$	(b) $xy = 0$	$x = 0$
(c) $x \neq 3$ or $x \neq -3$	$x^2 = 9$	(d) n is even	$n^2 \geq 0$
(e) n is odd	n is prime	(f) $x > 8$	$x \leq 8$
(g) $x < -6$	$x < -7$	(h) $x > 3$	$x^2 > 9$
(i) $5 \leq x < 6$	$5 < x \leq 6$	(j) $x^2 + x - 12 = 0$	$x = 3$ or -4
(k) $x^2 > x$	$x > 1$	(l) $x^2 > y^2$	$x > y > 0$

Solution

(a) $x = 1 \Rightarrow x^2 = 1$, i.e. $p \Rightarrow q$, but $p \not\Leftarrow q$ as $x = -1$ is the exception.

(b) $p \Leftarrow q$ because $xy = 0$ means that either $x = 0$ or $y = 0$.

(c) $\bar{p}:x = 3$ or $x = -3$ in which case $x^2 = 9$; also if $x^2 = 9$, $x = \pm 3$, so $\bar{p} \Leftrightarrow q$.

(d) For any $x \in \mathbb{R}$, $x^2 \geq 0$, so this is true for any integer n by default. However $n^2 \geq 0$ tells us nothing about n. Therefore $p \Rightarrow q$.

(e) The prime numbers exclude $n = 1$ and include $n = 2$, so $p \not\Rightarrow q$ and $p \not\Leftarrow q$.

(f) p and q are exactly opposite statements, so $\bar{p} = q$.

(g) Any number < -7 is automatically < -6, so $p \Leftarrow q$.

(h) If $x > 3$, then $x^2 > 3x > 9$ (because $x > 0$), so $p \Rightarrow q$. But if $x < -3$, $x^2 > 9$, so $p \not\Leftarrow q$.

(i) p: a real number x lies between 5 and 6, including 5 but not 6.
q: a real number lies between 5 and 6 including 6 but not 5.
Though very nearly the same, p and q are different statements, each possessing an exceptional case not covered by the other. So $p \not\Rightarrow q$, $p \not\Leftarrow q$.

(j) Solving $x^2 + x - 12$ gives $(x + 4)(x - 3) = 0$, i.e. $x = 3$ or -4. Conversely, if $x = 3$ or -4 then $x^2 + x - 12 = 0$, so $p \Leftrightarrow q$.

(k) If $x > 1$ then $x^2 > x$ because $x > 0$. However, if $x < 0$ then $x^2 > x$ automatically, therefore $p \Leftarrow q$ only.

(l) If $x > y > 0$ then $x > y$ leads to $x^2 > xy$ (multiplying by $x > 0$) and to $xy > y^2$ (multiplying by $y > 0$). Hence $x^2 > xy > y^2$ and therefore $x^2 > y^2$. However, $16 = (-4)^2 > (-3)^2 = 9$. But $-4 \not> -3$ and neither number is > 0. Hence $p \Leftarrow q$ but $p \not\Rightarrow q$. ∎

In this book use will be made of the concepts of proof so far described. We will rely upon arguments of the form developed here. We introduced propositional logic to formalise and sharpen up our arguments and to open the way into the symbolic logic which underpins computer science. Should you undertake such a study you will need further logical symbols but for the remainder of this book you will find that the development to date is more than enough.

Exercise 6.4

1 The following examples relate two propositions p and q. In each case determine whether $p \Rightarrow q, p \Leftarrow q, p \Leftrightarrow q$ or none of these; x and y are real numbers and n is a positive integer.

	p	q
(a)	$x^2 - 5x + 4 = 0$	$x = 1$
(b)	$x + 1 = 0$	$x = -1$
(c)	n is odd	n^2 is odd
(d)	$x = 2$ or $x = -3$	$x^2 + x - 6 = 0$
(e)	$x > 5$	$x > 4$

	p	q
(f)	$y^2 = x$	$y = \sqrt{x}$
(g)	$-3 < x < 2$	$-1 < x < 1$
(h)	$x + 3 = 0$	$x = 4$
(i)	$2 < x < 3$	$2 \leq x \leq 3$
(j)	$2 \leq x < 3$	$2 < x \leq 3$
(k)	$x < y$	$x^3 < y^3$
(l)	$x < y$	$x^4 < y^4$

2   Under what circumstances is it true that $x + y < (x + y)^2$?

3   For the following propositions, either establish the truth or give a counterexample to establish falsehood.

(a) $n^3 - 1$ is never a prime number, for any positive integer n

(b) $x < \sqrt{x^2 + 1}$, for any $x \in \mathbb{R}$ (c) $x^2 + y^2 < (x + y)^2$; $x, y > 0$

(d) $x^2 + y^2 < (x + y)^2$; $x, y \in \mathbb{R}$

(e)   A quadratic equation always has two equal roots

(f)   If $x > y$ and $y > 5$, then $x > 5$ (g) $x^2 - 5x + 6 > 0$; $x \in \mathbb{R}$

(h) $x = -7 \Rightarrow x^2 - 2x - 63 = 0$ (i) $x = -7 \Leftarrow x^2 - 2x - 63 = 0$

(j) $x^2 + 2x + 2 \geq 1$

4   (a)   By writing $n = 2m$, establish that the square of an even integer must be divisible by 4.

(b)   If n has a factor f, i.e. $n = fq$, prove that f^2 and q^2 are factors of n^2.

5*   It is known that certain numbers, e.g. $\sqrt{2}$, are not rational. For $\sqrt{2}$ we prove the result by writing

$$\sqrt{2} = \frac{p}{q}$$

where p and q are integers with no common factor and one of them at most is even. Square p and use the result in Question 4(a) to obtain a contradiction.

6   If two numbers a and b are multiplied together and the product $ab = 0$, then either a or b, or both, must be zero. Factorise $x^3 - y^3$ and use this property to prove that no two different numbers x and y can have the same cube.

7 Many integers of the form

$$N = n^2 + n + 41$$

are prime. By taking $n = 40$ obtain a contradiction to the hypothesis that they are all prime. Is there another value of n which makes n composite?

8 Use the method of induction to prove the following (n is an integer, $x \in \mathbb{R}$).

(a) $2^n > n,\ n \geq 1$

(b) $(1 + x)^n > 1 + nx,\ n \geq 1,\ x > 0$

(c) $n^n > n!,\ n > 1$

(d) $1^3 + 2^3 + 3^3 + \cdots + n^3 = \left(\dfrac{1}{2} n(n + 1) \right)^2 = (1 + 2 + 3 + \cdots + n)^2$

9 Prove that

$$\frac{1 - x^n}{1 - x} = 1 + x + x^2 + \cdots + x^{n-1}.$$

Multiply up by the denominator and use the method of induction.

10 Establish that

$$\frac{1}{r(r + 1)} = \frac{1}{r} - \frac{1}{r + 1}$$

Hence obtain a compact form for

$$\frac{1}{1.2} + \frac{1}{2.3} + \cdots + \frac{1}{(n - 1)n}$$

proving the result the method of induction.

11* If N is the smallest integer for which $\dfrac{1}{2^n} < \dfrac{1}{2} \times 10^{-m}$, write $\dfrac{1}{2^n} = \dfrac{1}{2^{n-N}} \times \dfrac{1}{2^N}$, and prove that $\dfrac{1}{2^n} < \dfrac{1}{2} \times 10^{-m}$ for any $n \geq N$.

12* An arithmetic series of the form $a + (a + d) + (a + 2d) + \cdots + (a + (n - 1)d)$ has the sum

$$S_n = \frac{n}{2} [2a + (n - 1)d].$$

Prove this result by the method of induction. Hence determine the sum of the following series:

(a) $2 + 5 + 8 + \cdots + 32$

(b) $41 + 36 + 31 + \cdots + 6$

(c) $5 + 1 - 3 - 7 - \cdots - 39.$

SUMMARY

- **Axiom:** a fundamental statement which is assumed to be true
- **Theorem:** a statement which can be derived from axioms
- **Corollary:** a statement which follows from a theorem
- **Pythagoras' theorem:** in a right-angled triangle the square of the length of the hypotenuse is equal to the sum of the squares of the lengths of the other two sides
- **Triangle theorems:**

    The bisectors of the angles meet at a point
    The perpendicular bisectors of the sides meet at a point
    The medians meet at a point
    The altitudes meet at a point

- **Circle theorems:**

    The angle subtended at the centre by an arc is
    twice the angle it subtends at the circumference

    The angle between a tangent to a circle and
    a chord to the point of contact is equal to any
    angle in the alternate segment

- **Sufficient condition:** p is a sufficient condition for q if q is true when p is true
- **Necessary condition:** p is a necessary condition for q if p is true when q is true
- **Methods of proof:**

    Reductio ad absurdum
    mathematical induction
    negation

Answers

Exercise 6.1

1 $(N - 1)(N + 1) \equiv N^2 - 1$

2 $N^3 - 1 \equiv (N - 1)(N^2 + N + 1)$; with $N = 2$, $N - 1 = 1$ is a trivial factor and $7 = 2^3 - 1$ is prime

3 (a) Every second integer is divisible by 2, every third by 3 and every fourth integer is divisible by 4.

    (b) 120.

4 All except (d)

5 4.8 m

6 (a) $\sqrt{105} = 10.247$ (b) $\sqrt{181} = 13.453$

7 (a) $45°$ (b) $\sqrt{2}$

8 $AC = 6.798$, $AD = 8.152$, $AE = 7.619$

10 $\sqrt{2} + 1 \pm \sqrt{2(\sqrt{2} + 1)} = 4.6116$ or 0.2168

Exercise 6.2

2 (c) H must be the circumcentre of triangle PQR

4 (d) $90°$, a right angle

5 $YQ = 2.25$

8 (a) $70°$ (b) $2\sqrt{2}$ (c) $57°$

    (d) $65°$ (e) $60°$

Exercise 6.3

1 (b) $\dfrac{4}{7} < \dfrac{3}{5} < \dfrac{2}{3} < \dfrac{8}{11}$

3 6

6　$\dfrac{2}{3} < \dfrac{x+2}{x+3} < 1$, if $x > 0$

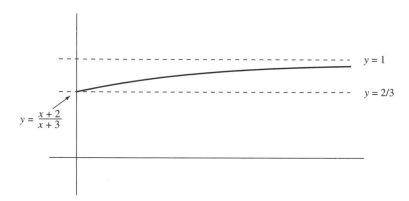

Exercise 6.4

1　(a)　$p \Leftarrow q$　　(b)　$p \Leftrightarrow q$　　(c)　$p \Leftrightarrow q$　　(d)　$p \Leftrightarrow q$

　　(e)　$p \Rightarrow q$　　(f)　$p \Leftarrow q$　　(g)　$p \Rightarrow q$　　(h)　neither

　　(i)　$p \Rightarrow q$　　(j)　neither　　(k)　$p \Leftrightarrow q$　　(l)　neither

2　$x + y > 1$ or $x + y < 0$

3　(a)　False: $n = 2$ gives $n^3 - 1 = 7$, which is prime

　　(b)　True

　　(c)　True

　　(d)　False: if $x = 1$, $y = -2$, LHS is 5, RHS is 1

　　(e)　False: $x^2 + x - 2 = 0$ has roots $x = 1$ and $x = -2$

　　(f)　True

　　(g)　False: LHS ≤ 0 for $2 \leq x \leq 3$

　　(h)　True

　　(i)　False: $x^2 - 2x - 63 = 0$ has roots $x = -7$ and $x = 9$

　　(j)　True

7　Yes, $n = 41$

10　$1 - \dfrac{1}{n} = \dfrac{n-1}{n}$

12　(a)　187　　　　　　　(b)　188　　　　　　　(c)　-204

7 TRIGONOMETRY

INTRODUCTION

The name *trigonometry* derives from the Greek words *trigonon* 'triangle' and *metron* 'measurement'. It is the basis of surveying and navigation. Indeed, it pervades every aspect of our lives. When an orienteer uses a map, when a toolmaker consults an engineering blueprint, when NASA engineers program a computer on a space probe, each of them relies upon trigonometry.

OBJECTIVES

After working through this chapter you should be able to

- define the sine, cosine and tangent of an acute angle
- state and use the fundamental identities derived from Pythagoras' theorem
- define the three reciprocal trigonometric ratios, secant, cosecant and cotangent
- state and apply the CAST rule
- relate the trigonometric ratios of an angle to those of its complement and its supplement
- state and use the sine rule for a triangle
- state and use the cosine rule for a triangle
- calculate the area of a triangle from the lengths of two sides and the included angle
- solve a triangle given sufficient information about its sides and angles
- calculate the angle between a line and a plane
- calculate the angle between two planes
- find the general solution of simple equations involving trigonometric ratios

7.1 TANGENT, SINE AND COSINE

Since the time of Thales in the sixth century BC the problem of estimating the height of a tall and distant object has been solved by the use of similar triangles. Figure 7.1(a) depicts the essential idea.

PQ represents a tower which casts a shadow QC on level ground. AB is a stick placed vertically into the ground so that the points C, A and P lie on a straight line. Triangles $\dfrac{PQC}{ABC}$ are similar, so

$$\frac{PQ}{AB} = \frac{QC}{BC}$$

from which the height PQ can be calculated (we used the term *estimating* earlier because measurements of AB, BC and QC will be somewhat inaccurate).

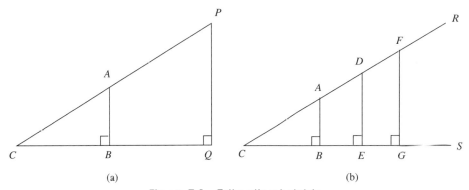

Figure 7.1 Estimating heights

It follows from the similarity of the triangles that $\dfrac{AB}{BC} = \dfrac{PQ}{QC}$. Figure 7.1(b) shows an angle $R\hat{C}S$ and three line segments AB, DE, FG perpendicular to CS. Using similar triangles we see that the ratios $\dfrac{AB}{BC}$, $\dfrac{DE}{CE}$ and $\dfrac{FG}{CG}$ are all equal.

This constant ratio is a property of the angle and is called the **tangent** of the angle, or **tan** for short (this is the name of the button on your calculator).

In the right-angled triangle shown in Figure 7.2 we focus attention on the angle $A\hat{C}B$ or, more simply, $\hat{C}$. We may write

$$\tan \hat{C} = \frac{AB}{CB} = \frac{\text{opposite}}{\text{adjacent}}$$

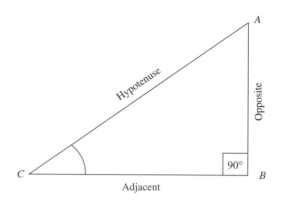

Figure 7.2 Definition of tangent of an angle

where we have labelled the side opposite the angle $\hat{C}$ as **opposite**; the side opposite the right angle is, as usual, the **hypotenuse** and the remaining side is the **adjacent**.

Reference back to Figure 7.1 allows the definition of two other constant ratios associated with an angle. From Figure 7.1(a) notice that $\dfrac{AB}{CA} = \dfrac{PQ}{CP}$ and that $\dfrac{CB}{CA} = \dfrac{CQ}{CP}$. These ratios are respectively the **sine** and **cosine** of the angle $\hat{C}$, written **sin** and **cos** for short (again, these are the names of the appropriate buttons on your calculator).

From Figure 7.2 we may write

$$\sin \hat{C} = \frac{AB}{CA} = \frac{\text{opposite}}{\text{hypotenuse}} \qquad \cos \hat{C} = \frac{CB}{AB} = \frac{\text{adjacent}}{\text{hypotenuse}}$$

Examples

1. At a point 50 m horizontally from the foot of a tall building, the angle of elevation of the top of the tower is 35°; what is the height of the building?

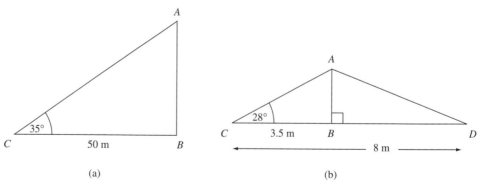

(a)   (b)

Figure 7.3 Diagrams for (a) the first example and (b) the second example

In Figure 7.3(a) AB represents the tower and $A\hat{C}B$ is the angle of elevation. We need to find the length AB.
Now

$$\frac{AB}{CB} = \tan\hat{C}, \text{ so } AB = CB\tan\hat{C}$$
$$= 50 \times 0.7002$$
$$= 35.0 \text{ m}$$

2. Figure 7.3(b) represents a cross-section of a roof in which the span CD is 8 m and the angle $A\hat{C}B$ is 28°. If CB is 3.5 m calculate the rise AB and the angle $A\hat{D}B$.

   From triangle ABC, $\tan\hat{C} = \dfrac{AB}{BC}$ so that $AB = BC\tan\hat{C} = 3.5\tan 28° = 1.86$ m.
   Also

$$\tan A\hat{D}B = \frac{AB}{BD} = \frac{1.86}{4.5} = 0.4136.$$

   Hence $A\hat{D}B = 22.5°$ (using INV TAN).

3. A 25 m ladder has to reach a window 22 m from the ground (which is assumed level). At what angle to the horizontal should the ladder be set and how far from the wall should it be placed?

   In Figure 7.4(a) AC is the ladder and AB is the window. The required angle is $A\hat{C}B$.

   Then $\qquad \sin A\hat{C}B = \dfrac{AB}{CA} = \dfrac{22}{25} = 0.88$, so $A\hat{C}B = 61.6°$ (using INV SIN).

   The required distance is $CB = AC\cos 61.6° = 25 \times 0.474\,97 = 11.87$ m.

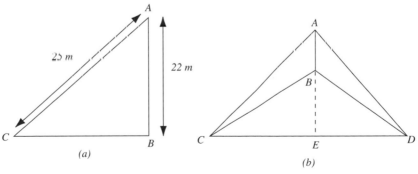

Figure 7.4 Diagrams for (a) the third example and (b) the fourth example

4. Figure 7.4(b) represents a cross-section of a symmetrical frame with $CB = BD = 10$ m, $BE = 3$ m and $A\hat{C}B = 20°$. Find the lengths CA and AB.

The first step in finding AC is to find $B\hat{C}E$ and then $A\hat{C}E$. Now

$$\sin B\hat{C}E = \frac{BE}{CB} = \frac{3}{10} \text{ so that } B\hat{C}E = 17.46°.$$

Hence $A\hat{C}E = A\hat{C}B + B\hat{C}E = 20° + 17.46° = 37.46°$.
Now $CE = CB \cos B\hat{C}E = 10 \cos 17.46° = 9.54$ m. Then

$$\frac{CE}{CA} = \cos A\hat{C}E \quad \text{so that } CA = \frac{CE}{\cos A\hat{C}E} = \frac{9.54}{\cos 37.46°} = 12.02 \text{ m.}$$

Finally

$$AB = AE - BE$$
$$= CA \sin A\hat{C}E - 3$$
$$= 7.31 - 3$$
$$= 4.31 \text{ m}$$

Graphs of sine, cosine and tangent

Figure 7.5 shows the effects on the values of sine, cosine and tangent of increasing the angle from $0°$ to $90°$. The point C is fixed and A moves along the arc of a circle.

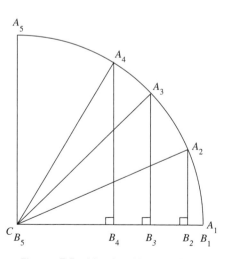

Figure 7.5 Varying the angle

As the angle $A\hat{C}B$ increases, the lengths CA_1, CA_2, CA_3, CA_4, remain constant (equal to r say) but the heights $A_1B_1, A_2B_2, A_3B_3, A_4B_4, A_5B_5$ increase to r, hence the sine of the angle increases from 0 to 1. The lengths $CB_1, CB_2, CB_3, CB_4, CB_5$ decrease from r to zero. Hence the cosine of the angle decreases from 1 to 0. Since the length of the opposite side is increasing and the length of the adjacent side is decreasing, the tangent of the angle is increasing—without limit.

Figure 7.6 shows the graphs of sine, cosine and tangent.

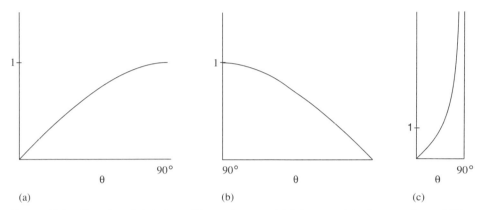

Figure 7.6 Graphs of (a) sine, (b) cosine and (c) tangent between 0 and 90°

Some useful results

Figure 7.7(a) depicts a right-angled isosceles triangle. If the lengths AB and CB are 1 then from Pythagoras' theorem we deduce that the hypotenuse has length $\sqrt{2}$. It follows using $A\hat{C}B$ as an example that

$$\sin 45° = \frac{1}{\sqrt{2}}, \quad \cos 45° = \frac{1}{\sqrt{2}}, \quad \tan 45° = 1$$

Figure 7.7(b) depicts an equilateral triangle ACD. B is the midpoint of CD and AB is a median, the perpendicular bisector of CD and the bisector of $C\hat{A}D$. If the lengths of the sides of the triangle ACD are 2, then the length CB is 1 and, via Pythagoras' theorem, the length of AB is $\sqrt{3}$.

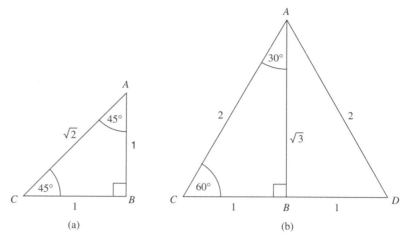

Figure 7.7 Ratios of special angles

Using triangle ACB and the angle $A\hat{C}B$, we see that

$$\sin 60° = \frac{\sqrt{3}}{2}, \quad \cos 60° = \frac{1}{2}, \quad \tan 60° = \sqrt{3}$$

Using $C\hat{A}B$ we obtain

$$\sin 30° = \frac{1}{2}, \quad \cos 30° = \frac{\sqrt{3}}{2}, \quad \tan 30° = \frac{1}{\sqrt{3}}$$

It is no coincidence that $\sin 60° = \cos 30°$ and $\cos 60° = \sin 30°$. Consider Figure 7.8 where θ is measured in degrees:

$$\sin \theta = \frac{AB}{CA} = \cos(90° - \theta) \quad \text{and} \quad \cos \theta = \frac{CB}{CA} = \sin(90° - \theta)$$

We highlight these important results:

$$\sin(90° - \theta) = \cos\theta, \quad \cos(90° - \theta) = \sin\theta$$

When the sum of two angles is 90° the angles are called **complementary**, hence we may think of the cosine as the complementary **sine**.

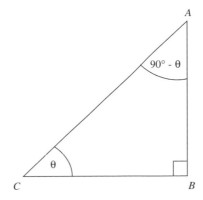

Figure 7.8 Complementary angles

Two identities

Reference to Figure 7.8 shows that

$$\sin\theta = \frac{AB}{CA}, \quad \cos\theta = \frac{CB}{CA}, \quad \tan\theta = \frac{AB}{CB}$$

so that

$$\tan\theta \equiv \frac{\sin\theta}{\cos\theta}$$

Furthermore, since $AB^2 + CB^2 = CA^2$, dividing through by CA^2 gives

$$\left(\frac{AB}{CA}\right)^2 + \left(\frac{CB}{CA}\right)^2 = 1$$

or

$$\sin^2 \theta + \cos^2 \theta \equiv 1$$

These results are true for any angle, not simply angles between $0°$ and $90°$.

Ratios for obtuse angles

We saw in Chapter 5 that some triangles contain obtuse angles, i.e. angles of magnitude greater than $90°$ but less than $180°$. In the next section we shall need to calculate sines and cosines of these angles. If you use your pocket calculator you will find that $\sin 150° = 0.5$ and $\cos 120° = -0.5$. What do these results mean? In Section 7.4 we consider ratios for angles of any magnitude, but for the moment we concentrate on acute and obtuse angles only.

In Figure 7.9 the point P lies on a circle of radius 1 unit, centred at O. As P moves anticlockwise from A to B the angle θ increases from $0°$ to $180°$.

In Figure 7.9(a) the height of P above the x-axis, $PN = OP \sin \theta = \sin \theta$ and the distance $ON = OP \cos \theta = \cos \theta$, so the coordinates of P are $(\cos \theta, \sin \theta)$.

Now consider Figure 7.9(b) where the angle $A\hat{O}P$ is obtuse. Clearly $PN = OP \sin \phi = \sin \phi$ and $ON = OP \cos \phi = \cos \phi$. The coordinates of P are $(\cos \phi, \sin \phi)$, i.e. $(-\cos(180° - \theta), \sin(180° - \theta))$.

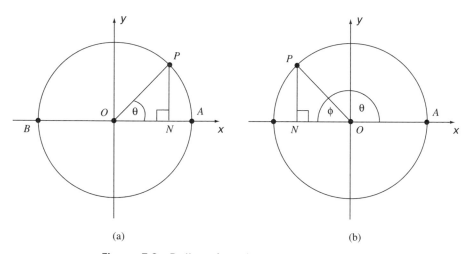

(a)   (b)

Figure 7.9 Ratios of acute and obtuse angles

The *x*-coordinate is negative because *P* is to the left of the *x*-axis. If we are being consistent we should expect the coordinate of *P* to be $(\cos\theta, \sin\theta)$. We therefore notice that the sine of an obtuse angle is positive and the cosine is negative. If we *define* the tangent to be the ratio sine:cosine then the tangent of an obtuse angle will also be negative. In fact we can make the following general definition

$$\tan\theta \equiv \frac{\sin\theta}{\cos\theta}$$

since this definition is completely consistent with our earlier definition for an acute angle; see Figure 7.2.

We have the following definitions for an acute angle θ:

$$\sin(180° - \theta) \equiv \sin\theta \qquad \cos(180° - \theta) = -\cos\theta$$
$$\tan(180° - \theta) \equiv -\tan\theta$$

Exercise 7.1

1 Convert the following angles from degrees to radians:

(a)   60°   (b)   135°   (c)   210°

(d)   300°   (e)   144°   (f)   405°

(g)   900°

2 Convert the following angles from radians to degrees:

(a) $(\pi/5)^c$ (b) $(3\pi)^c$ (c) $(4\pi/9)^c$

(d) $(7\pi/36)^c$ (e) $(11\pi/2)^c$ (f) $(41\pi/4)^c$

(g) $(-3\pi/2)^c$

 **3** Express the following angles in radians:

(a)   46.281°   (b)   97.929°   (c)   305.06°

Express the following angles in degrees:

(d) 1^c (e) 1.646^c (f) 4.844^c

4   Using your calculator, determine the sine, cosine and tangent of the angles in Questions 1, 2 and 3.

5   Given θ to be an acute angle, the accompanying diagram shows that

$$\sin\theta = \frac{y}{r}, \quad \cos\theta = \frac{x}{r}, \quad \tan\theta = \frac{y}{x}.$$

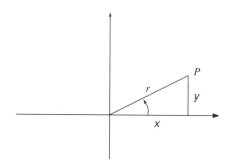 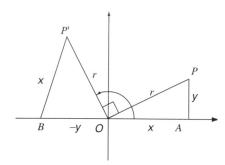

Adding $\pi/2$, as in the second diagram, we can find the sine, cosine and tangent of $(\theta + \pi/2)$, by using the sign convention with the supplementary angle $P'\hat{O}B$, i.e.

$$\sin(\theta + \pi/2) = \frac{x}{r} = \cos\theta$$

$$\cos(\theta + \pi/2) = \frac{-y}{r} = -\sin\theta$$

$$\tan(\theta + \pi/2) = \frac{x}{-y} = -\frac{1}{\tan\theta}$$

Repeat the process to find the sin, cos and tan of

(a) $\theta + \pi$ (b) $\theta + 2\pi$ (c) $-\theta$

(d) $\pi/2 - \theta$ (e) $\pi - \theta$ (f) $\theta + 3\pi/2$

Give your answers in the form $\cos(\theta + \pi/2) = -\sin\theta$, etc.

6   Use your calculator to determine values of $\sin\theta$ at $10°$ intervals between $0°$ and $180°$ (π^c). Use the results of the properties in parts (a), (b) and (c) of Question 5 to extend the graph to $-3\pi \le \theta \le 3\pi$. Do the same for $\cos\theta$ and $\tan\theta$.

7 Depict the following angles on a diagram:

(a) $382°$ (b) $-104°$ (c) $510°$

and write down the equivalent angle between $0°$ and $360°$ with the same sin, cos and tan.

8 Using the values $\sin 30° = \dfrac{1}{2} = \cos 60°$, etc., show that

(a) $\sin 60° \cos 30° + \cos 60° \sin 30° = 1$

(b) $\sin^2 60° + \dfrac{1}{2}\sin^2 45° = 1$

(c) $(\tan 45° + \sin 60°)(\cos 30° - \tan 45°) + \sin^4 45° = 0$

9 In the illustrated Pythagorean triangle, sides 3, 4, 5, find θ by pressing INV SIN or SIN^{-1} on the calculator. What is ϕ? Do the same for the Pythagorean triangles whose sides are 5, 12, 13 and 8, 15, 17.

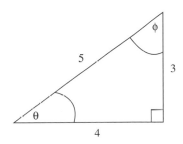

10 Use the identity $\sin^2 \theta + \cos^2 \theta = 1$ to establish the following identities:

(a) $1 - \cos \theta \equiv \dfrac{\sin^2 \theta}{1 + \cos \theta}$

(b) $(1 + \sin \theta - \cos \theta)(1 - \sin \theta + \cos \theta) \equiv 2 \sin \theta \cos \theta$

(c) $(1 - \sin^2 \theta)(1 + \tan^2 \theta) \equiv 1$

(d) $\dfrac{5 \sin \theta - 3}{4 + 5 \cos \theta} \equiv \dfrac{4 - 5 \cos \theta}{5 \sin \theta + 3}$

(e) $(\tan\theta - \sin\theta)(\tan\theta + \sin\theta) \equiv \dfrac{\sin^4\theta}{\cos^2\theta}$

(f) $\sin^3\theta - \cos^3\theta \equiv (\sin\theta - \cos\theta)(1 + \sin\theta\cos\theta)$

7.2 SOLUTION OF TRIANGLES

Given certain information about a triangle, in the form of the lengths of one or more of its sides and the value of one or more of its angles, it may be possible to deduce the unknown sides and angles. The special case of a right-angled triangle can be solved using the information in the previous section. Here we develop some rules for a general triangle. First, we recall a result from Chapter 5.

> The sum of the angles of a triangle is 180°.

Now we develop some rules to help us in our task.

Sine rule

In a triangle *the lengths of the sides are proportional to the sines of the angles opposite them.* We prove this result by considering separately the cases of an acute-angled triangle and an obtuse-angled triangle.

In Figure 7.10(a) AM is drawn perpendicular to BC and in Figure 7.10(b) AM is perpendicular to BC produced.

In Figure 7.10(a) we find from triangle ABM that $AM = c\sin\hat{B}$ and from triangle ACM, $AM = b\sin\hat{C}$, so

$$c\sin\hat{B} = b\sin\hat{C} \tag{1}$$

In the case of Figure 7.10(b), since $A\hat{C}B$ and $A\hat{C}M$ are supplementary angles, $\sin A\hat{C}M = \sin A\hat{C}B = \sin\hat{C}$, so $AM = c\sin\hat{B} = b\sin\hat{C}$ as before.

Equation (1) can be divided through by bc to give

$$\frac{\sin\hat{B}}{b} = \frac{\sin\hat{C}}{c}$$

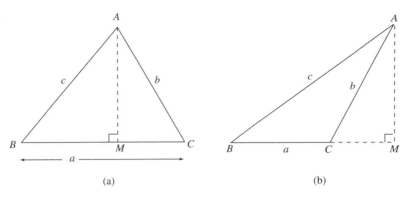

Figure 7.10 Sine rule

Similarly, by drawing the perpendicular from B to AC, we can show that

$$\frac{\sin \hat{C}}{c} = \frac{\sin \hat{A}}{a}$$

Pooling these results we have the **sine formula** or **sine rule**:

$$\frac{\sin \hat{A}}{a} = \frac{\sin \hat{B}}{b} = \frac{\sin \hat{C}}{c}$$

By inverting each fraction, we obtain another version of the formula which may be more straightforward to apply in some circumstances:

$$\frac{a}{\sin \hat{A}} = \frac{b}{\sin \hat{B}} = \frac{c}{\sin \hat{C}}$$

Example

In the triangle ABC, $\hat{C} = 47°$, $\hat{B} = 34°$ and $a = 125$ mm. Calculate b and c.
First, we use the angle sum to determine the third angle. Hence

$$\hat{A} = 180° - 47° - 34° = 99°$$

Applying the sine rule

$$\frac{125}{\sin 99°} = \frac{b}{\sin 34°} = \frac{c}{\sin 47°}.$$

Then

$$b = 125 \sin 34° / \sin 99° = 70.8 \text{ mm} \quad \text{and} \quad c = 125 \sin 47° / \sin 99° = 92.6 \text{ mm}$$

(You might like to draw carefully the triangle from the information given and see whether the answers above are sensible.) ∎

Cosine rule

As with the sine rule we consider two cases, they are illustrated in Figure 7.11.
  Using Pythagoras' theorem on triangle ACM,

$$(AM)^2 = (AC)^2 - (CM)^2 = b^2 - x^2$$

Using Pythagoras' theorem on triangle ABM,

$$(AM)^2 = (AB)^2 - (BM)^2$$

From Figure 7.11(a) this becomes

$$(AM)^2 = c^2 - (a - x)^2$$

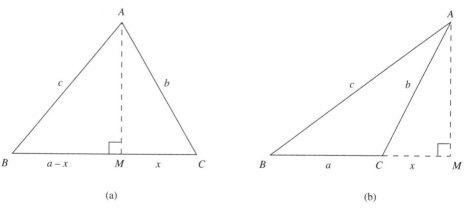

(a) (b)

Figure 7.11 Cosine rule

Hence

$$b^2 - x^2 = c^2 - (a - x)^2$$
$$= c^2 - a^2 + 2ax - x^2$$

so that $\qquad c^2 = a^2 + b^2 - 2ax$

From triangle ACM,

$$x = b \cos \hat{C}$$

Then

$$c^2 = a^2 + b^2 - 2ab \cos \hat{C}$$

Now we consider the case of an obtuse angle.
    Using Figure 7.11(b)

$$(AM)^2 = (AB)^2 - (BM)^2 = c^2 - (a + x)^2$$

Hence

$$b^2 - x^2 = c^2 - (a + x)^2$$
$$= c^2 - a^2 - 2ax - x^2$$

so that $\qquad c^2 = a^2 + b^2 + 2ax$

From triangle ACM,

$$x = b \cos A\hat{C}M = b \cos(180° - \hat{C}) = -b \cos \hat{C}$$

Hence

$$c^2 = a^2 + b^2 - 2ab \cos \hat{C}$$

Therefore, the same result is obtained whether or not the triangle contains an obtuse angle.
    We chose to draw a perpendicular from the vertex at A to the opposite side; we could have chosen B or C as the vertex and we would have developed two similar formulae. We present all three options as a summary of the **cosine rule**:

$$a^2 = b^2 + c^2 - 2bc \cos \hat{A}$$
$$b^2 = c^2 + a^2 - 2ca \cos \hat{B}$$
$$c^2 = a^2 + b^2 - 2ab \cos \hat{C}$$

The cosine rule as stated above can be used to calculate one side of a triangle given the two other sides and the angle between them. It can also be used to calculate an angle given the three sides of a triangle. The formulae above are then rearranged as follows:

$$\cos \hat{A} = \frac{b^2 + c^2 - a^2}{2bc}$$
$$\cos \hat{B} = \frac{c^2 + a^2 - b^2}{2ca}$$
$$\cos \hat{C} = \frac{a^2 + b^2 - c^2}{2ab}$$

Examples

1. The sides of triangle ABC are $a = 120$ mm, $b = 135$ mm, $c = 200$ mm; find the angles $\hat{A}$ and $\hat{C}$. We use the formulae

$$\cos \hat{A} = \frac{b^2 + c^2 - a^2}{2ac} = \frac{18\,225 + 40\,000 - 14\,400}{2 \times 135 \times 200} = 0.8116$$

therefore $\hat{A} = 45.75°$ (using INV COS).

Similarly $\cos \hat{C} = \dfrac{14\,400 + 18\,225 - 40\,000}{2 \times 120 \times 135} = -0.2276$

therefore $\hat{C} = 103.2°$ (obtuse, because the cosine is negative)

2. In triangle ABC, $\hat{A} = 26.5°$, $b = 130$ mm and $c = 220$ mm. Find the length of the third side. Start with

$$a^2 = b^2 + c^2 - 2bc \cos A$$

then $a^2 = 16\,900 + 48\,400 - 2 \times 130 \times 220 \times \cos 26.5° = 14\,109.75$

Therefore $a = 118$ mm   ∎

Solution of triangles

It is useful to note that the largest angle in a triangle is opposite the longest side and the smallest angle is opposite the shortest side. There are four cases to consider.

Three sides are given

(i)   We first find the largest angle using the cosine rule.

(ii) Then we find a second angle using the sine rule.

(iii) Finally we use the angle sum property to find the third angle.

There are two points to note. First, we advise finding the largest angle first because the cosine of an obtuse angle is negative and if the largest angle is obtuse this presents no problem; the other angles must be acute and knowing the sine of an acute angle determines that angle uniquely. Second, if we wish to have a check on accuracy we can replace (iii) by the use of the sine rule then apply the angle sum property as the check.

Two sides and the included angle are given

(i)   Calculate the third side using the cosine rule.

(ii) Find the smaller of the unknown angles using the sine rule.

(iii) Use the angle sum property to determine the third angle.

Again, there are two points to note. First, we find the smaller angle at step (ii) because it must be acute, hence it can be determined from knowledge of its sine. Second, we can employ the sine rule a second time at step (iii) and use the angle sum property as a check. However, note the following example.

Example

In triangle ABC, $\hat{A} = 20°$, $b = 4$, $c = 8$. Find a, $\hat{B}$, $\hat{C}$.

(i)   We use

$$a^2 = b^2 + c^2 - 2bc \cos \hat{A}$$
$$= 16 + 64 - 2 \times 4 \times 8 \cos 20°$$
$$= 19.86$$

therefore $a = 4.456$.

(ii) The smaller angle is opposite the smaller of sides b and c, so

$$\frac{\sin \hat{B}}{b} = \frac{\sin \hat{A}}{a}$$

so that $\quad \sin \hat{B} = \dfrac{4 \sin 20°}{4.456} = 0.307$

therefore $\quad \hat{B} = 17.88°$

(iii) If we use the sine rule again

$$\sin \hat{C} = \frac{c \sin \hat{A}}{a} = \frac{8 \times \sin 20°}{4.456} = 0.614.$$

INV SIN on the calculator suggests that $\hat{C} = 37.88$. This is clearly wrong since $\hat{A} + \hat{B} + \hat{C} \neq 180°$ with this value of $\hat{C}$. The use of the angle sum property gives

$$\hat{C} = 180° - 20° - 17.88° = 142.12°$$

The problem has arisen because $\sin 142.12° = \sin 37.88°$ as a consequence of the fact that $142.12°$ and $37.88°$ are supplementary. This ambiguity occurs again in the fourth case. ∎

Two angles and one side are given

(i) Use the angle sum property to find the third angle.

(ii) Use the sine rule to find one of the remaining sides.

(iii) Use the sine rule again to find the third side.

Two sides and an angle opposite one of them are given

This is known as the **ambiguous case** since the information can lead to two possible triangles.

Consider Figure 7.12 where we have assumed that c, b and $\hat{B}$ are the given quantities.

To construct the triangle from the information given, we draw AB to the required length and a line from B to form the given angle $A\hat{B}C$. From A we draw the arc of a circle of radius b and where it intersects the second line is the third vertex C.

Figure 7.12(a) shows that no triangle can be formed if b is not large enough; in fact, Figure 7.12(b) shows that the shortest possible length for b is $c \sin \hat{B}$. Since this case represents a right-angled triangle, it is easy to find the remaining side and angles.

Figure 7.12(c) seems to indicate that two triangles are possible but one of them requires the third vertex to lie on the wrong side of AB and is inadmissible. This happens when $b > c$. The ambiguous case is shown in Figure 7.12(d) and arises when $c \sin \hat{B} < b < c$. Two examples are now given to illustrate the cases depicted in the Figure 7.12(c) and (d).

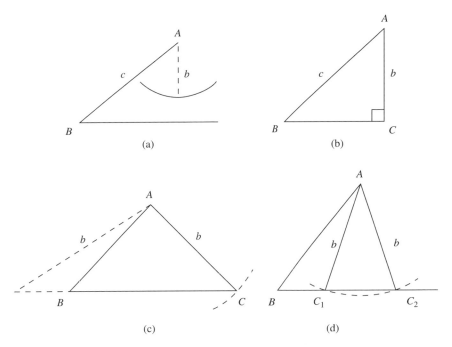

Figure 7.12 The ambiguous case

Examples

1. Solve the triangle ABC if $a = 24$, $c = 25$ and $\hat{C} = 66°$.

   (i)   Using the sine rule

   $$\frac{\sin \hat{A}}{a} = \frac{\sin \hat{C}}{c}$$

   so that $\sin \hat{A} = \dfrac{a \sin \hat{C}}{c} - \dfrac{24 \sin 66°}{25} = 0.8770$

   therefore $\hat{A} = 61.28°$

   The other option $\hat{A} = 180° - 61.28° = 118.72°$ is not possible since $\hat{A} + \hat{C} > 180°$.

   (ii)   The third angle, $\hat{B} = 180° - 66° - 61.28° = 52.72°$.

(iii) Using the sine rule again

$$\frac{b}{\sin \hat{B}} = \frac{c}{\sin \hat{C}}$$

therefore $b = c \dfrac{\sin \hat{B}}{\sin \hat{C}} = \dfrac{25 \sin 52.72°}{\sin 66°} = 21.8.$

Note that we used the function $\dfrac{c}{\sin \hat{C}}$, in preference to $\dfrac{a}{\sin \hat{A}}$, since it uses information given, not estimates derived from that information. Note also that we *could* have used the cosine rule to determine b.

2. Solve the triangle ABC if $a = 7$, $c = 6$ and $\hat{C} = 29.45°$.

(i)   Using the sine rule

$$\sin \hat{A} = \frac{a \sin \hat{C}}{c} = \frac{7 \sin 29.45°}{6} = 0.5736.$$

Therefore $\hat{A} = 35.0°$ or $145°$

Each of these answers is possible. First we take $\hat{A} = 35.0°$.

(ii) The third angle, $\hat{B} = 180° - 29.45° - 35.0 = 115.55°$.

(iii) The sine rule gives

$$b = \frac{c \sin \hat{B}}{\sin \hat{C}} = \frac{6 \sin 115.5°}{\sin 29.45°} = 11.0.$$

Second we take $\hat{A} = 145.0°$.

(ii) The third angle, $\hat{B} = 180° - 29.45° - 145.0° = 5.55°$.

(iii) The sine rule gives

$$b = \frac{6 \sin 5.55°}{\sin 29.45°} = 1.18.$$

In summary the two results are

$$\hat{A} = 35.0, \quad \hat{B} = 115.55°, \quad b = 11.0 \quad \text{or}$$
$$\hat{A} = 145.0, \quad \hat{B} = 5.55°, \quad b = 1.18.$$

Draw the two triangles to convince yourself that they are both solutions.

Area of a triangle

In Figure 7.13 we show a triangle ABC and an altitude AD formed by drawing a line from A perpendicular to BC or BC produced, as in diagram (b).

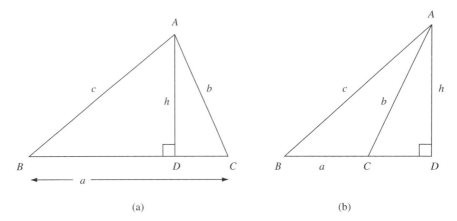

(a)   (b)

Figure 7.13 Area of a triangle

As we saw in Chapter 5 the area of either triangle, denoted $\triangle$, is given by

$$\triangle = \frac{1}{2} \times \text{base} \times \text{height}$$

so

$$\triangle = \frac{1}{2} ah$$

This result was obtained by noting that the area of a triangle is half the area of a rectangle with sides equal to the triangle's base and height. Since the altitude is seldom given, but usually has to be derived, it is more convenient to employ a formula which involves only the angles and sides of a triangle.

Reference to Figure 7.13 shows that

$$h = c \sin \hat{B}$$

Hence

$$\triangle = \frac{1}{2} ac \sin \hat{B},$$

In Figure 7.13(a) we see that

$$h = b \sin \hat{C}$$

and we could write

$$\triangle = \frac{1}{2} ab \sin \hat{C}$$

In Figure 7.13(b), however,

$$h = b \sin A\hat{C}D = b \sin(180° - A\hat{C}B) = b \sin(180° - \hat{C}) = b \sin \hat{C}$$

and the result is the same, showing that it does not matter whether the given angle is acute or obtuse.

It can also be shown that

$$\triangle = \frac{1}{2} bc \sin \hat{A}$$

Hence the area of a triangle is given by

$$\triangle = \frac{1}{2} ab \sin \hat{C} = \frac{1}{2} bc \sin \hat{A} = \frac{1}{2} ac \sin \hat{B}$$

A third way of finding the area of a triangle can be used if all three sides of a triangle are given. We quote the result without proof.

$$\triangle = \sqrt{s(s-a)(s-b)(s-c)} \quad \text{where} \quad s = \frac{1}{2}(a+b+c)$$

Examples

1. Find the area of triangle ABC where $a = 3.1$, $b = 39$, $\hat{C} = 52°$.

$$\triangle = \frac{1}{2} \times 3.1 \times 39 \times \sin 52° = 47.64.$$

2. Find the area of the triangle with sides 57.4, 70.8, 103.6. For convenience let $a = 57.4$, $b = 70.8$, $c = 103.6$. First we calculate the **semi-perimeter**, s:

$$s = \frac{1}{2}(57.4 + 70.8 + 103.6) = 115.9.$$

Then

$$\triangle = \sqrt{(115.9)(115.9 - 57.4)(115.9 - 70.8)(115.9 - 103.6)}$$
$$= \sqrt{(115.9)(58.5)(45.1)(12.3)}$$
$$= 1939. \qquad \blacksquare$$

Exercise 7.2

 1 Use the sine rule to determine the unknown sides and angles of the triangles given below.

(a) $a = 4$, $b = 6$, $\hat{A} = 37°$ (B acute)   (b) $b = 7$, $c = 5$, $\hat{B} = 57°$

(c) $\hat{B} = 82°$, $\hat{A} = 41°$, $c = 8$ (d) $\hat{A} = 117°$, $\hat{B} = 11°$, $a = 6$

(e) $\hat{A} = 90°$, $b = 7$, $c = 24$ (f) $\hat{A} = \hat{B}$, $\hat{C} = 40°$, $c = 5$

 2 Use the cosine rule to determine the unknown sides and angles of the triangles given below.

(a) $a = 4$, $b = 5$, $c = 6$ (b) $a = 10$, $b = 7$, $c = 6$

(c) $\hat{A} = 47°$, $b = 3$, $c = 4$ (d) $a = 6$, $b = 8$, $\hat{c} = 121°$

 3 The accompanying figure shows $AB = 6$ cm and the line AG drawn at $40°$ to AB. At the point B an arc of radius 4 cm is drawn. This meets AG at the points C_1 and C_2 respectively. Now solve the triangles ABC_1 and ABC_2 given $A = 40°$, $a = 4$, $c = 6$, noting that there are two possible solutions.

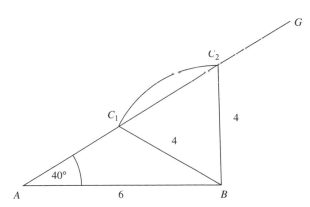

4 Determine all possible triangle solutions, where possible, for the following data sets:

(a) $a = 3.5$, $b = 5$, $c = 3$ (b) $\hat{A} = 34°$, $\hat{B} = 42°$, $c = 21.63$

(c) $b = 6$, $c = 8$, $\hat{B} = 41°$ (d) $b = 4$, $c = 9$, $\hat{B} = 37°$

 5 Find the lengths of the sides of a parallelogram whose diagonals are of length 3 and 2 and meet at an angle of 42°.

 6 (a)   If $\triangle ABC$ is obtuse prove that

$$R = \frac{a}{2 \sin \hat{A}} = \frac{b}{2 \sin \hat{B}} = \frac{c}{2 \sin \hat{C}}$$

where R is the radius of the circumscribed circle.

(b)   Determine R for the triangles in Question 4. What do you observe when two possible triangle solutions exist?

 7 Determine the area of each triangle in Question 4.

8 (a)   Find the points where the straight line $y = 2x - 3$ crosses the coordinate axes.

(b)   Determine the area of the triangle enclosed by the crossing points and the origin.

(c)   What are the angles in the triangle?

9 Use Pythagoras' theorem in the right-angled triangle shown in the figure to prove that the distance between (x_1, y_1) and (x_2, y_2) is

$$\{(x_2 - x_1)^2 + (y_2 - y_1)^2\}^{1/2}$$

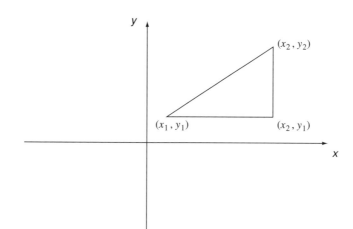

Find the distance between

(a)   (3, 4) and (1, 0)   (b)   (−2, 3) and (3, 4)

(c)   (1, 0) and (−2, 3).

Hence find the area of the triangle formed by (3, 4), (1, 0) and (−2, 3).

10   Draw a graphical picture of the triangle in Question 9.

(a)   Determine the equation of the three straight lines which make up its sides in the form $y = mx + c$.

(b)   Using INV TAN on a calculator, find the angle of intersection of each straight line with the x-axis.

(c)   Check that the angles in (b) match the angles in the triangle.

11   For the general scalene triangle whose sides are a, b, c we know from the cosine rule that

$$\cos \hat{A} = \frac{b^2 + c^2 - a^2}{2bc}.$$

By defining $s = \frac{1}{2}(a + b + c)$, the **semi-perimeter**, it can be proved that

$$\sin \hat{A} = \frac{2\{s(s - a)(s - b)(s - c)\}^{1/2}}{bc}$$

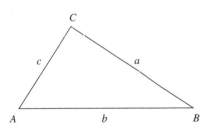

(a)   Deduce that the area of the triangle is

$$\triangle = \{s(s - a)(s - b)(s - c)\}^{1/2}.$$

(b)   Verify the result for the triangle in Question 9.

12 Verify that the area of the triangle AIB is equal to $\frac{1}{2}rs$, where r is the radius of the inscribed circle. Using the notation of Question 11, deduce that for the triangle ABC, $\triangle = rs$.

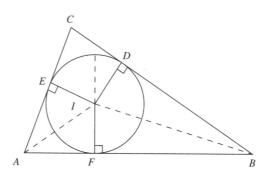

13 (a) Prove that the radius of the circumscribed circle R is

$$R = \frac{abc}{4\triangle}$$

(b) For an equilateral triangle determine the ratio r/R.

14 Using the standard notation in the triangle, prove the following results:

(a) $a(\sin \hat{B} - \sin \hat{C}) + b(\sin \hat{C} - \sin \hat{A}) + c(\sin \hat{A} - \sin \hat{B}) = 0$

(b) $\dfrac{\cos \hat{A}}{a} + \dfrac{\cos \hat{B}}{b} + \dfrac{\cos \hat{C}}{c} = \dfrac{a^2 + b^2 + c^2}{2abc}$

(c) $a(c \cos \hat{B} - b \cos \hat{C}) = c^2 - b^2$

(d) $\dfrac{a \sin \hat{A} + b \sin \hat{B}}{c \sin \hat{C}} = \dfrac{a^2 + b^2}{c^2}$

(e) $\dfrac{b \cos \hat{C} + c \cos \hat{B}}{a \cos \hat{C} + c \cos \hat{A}} = \dfrac{\sin \hat{A}}{\sin \hat{B}}$.

15 The triangle ABC has vertices at (x_1, y_1), (x_2, y_2) and (x_3, y_3) as shown.

(a) Write down the areas of the parallelograms, $ALMC$, $CMNB$ and $ALNB$.
(b) Deduce that the area of triangle ABC is given by

$$\triangle = \frac{1}{2}\{x_1(y_2 - y_3) + x_2(y_3 - y_1) + x_3(y_1 - y_2)\}.$$

(c) What conclusion could you draw if $\triangle = 0$?

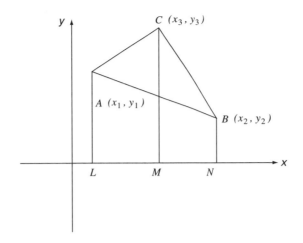

16 Determine the areas of the triangles whose vertices are

(a) $(-3, 2), (2, 3), (-1, 4)$

(b) $(0, 0), (2, 1), (-2, 4)$

(c) $(a, 0), (b, b), (4b - a, 4b)$ where $b > a > 0$

(d) $(-3, 7), (-1, 4), (3, -2)$.

17 A quadrilateral is formed by the points $(x_1, y_1), (x_2, y_2), (x_3, y_3)$ and (x_4, y_4) as shown in the diagram.

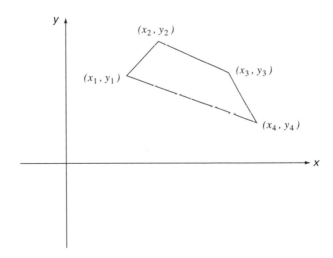

(a)   Its area is given by

$$\frac{1}{2}|x_1(y_2 - y_3) + x_2(y_3 - y_1) + x_3(y_1 - y_2)|$$

$$+ \frac{1}{2}|x_1(y_3 - y_4) + x_3(y_4 - y_1) + x_4(y_1 - y_3)|$$

Why are the modulus signs so important?

(b)   By drawing a simple sketch, but without giving detail, explain how you would find the area of a pentagon knowing the coordinates of its vertices.

(c)   A quadrilateral is formed by the points $(6, -2)$, $(-2, -1)$, $(3, 5)$ and $(6, 3)$. Sketch it on graph paper and determine its area.

18   By solving the triangles ABD and BCD determine the area of the quadrilateral shown in the diagram.

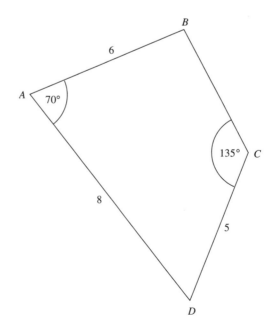

19 How far is P from O?

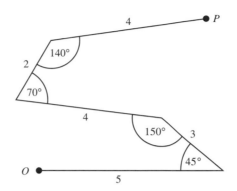

20 Determine the distance AC to the nearest whole number.

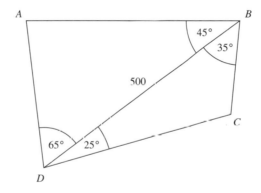

7.3 DISTANCES AND ANGLES IN SOLIDS

Problems which are concerned with solid figures are usually solved by selecting suitable right-angled triangles. A good starting-point is the drawing of a clear diagram. Some useful tips are given.

(i) Keep the diagram simple; draw two or more diagrams if necessary.

(ii) Show a three-dimensional solid from a point of view which is not head-on but from a point above and to one side.

(iii) Indicate hidden edges by dotted lines.

In the examples which follow, these and other aspects of good practice will be used.

The angle between a line and a plane

The angle between a line and a plane is the angle between the line and its projection on the plane.

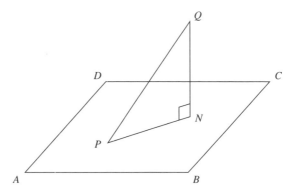

Figure 7.14 Angle between line and plane

In Figure 7.14 the line segment PQ meets the plane $ABCD$ at P. QN is the perpendicular from Q to the plane. PN is the projection of PQ on the plane and PQN is a right-angled triangle. The angle $Q\hat{P}N$ is the angle between the line and the plane.

Note that a line which is perpendicular to a plane is perpendicular to every line in the plane.

The angle between two planes

Two planes which are not parallel meet in a straight line, known as the **common line**.

To determine the angle between two planes, draw a line in each plane from a point on the common line so that the two lines drawn are perpendicular to the common line. The angle between these two lines is the angle between the planes. In Figure 7.15 the common line is AB and the two lines perpendicular to AB are OP and OQ. The required angle is $P\hat{O}Q$.

If one of the planes is horizontal, say the plane containing AOB and Q, then the line OP is a **line of greatest slope** in the inclined plane. It indicates the steepest path up the

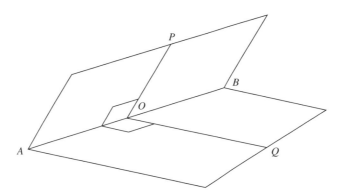

Figure 7.15 Angle between two planes

inclined plane from O; wherever we start on AB the line of greatest slope is parallel to OP.

Note that if two planes are perpendicular then it is not the case that each line in one plane is perpendicular to each line in the other. See for instance Figure 7.16; the lines AD and BG are not perpendicular.

Examples

1. Figure 7.16 shows a cuboid. Find

   (i)   the length of the diagonal AG
   (ii) the angle $D\hat{A}G$
   (iii) the angle between the line AG and the plane $ABCD$
   (iv) the angle between the planes DAG and $ABCD$

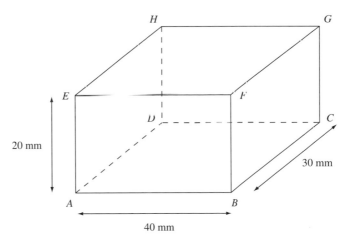

Figure 7.16 Diagram for the cuboid example

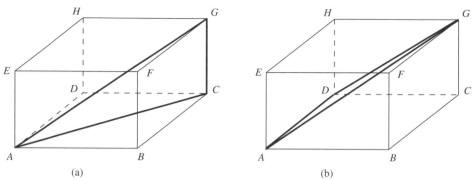

Figure 7.17 Construction for the cuboid example

In Figure 7.17(a) we have highlighted the triangle ACG

(i) First we use Pythagoras' theorem on the triangle ABC:

$$(AC)^2 = (AB)^2 + (BC)^2 = 1600 + 900 = 2500$$

Hence $AC = 50$ mm.

Now we apply Pythagoras' theorem to the triangle ACG:

$$(AG)^2 = (AC)^2 + (CG)^2 = 2500 + 400 = 2900$$

Hence $AG = \sqrt{2900} = 53.85$ mm.

Although it was easy to do so in this example, it was not necessary to evaluate AC itself; its square was all that was required.

(ii) In Figure 7.17(b) we have highlighted triangle ADG; it has a right angle at D. The length DG is found via Pythagoras' theorem to be

$$\sqrt{(AG)^2 - (AD)^2} = \sqrt{2900 - 900} = \sqrt{2000} = 44.72 \text{ mm}$$

The angle $D\hat{A}G$ can be found by first calculating its sine, its cosine or its tangent. We shall use the cosine:

$$\cos D\hat{A}G = \frac{DA}{AG} = \frac{30}{44.72} = 0.6708$$

Hence $D\hat{A}G = 47.9°$.

(iii) The specified angle is found by considering triangle AGC; C is the foot of the perpendicular from G to the plane $ABCD$.

Hence the angle we require is $G\hat{A}C$. Now

$$\tan G\hat{A}C = \frac{GC}{AC} = \frac{20}{50} = 0.4$$

so that $G\hat{A}C = 21.8°$.

(iv) The common line of the planes DAG and $ABCD$ is AD. Suitable perpendicular lines to it are GD and DC and the required angle is $G\hat{D}C$. From the right-angled triangle GDC it follows that $\tan G\hat{D}C = \dfrac{20}{40} = 0.5$, therefore $G\hat{D}C = 26.6°$.

2. Figure 7.18 shows a pyramid on a square base of side 120 mm. The sloping edges of the pyramid are of length 200 mm. G is vertically below P. Find

(i)   the altitude of the pyramid

(ii) the angle $P\hat{E}G$

(iii) the angle $P\hat{B}E$

(iv) the complete surface area of the pyramid

(i)   The altitude of the pyramid is the length $PG \times GE = \dfrac{1}{2} \times 120 = 60$ mm. From the right-angled triangle CGE we calculate

$$(GC)^2 = (GE)^2 + (EC)^2 = (60)^2 + (60)^2 = 7200$$

Therefore $GC = \sqrt{7200} = 60\sqrt{2}$ mm.
From the right-angled triangle PGC we calculate

$$(PG)^2 = (PC)^2 - (GC)^2 = 40\,000 - 7200 = 32\,800$$

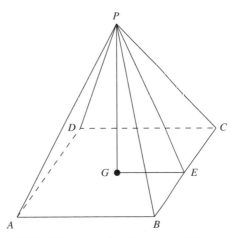

Figure 7.18 Diagram for the pyramid example

therefore $PG = 181.1$ mm.

(ii) From triangle PGE, $\tan P\hat{E}G = \dfrac{181.1}{60}$ so that $P\hat{E}G = 71.7°$.

(iii) In the right-angled triangle PBE, $PB = 200$ mm and $BE = \dfrac{1}{2} \times 120 = 60$ mm. Hence

$$\cos P\hat{B}E = \frac{BE}{BP} = \frac{60}{200}$$
$$P\hat{B}E = 72.5°$$

(iv) First we find PE from triangle PBE:

$$(PE)^2 = (PB)^2 - (BE)^2 = 40\,000 - 3600 = 36\,400$$

so that $PE = 190.79$ mm

We could have found PE from triangle PGE:

$$(PE)^2 = (PG)^2 + (GE)^2 = 32\,800 + 3600 = 36\,400 \quad \text{(as before)}$$

The area of triangle PBC is

$$\frac{1}{2} PE \times BC = \frac{1}{2} \times 190.79 \times 120 = 11\,447.3 \text{ mm}^2$$

The total surface area is

$$4 \times \text{area of } PBC + \text{area of base} = 4 \times 11\,447.3 + 14\,400$$
$$= 60\,189 \text{ mm}^2$$

3. A plane is inclined at $30°$ to the horizontal. A line in this plane makes an angle of $25°$ with the line of greatest slope. Find the angle which the first line makes with the horizontal.

Refer to Figure 7.19(a). The line BE makes an angle $25°$ with the line of greatest slope BC, which makes an angle $30°$ with the horizontal, so

Then $\dfrac{BC}{BE} = \cos 25°$ and $BE = \dfrac{BC}{\cos 25°}$

Let the perpendicular from M meet the plane $ABCD$ at C; then $\dfrac{CM}{BC} = \sin 30°$ so that $CM = BC \sin 30°$.

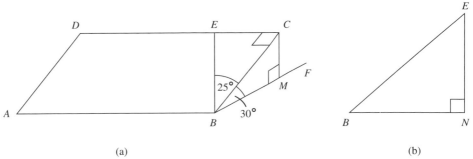

Figure 7.19 Diagrams for the plane example

This is also the height of E above the horizontal plane through A, B and M. If N is the foot of the perpendicular from E to the plane $ABCD$ then, as we see from Figure 7.19(b),

$$\sin E\hat{B}N = \frac{EN}{BE} = \frac{CM}{BE} = \frac{BC\sin 30°}{BC}\cos 25°$$
$$= \sin 30° \cos 25°$$
$$= 0.4532.$$

Then $E\hat{B}N = 26.9°$. ∎

Exercise 7.3

1 A car travelling north at 100 km hr^{-1} along a straight road first observes a tall radio mast on a bearing 040°. Twelve minutes later the bearing is 110°. How far is the mast to the east of the road?

2 A tall chimney at the end of a horizontal street subtends an angle of 10° to an observer on the pavement. By walking 100 m towards the chimney the observer notices that the elevation has risen to 15°. How tall is the chimney?

3 A block of flats 50 m tall lies east–west. Find the length of shadow cast at midday

    (a)   in summer, solar elevation 62°   (b)   in winter, solar elevation 15°

4 A fan belt is required to revolve around two cylindrical flywheels of radii 2 cm and 5 cm whose centres are 12 cm apart. How long is it?

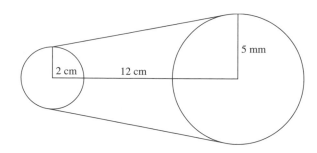

5   A cylindrical log is to be pulped at a sawmill in four equal volumes. The first stage is to slice it horizontally along its length. The slicing cross-section looks like the first diagram so that corresponding shaded areas are equal. Now refer to the second diagram. The upper and lower slicing levels are determined by arranging θ so that the area shaded is $\frac{1}{4}\pi R^2$, where R is the radius of the log. Prove that θ must satisfy the equation

$$2\theta - \pi = 2\sin\theta$$

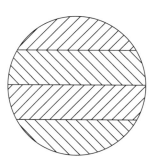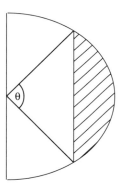

6   A steel beam 20 m long and density 8.5 gm cm^{-3} is being used to refit a building with a new pitched roof. The cross-section of the beam is specified to be of the form below, with distances in centimetres. Determine the mass of the beam to the nearest kilogram.

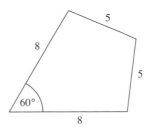

 7 *A* and *B* are two points on level ground, with *B* 2.5 km south of *A*. At a given instant an aeroplane is observed at a bearing 120° from *A* and 050° from *B*. The angle of elevation is 35° from *A*. Find the height of the aeroplane and its angle of elevation from *B*.

 8 As part of an accident emergency procedure, following an underground gas leak, the area within a radius 100 m from the leak must be cleared at ground level. For example the point *P*, 60 m east and 80 m north of the leak is on the edge of the clearance area (or cordon).

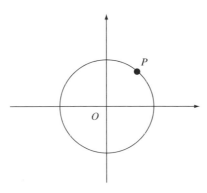

(a) A building 90 m south of the site of a detected leak lies east–west. What length of its frontage is exposed at ground level?

(b) At what level up the building might the front be considered safe if the upward effects of the blast are ignored?

(c) If (x, y) are the coordinates of a general point on the circle, what equation do x and y satisfy?

 9 A cuboid has the dimensions shown.

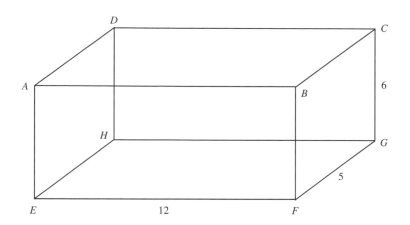

Determine

(a)   the angle *EC* makes with the horizontal

(b)   the acute angle between *EC* and *DF.*

10   The plane containing the triangle *ABC* is at 45° to the horizontal. *AB* is in the direction of steepest slope and *BC* is horizontal. Determine the angle *AC* makes with the horizontal plane.

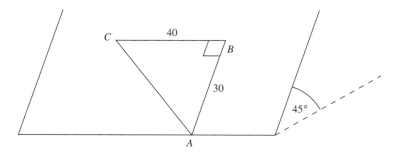

11   Three sonar detectors monitor fish movement from the bottom of a deep lake of 500 m depth. They are arranged in an equilateral triangle and each has an effective detection range of 1 km. How far apart can they be placed to ensure total coverage of breadth and depth within that triangle?

12   For a square pyramid of side 10 m and height 15 m, determine

(a)   the total volume,

(b)   the total surface area, excluding the base

(c)   the angles of slopes of both edges and inclined planes.

13   A ship passing a cape in good visibility observes a large mountain due north on shore. Two hours later and 40 km NW of the original position, the ship observes the mountain again, this time due east at an elevation to the horizontal of 5°. How far away from the ship on the later observation is the base of the mountain and how tall is it?

 14   Two pitched roofs meet at the corner of a building at right angles. *AB* represents the gulley between them at the inward corners. If each roof has a pitch of 15°, what is the angle of slope of the gully?

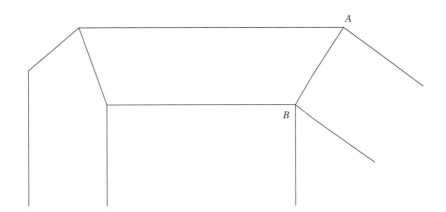

15   Find the distances between the following pairs of points:

(a)   (3, 1, 5) and (2, 4, 7)   (b)   (−1, −5, 3) and (6, 5, −5)

(c)   (4, 7, 11) and (−1, −9, 16).

16   Determine the distance of each of the points (0, 0, 0), (1, 0, 3), (4, 1, −1) from (2, 2, 1). What conclusion do you draw?

17   What is the volume of the tetrahedron formed by the points (1, 1, 0), (2, 1, −1), (0, 0, 0) and (6, 0, 0)?

 18   A tetrahedron is formed by the points *A*(3, 0, 0), *B*(−1, −2, 0), *C*(−1, 2, 0) and *D*(0, 0, 4). Given that the plane $z = 0$ is the base, determine the angle of slope of the plane formed by *B*, *C*, *D* with the horizontal.

7.4   RATIOS OF ANY ANGLE AND PERIODIC MODELLING

In Section 7.2 we extended the definition of sine, cosine and tangent to include obtuse angles. We continue that extension to embrace angles of any magnitude.

In Figure 7.20 two positions of the rotating line OP on a circle of radius 1 are shown so that P appears in the third and fourth quadrants.

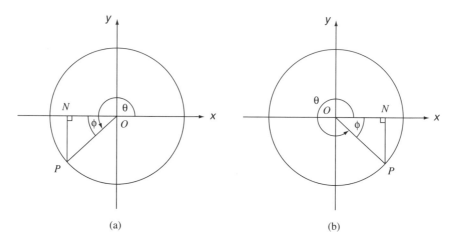

Figure 7.20   Ratios of angles greater than 180°

In Figure 7.20(a) $PN = OP \sin \phi = \sin \phi$ and $ON = OP \cos \phi = \cos \phi$. The coordinates of P are $(-\cos \phi, -\sin \phi)$, i.e. $(-\cos(\theta - 180°), -\sin(\theta - 180°))$.

We have defined the coordinates of P to be $(\cos \theta, \sin \theta)$ and therefore both the sine and the cosine of an angle between 180° and 270° are negative. However, since the tangent has been defined as the ratio of sine to cosine, in this quadrant it is positive. We therefore have the following results for an angle θ between 180° and 270°:

$$\sin \theta \equiv -\sin(\theta - 180°) \quad \cos \theta \equiv -\cos(\theta - 180°)$$
$$\tan \theta \equiv \tan(\theta - 180°)$$

Similarly, referring to Figure 7.20(b), we find that

$$\sin \theta \equiv -\sin(360° - \theta) \quad \cos \theta \equiv \cos(360° - \theta)$$
$$\tan \theta \equiv -\tan(360° - \theta)$$

Example

Find the three trigonometric ratios for the angles
(i) 240°   (ii) 330°

Solution

(i) $\theta = 240°$ is in the third quadrant and $\theta - 180° = 60°$. Hence

$$\sin 240° = \sin(240° - 180°) = -\sin 60°$$
$$\cos 240° = -\cos 60°, \ \tan 240° = \tan 60°$$

(ii) $\theta = 330°$ is in the fourth quadrant and $360° - \theta = 30°$, so

$$\sin 330° = \sin(360° - 330°) = -\sin 30°$$
$$\cos 330° = \cos 30°, \ \tan 330° = -\tan 30°$$ ■

We should make a summary of our results so far.

The ratios which are positive in each quadrant are given by the following rule (sometimes known as the CAST rule).

S (sin)	A (all)
T (tan)	C (cos)

Otherwise the ratio is negative in sign.

As an example, the angle $210°$ is in the third quadrant and has a positive tangent but negative sine and cosine.

The magnitude of the trigonometric ratios are best calculated using the acute angle formed with the horizontal axis. For example, in the second quadrant the cosine of an angle is negative and the magnitude of $\cos 135° = -$the magnitude of $\cos 45° = -\dfrac{1}{\sqrt{2}} \simeq -0.707$; see Figure 7.21.

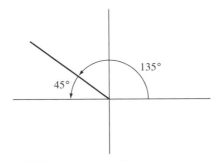

Figure 7.21   Angle for the second quadrant

Examples

First Quadrant

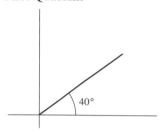

Sign	Value
$\sin 40° = +$	$\sin 40° = 0.643$
$\cos 40° = +$	$\cos 40° = 0.766$
$\tan 40° = +$	$\tan 40° = 0.839$

Second Quadrant

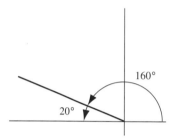

Sign	Value
$\sin 160° = +$	$\sin 20° = 0.342$
$\cos 160° = -$	$-\cos 20° = -0.940$
$\tan 160° = -$	$-\tan 20° = -0.364$

Third Quadrant

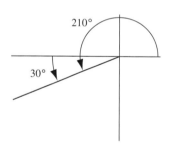

Sign	Value
$\sin 210° = -$	$-\sin 30° = -0.500$
$\cos 210° = -$	$-\cos 30° = -0.866$
$\tan 210° = +$	$\tan 30° = 0.577$

Fourth Quadrant

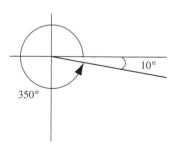

Sign	Value
$\sin 350° = -$	$-\sin 10° = -0.174$
$\cos 350° = +$	$\cos 10° = 0.985$
$\tan 350° = -$	$-\tan 10° = -0.176$

A useful generalisation of the above examples is provided by Figure 7.22 and the formulae enclosed in the following display.

$$\sin(180° - \theta) \equiv \sin \theta \qquad \cos(180° - \theta) \equiv -\cos \theta \qquad \tan(180° - \theta) \equiv -\tan \theta$$
$$\sin(180° + \theta) \equiv -\sin \theta \qquad \cos(180° + \theta) \equiv -\cos \theta \qquad \tan(180° + \theta) \equiv \tan \theta$$
$$\sin(360° - \theta) \equiv -\sin \theta \qquad \cos(360° - \theta) \equiv \cos \theta \qquad \tan(360° - \theta) = -\tan \theta$$

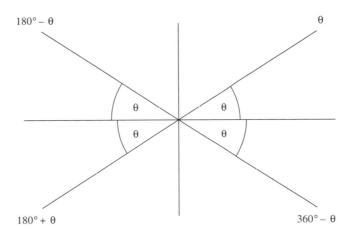

Figure 7.22 Related angles

Example

Find

(a) $\sin 300°$ (b) $\tan 200°$

(a) The angle $300°$ is in the fourth quadrant and can be written as $(360° - 60°)$. Hence $\sin 300° = -\sin 60°$.

(b) The angle $200°$ is in the third quadrant and can be written as $(180° + 20°)$. Hence $\tan 200° = +\tan 20°$. ■

Ratios of angles greater than 360°

If we return to the rotating line OP in Figure 7.20 we may imagine rotating machines where the line has gone through more than one revolution.

If we compare the positions of OP after a rotation of $390°$ with that of $30°$ we will find that they are identical since $390° = 360° + 30°$, i.e. one full revolution plus $30°$. Indeed a rotation of $750°$ produces an identical result since $750° = 360° + 360° + 30°$, i.e. two complete revolutions plus 30°. Clearly each of the ratios for these three angles has the same value at all three angles, e.g. $\sin 750° = \sin 390° = \sin 30°$.

We may generalise this result to say that

$$\sin(360° + \theta) \equiv \sin\theta, \quad \cos(360° + \theta) \equiv \cos\theta, \qquad \tan(360° + \theta) \equiv \tan\theta$$

In this context angles are often measured in radians and the equivalent results are as follows:

$$\sin(2\pi + \theta) \equiv \sin\theta, \quad \cos(2\pi + \theta) \equiv \cos\theta, \qquad \tan(2\pi + \theta) \equiv \tan\theta$$

Example

Find

(i) $\cos 780°$ (ii) $\tan 600°$ (iii) $\sin\dfrac{11\pi}{4}$

(i) $780° = 720° + 60°$ therefore $\cos 780° = \cos 60° = 0.5$

(ii) $600° = 360° + 240°$ therefore $\tan 600° = \tan 240°$

  Since $240° = 180° + 60°$, $\tan 600° = +\tan 60° = \sqrt{3}$

(iii) $\dfrac{11\pi}{4} = 2\pi + \dfrac{3\pi}{4}$ therefore $\sin\dfrac{11\pi}{4} = \sin\dfrac{3\pi}{4}$

  Since $\dfrac{3\pi}{4} = \pi - \dfrac{\pi}{4}$, $\sin\dfrac{11\pi}{4} = +\sin\dfrac{\pi}{4} = \dfrac{1}{\sqrt{2}}$.   ■

Ratios of negative angles

If θ measured anticlockwise from the positive x-axis is a *positive* angle then θ measured clockwise from that position is a *negative* angle. Figure 7.23 illustrates the situation for an acute negative angle.

Now OP is in exactly the position it would have if it were to rotate anticlockwise from OA through an angle $(360° - \theta)$. Hence

$$\sin(-\theta) \equiv \sin(360° - \theta) \equiv -\sin\theta$$
$$\cos(-\theta) \equiv \cos(360° - \theta) \equiv \cos\theta$$
$$\tan(-\theta) \equiv \tan(360° - \theta) \equiv -\tan\theta$$

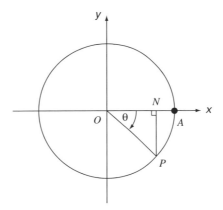

Figure 7.23 Negative angles

The conclusions are true whatever the magnitude of the negative angle, and we may state the following:

$$\sin(-\theta) \equiv -\sin\theta, \qquad \cos(-\theta) \equiv \cos\theta, \qquad \tan(-\theta) \equiv -\tan\theta$$

Example

Find
(i) $\sin(-200°)$ (ii) $\tan(-300°)$

(i) $\sin(-200°) = -\sin 200° = -\sin(180° + 20°) = -(-\sin 20°) = +\sin 20°$. Note that $-200°$ occupies the same position as $160°$ and
$\sin 160° = \sin(180° - 20°) = \sin 20°$.

(ii) $\tan(-300°) = -\tan 300° = -\tan(360° - 60°) = \tan 60°$.

Graphs of the ratios

We now display the graphs of $\sin\theta$ and $\cos\theta$ for values of θ between $-180°$ and $+540°$. We calculated the values for several angles in the range $0°$ to $90°$ and used the properties of the ratio of angles in other quadrants to extend the graphs. Figure 7.24 shows $\sin\theta$ and Figure 7.25 shows $\cos\theta$. Notice that we have marked the scales in Figure 7.21 both in degrees and in radians for comparison.

Notice how the pattern repeats every $360°$ or 2π radians. This feature is called **periodic** behaviour and the **period** is $360°$ or 2π radians. Because of the shape of the graphs, periodic phenomena, e.g. oscillations, can often be expressed in terms of sines and cosines.

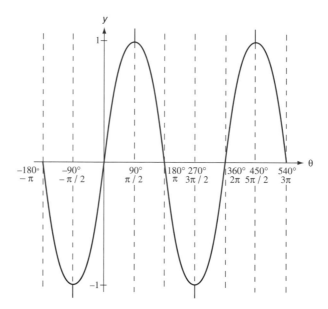

Figure 7.24 Sine of any angle

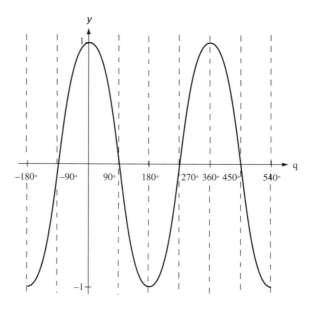

Figure 7.25 Cosine of any angle

The graph of $\tan \theta$ has a surprise for us. It is shown in Figure 7.26.

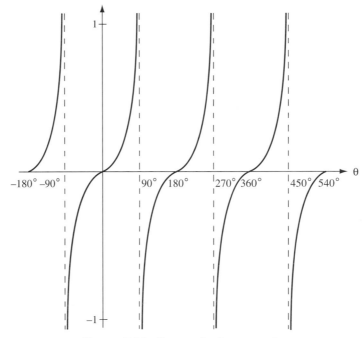

Figure 7.26   Tangent of any angle

The graph shows that $\tan \theta$ is also periodic, but the smallest interval over which the pattern repeats is 180° or π radians, even though it is the ratio of two quantities which have a period of 360° or 2π radians.

Furthermore, the graphs of sine and cosine oscillate smoothly between the values -1 and $+1$ but the tangent has violent breaks in its graph every 180°. Each of its **branches** can be moved 180° to the left or to the right to coincide with its neighbour.

Solutions of trigonometric equations

tan θ = p

Example

Suppose that we wish to solve the equation

$$\tan \theta = 1$$

If we input 1 to a pocket calculator and press INV TAN, we obtain $\theta = 45°$ or, in radian mode $\theta \simeq 0.7854$, which is an approximation to $\theta = \dfrac{\pi}{4}$. However, if we look at Figure

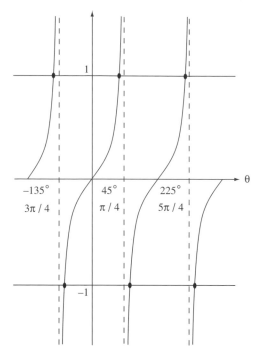

Figure 7.27 $\tan\theta = 1$ or -1

7.27 we see there are other solutions of the equation, i.e. there are several angles whose tangent is 1. In fact there are infinitely many, both to the left of 45° and to its right.

We consider two cases.

Solutions between 0° and 360° (0 and 2π radians)

There is a second solution, namely $\theta = 225°$ $\left(\text{or } \dfrac{5\pi}{4} \text{ radians}\right)$ found by adding 180° (or π radians) to the calculator solution.

All solutions of the equation

If we add or subtract multiples of 180° (or π radians) to the calculator solution then we generate all the possible solutions. Here are some of them:

$$\ldots, -315°, \quad -135°, \quad 45°, \quad 225°, \quad 405°, \quad 585°, \ldots$$

or in radians

$$\ldots, -\frac{7\pi}{4}, \quad -\frac{3\pi}{4}, \quad \frac{\pi}{4}, \quad \frac{5\pi}{4}, \quad \frac{9\pi}{4}, \quad \frac{13\pi}{4}, \ldots$$ ■

In radian mode a compact formula can be used to cover all solutions. It is

$$n\pi + \theta_0, \qquad n = 0, \quad \pm 1, \quad \pm 2, \ldots$$

where θ_0 is the calculator solution of the equation.

Note that if we wish to solve the equation $\tan \theta = -1$ the calculator solution is $-45°$ and the solutions in the range $0° \leq \theta \leq 360°$ can be found by adding $180°$ twice to give first $\theta = 135°$ then $\theta = 315°$. Other solutions can be found by repeatedly adding or subtracting $180°$. The formula applies generally to the equation $\tan \theta = p$, whether p is positive or negative.

cos $\theta = p$, $-1 \leq p \leq 1$

Figure 7.28 shows the graph of $\cos \theta$ between $\theta = 0$ and $\theta = 360°$.

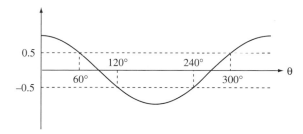

Figure 7.28 $\cos \theta = 0.5$ or -0.5

Example

Suppose we wish to solve the following equations:
(i) $\cos \theta = 0.5$
(ii) $\cos \theta = -0.5$

(i)   The calculator solution is $\theta = 60°$ or $\dfrac{\pi}{3}$ radians. In the range $0 \leq \theta < 360°$ there is

a second solution, $\theta = 360° - 60° = 300°$ or $\theta = 2\pi - \dfrac{\pi}{3} = \dfrac{5\pi}{3}$ radians. Other solutions can be found by adding or subtracting multiples of $360°$ or 2π radians, as appropriate.

(ii)   The calculator solution of $\cos \theta = -0.5$ is $\theta = 120°$ $\left(\text{or} \dfrac{2\pi}{3} \text{radians} \right)$. This angle is of the form $180° - \phi$ where $\phi = 60°$; the second solution in the range $0 \leq \theta \leq 360°$ is of the form $180° + \phi$, i.e. $\theta = 240°$ $\left(\text{or} \dfrac{4\pi}{3} \text{radians} \right)$. Note that $240° = 360° - 120°$.

Again, other solutions can be found by adding or subtracting multiples of 360° (or 2π radians). A compact formula for the solutions in radians of the equation $\cos \theta = p$ is

$$2n\pi \pm \theta_0, \qquad n = 0, \ \pm 1, \ \pm 2, \ldots$$

where θ_0 is the calculator solution.

$sin \ \theta = p, \ -1 \le p \le 1$

Example

Find solutions of the following equations:
(i) $\sin \theta = 0.5$,
(ii) $\sin \theta = -0.5$

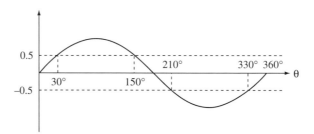

Figure 7.29 $\sin \theta = 0.5$ or -0.5

Figure 7.29 shows the graph of $\sin \theta$ for $0 \le \theta \le 360°$ (or $0 \le \theta \le 2\pi$ radians).

(i)   If we input 0.5 to a pocket calculator and press INV SIN, we obtain $\theta = 30°$ $\left(\text{or } \dfrac{\pi}{6}\right.$
radians$\left.\right)$. Looking at Figure 7.29 we see there is a second solution in the interval $0 \le \theta \le 360°$; it is the supplement of the first solution, i.e. $180° - 30° = 150°$
$\left(\text{or } \pi - \dfrac{\pi}{6} = \dfrac{5\pi}{6} \text{ radians}\right)$.

Other solutions are obtained by adding or subtracting multiples of 360° (or 2π radians) to each of these solutions. We therefore generate two sets of solutions:

$$\ldots, \quad -690°, \quad -300°, \quad 30°, \quad 390°, \quad 750°, \ldots$$

$$\text{and} \quad \ldots, \quad -570°, \quad -210°, \quad 150°, \quad 510°, \quad 870°, \ldots$$

In radian measure these solutions are

$$\ldots, \ -\frac{23\pi}{6}, \ -\frac{11\pi}{6}, \ \frac{\pi}{6}, \ \frac{17\pi}{6}, \ \frac{25\pi}{6}, \ \ldots$$

$$\text{and} \quad \ldots, \ -\frac{19\pi}{6}, \ -\frac{7\pi}{6}, \ \frac{5\pi}{6}, \ \frac{17\pi}{6}, \ \frac{29\pi}{6}, \ \ldots$$

(ii) A pocket calculator gives the solution of $\sin\theta = -0.5$ as $\theta = -30°$. If, as is often the case, we seek solutions in the interval $0° \le \theta \le 360°$ then we add $360°$ to this value to obtain $\theta = 330°$. The second solution is found by noting that $330° = 360° - 30° = 360° - \phi$ then calculating $180° + \phi = 180° + 30° = 210°$.

Other solutions are generated as before by adding or subtracting multiples of $360°$ to both solutions above. Hence the sets are

$$\ldots, \ -510°, \quad 150°, \quad 210°, \quad 570°, \quad 930°, \ \ldots$$

$$\text{and} \quad \ldots, \ -390°, \quad -30°, \quad 330°, \quad 690°, \quad 1050°, \ \ldots$$

or in radians

$$\ldots, \ -\frac{17\pi}{6}, \ -\frac{5\pi}{6}, \ \frac{7\pi}{6}, \ \frac{19\pi}{6}, \ \frac{31\pi}{6}, \ \ldots$$

$$\text{and} \quad \ldots, \ -\frac{13\pi}{6}, \ -\frac{\pi}{6}, \ \frac{11\pi}{6}, \ \frac{23\pi}{6}, \ \frac{35\pi}{6}, \ \ldots$$ ∎

Exercise 7.4

1 Given that the relationships $\sin(-x) = -\sin x$ and $\cos(-x) = \cos x$, define $\sin x$ and $\cos x$ respectively as *odd* and *even* functions, define the following as odd, even or neither.

(a) $\sin 2x$ (b) $\sin^2 x$ (c) $\sin(x + \pi/2)$

(d) $\cos^2 x$ (e) $\sin x + \sin 2x$ (f) $\sin(2x - 1)$

(g) $\tan^2 x$ (h) $\dfrac{2}{\sin x} + \dfrac{3}{\sin^3 x}$

2 On graphs directly below one another and in the range $0 \le x \le 2\pi$ draw

(a) $\sin x$ (b) $\sin 2x$ (c) $\sin^2 x$

What do you observe?

3 For $-\pi \leq x \leq \pi$ draw the graph of $y = \dfrac{1}{\cos x}$ and dot in the graph of $y = \cos x$.

4 Extend the graph of $y = \sin x$ to $-4\pi \leq x \leq 4\pi$ and draw in the line $y = \dfrac{1}{2}$.

    (a) For what values of x in $0 \leq x \leq 2\pi$ does $\sin x = \dfrac{1}{2}$?

    (b) Extend the result to the rest of the interval $-4\pi \leq x \leq 4\pi$.

    (c) What is the general solution for $\sin x = \dfrac{1}{2}$?

5 Repeat Question 4 for

    (a) $y = \cos x$ $\cos x = \dfrac{1}{2}$

    (b) $y = \tan x$ $\tan x = 1$

6 Solve $\sin(x\pi) = \dfrac{1}{2}$ for $0 \leq x \leq 2\pi$ by first setting $x\pi = z$ and obtaining solutions for z.

SUMMARY

- **Ratios of an angle:** in a right-angled triangle, opposite/hypotenuse is the **sine** of an angle, $\sin\theta$; adjacent/hypotenuse is its **cosine**, $\cos\theta$; sine/cosine is its **tangent**, $\tan\theta$.
- **Fundamental identity:** $\sin^2\theta + \cos^2\theta \equiv 1$.
- **Cast rule:** determines the sign of the trigonometric ratios in the four quadrants of the plane.
- **Useful results:**

$$\sin(90° - \theta) = \cos\theta$$
$$\cos(90° - \theta) = \sin\theta$$
$$\sin(180° - \theta) = \sin\theta$$
$$\sin(360° - \theta) = -\sin\theta$$

- **Sine rule:** in triangle ABC

$$\frac{a}{\sin A} = \frac{b}{\sin B} = \frac{c}{\sin C}$$

- **Cosine rule:** in triangle ABC

$$a^2 = b^2 + c^2 - 2bc\cos A$$

- **Area of a triangle:** triangle ABC with sides a, b. c has area $\frac{1}{2}bc\sin A$
- **Solution of a triangle:** triangles can be solved if we know

  (i)   three sides
  (ii) two sides and the included angle
  (iii) two angles and one side

The ambiguous case may occur if we are given two sides and a non-included angle.

- **Angle between a line and a plane:** the angle between the line and its projection on the plane.

- **Meeting of two planes:** unless they are parallel, two planes meet in a common line; the angle between the planes is the angle between their normals.

- **General solutions:**

$$\tan \theta = p \qquad\qquad \theta_0 \pm n\pi,$$
$$\cos \theta = p, \ |p| \le 1 \qquad 2n\pi \pm \theta_0$$
$$\sin \theta = p, \ |p| \le 1 \qquad \theta_0 \pm 2n\pi, \ \pi - \theta_0 \pm 2n\pi$$

where θ_0 is the value obtained from a calculator.

Answers

Exercise 7.1

1 (a) $\pi/3$ (b) $3\pi/4$ (c) $7\pi/6$

   (d) $5\pi/3$ (e) $4\pi/5$ (f) $9\pi/4$

   (g) 5π

2 (a) $36°$ (b) $540°$ (c) $80°$

   (d) $35°$ (e) $990°$ (f) $1845°$

   (g) $-270°$

3 (a) 0.8088^c (b) 1.7092^c (c) 5.324^c

   (d) $57.296°$ (e) $94.309°$ (f) $277.54°$

4 (selected answers)

   1(c) $\sin 210° = -\sin 30° = -\dfrac{1}{2}$ $\cos 210° = -\cos 30° = -\sqrt{3}/2$

       $\tan 210° = \tan 30° = \dfrac{1}{\sqrt{3}}$

   1(f) $405° = 360° + 45°$ $\sin 405° = \sin 45° = \dfrac{1}{\sqrt{2}}$, etc.

   1(g) $900° = 2 \times 360° + 180°$ $\sin 900° = \sin 180° = 0$

   2(a) $\sin\dfrac{\pi}{5} = 0.5878$ $\cos\dfrac{\pi}{5} = 0.8090$ $\tan\dfrac{\pi}{5} = 0.7265$

   2(e) $\dfrac{11\pi}{2} = 2 \times 2\pi + \dfrac{3\pi}{2}$ $\sin\dfrac{3\pi}{2} = -\sin\dfrac{\pi}{2} = -1$, etc.

   2(f) $\sin\dfrac{41\pi}{4} = \sin\left(10\pi + \dfrac{\pi}{4}\right) = \sin\dfrac{\pi}{4} = \dfrac{1}{\sqrt{2}}$, etc.

   2(g) $\sin\left(-\dfrac{3\pi}{2}\right) = -\sin\dfrac{3\pi}{2} = 1$, etc.

   3(c) $\sin 305.06° = -\sin(360° - 305.06°) = -\sin 54.95° = -0.8186$

       $\cos(360° - 305.06) = \cos 54.94° = 0.5744.$

5 (a) $-\sin\theta, -\cos\theta, \tan\theta$ (b) $\sin\theta, \cos\theta, \tan\theta$

   (c) $-\sin\theta, \cos\theta, -\tan\theta$ (d) $\cos\theta, \sin\theta, \dfrac{1}{\tan\theta}$

   (e) $\sin\theta, -\cos\theta, -\tan\theta$ (f) $-\cos\theta, \sin\theta, -\dfrac{1}{\tan\theta}$

6

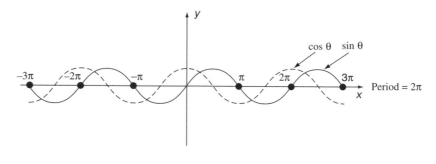

(a)

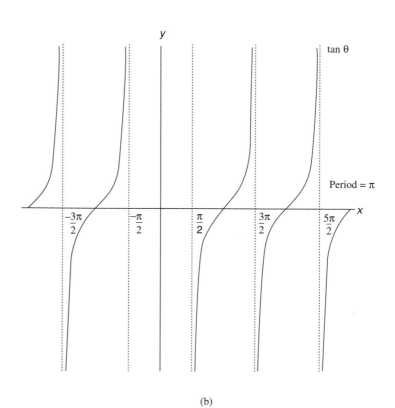

(b)

7 (a,b)   (c)

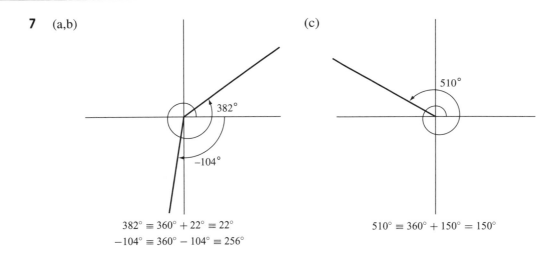

$382° \equiv 360° + 22° \equiv 22°$
$-104° \equiv 360° - 104° \equiv 256°$

$510° \equiv 360° + 150° = 150°$

9 (3, 4, 5), $\theta = 36.87°$, $\phi = 51.13°$; (5, 12, 13), $\theta = 22.62°$, $\phi = 67.38°$;
(8, 15, 17), $\theta = 28.07°$, $\phi = 61.93°$

Exercise 7.2

1 (a) $\hat{B} = 64.51°$, $\hat{C} = 78.49$, $c = 6.513$

(b) $\hat{C} = 36.8°$, $\hat{A} = 86.2°$, $a = 8.328$

(c) $\hat{C} = 57°$, $a = 6.26$, $b = 9.45$

(d) $\hat{C} = 52°$, $b = 1.28$, $c = 5.31$

(e) $\hat{B} = 16.26°$, $\hat{C} = 73.74°$, $a = 25$

(f) $\hat{A} = \hat{B} = 70°$, $a = b = 7.31$

2 (a) $\hat{A} = 41.40°$, $\hat{B} = 55.77°$, $\hat{C} = 82.82°$

(b) $\hat{A} = 100.29°$, $\hat{B} = 43.53°$, $\hat{C} = 36.18°$

(c) $a = 2.938$, $\hat{B} = 48.31°$, $\hat{C} = 84.69°$

(d) $c = 12.225$, $\hat{A} = 24.88°$, $\hat{B} = 34.12°$

3 (a) $C_1 = 105.38°$, $\hat{B} = 34.62°$, $b = 3.54$

(b) $C_2 = 74.62°$, $\hat{B} = 65.38°$, $b = 5.66$

4 (a) $\hat{A} = 43.53°$, $\hat{B} = 100.29°$, $\hat{C} = 36.18°$

(b) $a = 12.5$, $b = 14.9$, $\hat{C} = 104°$

(c) $\hat{A} = 77.98°$ or $20.02°$, $\hat{C} = 61.02°$ or $118.98°$, $a = 8.95$ or 3.13

(d) No triangle can exist

5 1.01, 2.34

6 4(a) 2.54   4(b) 11.14

4(c) 4.57 (same in both cases)

7 4(a) 5.165   4(b) 90.36   4(c) 6.215 or 23.48

8 (a) $(0, -3)$, $\left(\dfrac{3}{2}, 0\right)$ (b) $\dfrac{9}{4}$ (c) $27°, 63°, 90°$

9 (a) $2\sqrt{5}$ (b) $\sqrt{26}$ (c) $3\sqrt{2}$ Area $= 9$

10 BC: $y = 2(x - 1)$, $63.43°$, AB: $x + y = 1$, $-45°$, AC: $y = \dfrac{x + 17}{5}$, $11.31°$ $\hat{A} = 56.31°$, $\hat{B} = 71.57°$, $\hat{C} = 52.12°$

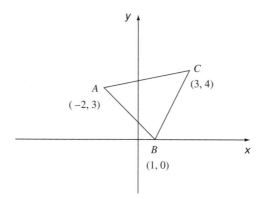

13 (b) $\dfrac{r}{R} = \dfrac{1}{2}$

15 (c) (x_1, y_1), (x_2, y_2), (x_3, y_3) are collinear

16 (a) 4   (b) 5   (c) ab (d) 0 (points are collinear)

17 (a) The area of the quadrilateral is the sum of the areas of the triangles. Magnitudes must be added and not allowed to cancel if quantities of the type $x_1(y_2 - y_3) + x_2(y_3 - y_1) + x_3(y_1 - y_2)$ work out negative.

(b)

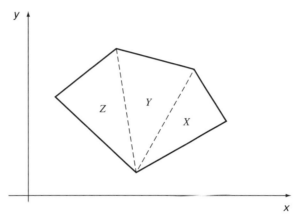

Sum of areas of triangles XYZ, for example

(c)

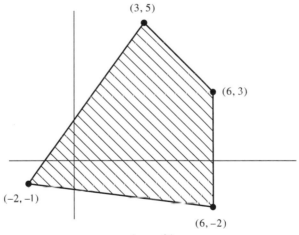

Area $= 34$

18 29.37

19 7.085

20 501

Exercise 7.3

1   12.86 km

2   51.57 m

3   (a)   26.59 m   (b)   186.60 m

4   46.74 cm

6   675 kg

7   1427 m, 31.78°

8   (a)   87.2 m   (b)   43.6 m   (c) $x^2 + y^2 = 100$

9   (a)   24.78°   (b)   45.24°

10   25.10°

11   1.5 km

12   (a)   500 m^3 (b)   316.2 m^2 (c)   64.76°, 71.56°

13   28.29 km, 2475 m

14   10.55°

15   (a)   3.742   (b)   14.59   (c)   17.49

16   Distance $= 3$ in each case. (2, 2, 1) is the centre of a sphere upon which all the points lie.

17   2 units

18   $\tan^{-1} 4 = 75.96°$

Exercise 7.4

1   (a)   odd   (b)   even
    (c) $\sin\left(x + \dfrac{\pi}{2}\right) = -\cos x$, so even   (d)   even
    (e)   odd   (f)   neither
    (g)   even   (h)   odd

2

(a)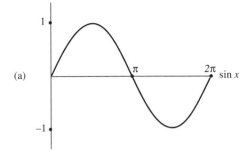

$\sin x$

(b)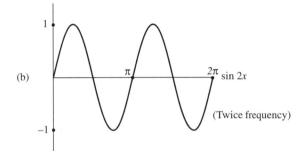

$\sin 2x$

(Twice frequency)

(c)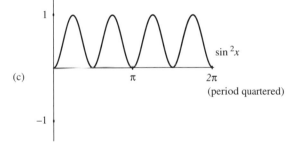

$\sin^2 x$

(period quartered)

3

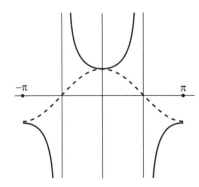

4

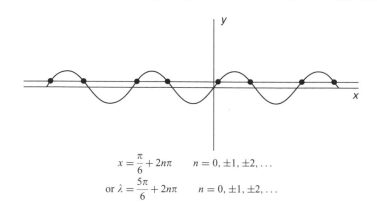

$$x = \frac{\pi}{6} + 2n\pi \qquad n = 0, \pm1, \pm2, \ldots$$
$$\text{or } \lambda = \frac{5\pi}{6} + 2n\pi \qquad n = 0, \pm1, \pm2, \ldots$$

5 (a)

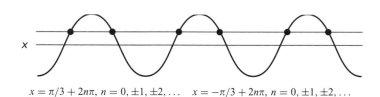

$$x = \pi/3 + 2n\pi, \, n = 0, \pm1, \pm2, \ldots \quad x = -\pi/3 + 2n\pi, \, n = 0, \pm1, \pm2, \ldots$$

(b)

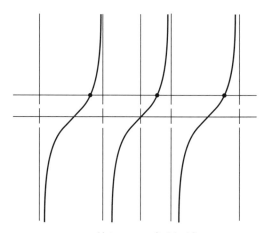

$$x = \pi/4 + n\pi, \, n = 0, \pm1, \pm2, \ldots$$

6 $z = \dfrac{\pi}{3}, \dfrac{2\pi}{3}, \dfrac{4\pi}{3}, \dfrac{5\pi}{3}, \left(z = \dfrac{\pi}{3} + 2n\pi \text{ or } z = \dfrac{2\pi}{3} + 2n\pi, \, n = 0, 1, 2, \ldots \right)$ i.e. $x = \dfrac{1}{3} + 2n$ or

$x = \dfrac{2}{3} + 2n, \, n = 0, 1, 2, \ldots$.

8 FURTHER ALGEBRA

INTRODUCTION

The seeds of algebra were sown in Chapter 2. The ideas of sequences and series underpin the important branch of mathematics known as calculus, which is introduced in Chapters 11 and 12. Two special classes of sequences, arithmetic and geometric, have applications when phenomena occur in a discrete rather than a continuous manner; for example, interest on a loan compounded monthly or the height reached at each bounce by an object which is colliding repeatedly with a horizontal surface. Polynomials have application in many areas of applied mathematics, including the automatic control of engineering systems.

OBJECTIVES

After working through this chapter you should be able to

- identify an arithmetic progression and its component parts
- find the sum of an arithmetic series
- identify a geometric progression and its component parts
- find the sum of a geometric series
- find the sum of an infinite geometric series, when it exists
- calculate the arithmetic mean of two numbers
- calculate the geometric mean of two numbers
- understand and use the sigma notation for series
- identify the terms associated with a polynomial expression
- evaluate a polynomial using nested multiplication
- know the remainder theorem for polynomials
- know the factor theorem for polynomials and use it to find factors
- deduce the main features of the graph of a simple rational function

8.1 ARITHMETIC AND GEOMETRIC PROGRESSIONS

A company opens an account to cover future repairs on its building. It allocates £800 in the first year and increases that amount by £50 per year for the next seven years. The amounts (in £) paid into the account in each of the eight years are 800, 850, 900, 950, 1000, 1050, 1100 and 1150.

This is an example of a **sequence**, which is a list of items (or **terms**) in a given order. The total amount (in £) paid into the account during the eight-year period is

$$800 + 850 + 900 + 950 + 1000 + 1050 + 1100 + 1150$$

This sum is an example of a **series**, which is simply the sum of the terms of a sequence. In this section we concentrate on two special examples of sequences in which the terms have a simple relationship to each other. In *Mathematics in Engineering and Science*, Chapter 10, the topics of sequences and series are dealt with further.

Arithmetic progressions

The sequence introduced above has the property that each term can be obtained from its predecessor by adding a fixed amount: in this case, 50. We say that the **common difference** between the terms is 50. The sequence is called an **arithmetic progression**, shortened to AP. There are four characteristic quantities of an AP: the first term, denoted by a, the common difference denoted by d, the number of terms denoted by n and the last term denoted by l.

In the example above, $a = 800$, $d = 50$, $n = 8$ and $l = 1150$. Note that $1150 = 800 + 7 \times 50$. In general,

$$l = a + (n - 1)d \tag{8.1}$$

The terms in the sequence can be written as $a, a + d, a + 2d, \ldots, a + 7d$. Knowing any three of the four characteristic equations it is possible to find the fourth from relationship (8.1).

Examples

1. In an AP, $a = 4$, $d = 3$, $l = 61$. From (8.1)

$$61 = 4 + (n - 1) \times 3$$

so that $\quad 3(n - 1) = 57$

and here $\quad\quad n = 20$.

(Write out the terms to verify this result)

2. In an AP, $a = 81$, $n = 25$, $l = 33$. From (8.1)

so that $33 = 81 + 24d$

$24d = -48$;   therefore $d = -2$

(Note that d can be negative in some cases. Again, write out the terms to check.)

3. The first three terms of an AP are 6, 13 and 20. Find a formula for the rth term and hence write down the 100th term. Here $a = 6$ and $d = 7$.

The first term is a, the second $a + d$, the third $a + 2d$ and the rth term is

$$a + (r - 1)d = 6 + (r - 1) \times 7$$
$$= 6 + 7r - 7$$
$$= 7r - 1$$

When $r = 100$, the term is $700 - 1 = 699$.   ■

Sum of an AP

In the example at the beginning of this section the sum of the AP can readily be found to be 7800, but it is tedious to compute. How much worse it would be to calculate the sum of the AP $6, 13, 20, \ldots, 699$. A short cut is needed. Returning to the first AP, let

$$S = 800 + 850 + 900 + 950 + 1000 + 1050 + 1100 + 1150$$

It should be obvious that we can write alternatively

$$S = 1150 + 1100 + 1050 + 1000 + 950 + 900 + 850 + 800$$

Adding the last two equations, we obtain

$$2S = 1950 + 1950 + 1950 + 1950 + 1950 + 1950 + 1950 + 1950$$
$$= 8 \times 1950$$
$$= 15\,600$$

so that $S = 7800$, as before. In general, if

$$S = a + [a + d] + [a + 2d] + \cdots + [a + (n - 1)d]$$

then

$$S = [a + (n - 1)d] + [a + (n - 2)d] + [a + (n - 3)d] + \cdots + a$$

and

$$2S = [2a + (n - 1)d] + [2a + (n - 1)d] + \cdots + [2a + (n - 1)d]$$
$$= n[2a + (n - 1)d]$$

so that

$$S = \frac{n}{2}[2a + (n - 1)d]$$

A more friendly formula is found by putting $l = a + (n - 1)d$ so that

$$S = \frac{n}{2}(a + l) \qquad\qquad (8.2)$$

Examples

1. Find the sum

$$3 + 5 + 7 + \cdots + 371$$

This is the sum of the AP

$$3, 5, 7, \ldots, 371$$

which has $a = 3$, $d = 2$, $l = 371$. The number of terms n is found from $n - 1 = (371 - 3)/2 = 184$, so $n = 185$.

The sum of an AP is called an **arithmetic series**. In general a series is the sum of a number of terms of a sequence. Hence

$$S = \frac{185}{2}(3 + 371) - \frac{185}{2} \times 374 = 34\,595$$

2. Find the sum of the first 20 terms of the AP whose first three terms are 8, 13 and 18. Here $a = 8$, $d = 5$, $n = 20$ so that $l = 8 + 19 \times 5 = 103$. Then

$$S = \frac{20}{2}(8 + 103) = 1110$$

3. The sum of the first 24 terms of an AP is 960. If the last term is 17 find the first term and the common difference between the terms. Using equation (8.2)

$$960 = \frac{24}{2}(a + 17) = 12(a + 17)$$

therefore $80 = a + 17$

so that $a = 63.$

Since $l = a + (n - 1)d$

then $17 = 63 + 23d$

so that $23d = -46$ and $d = -2.$ ■

Exercise 8.1

1 Find the common difference and the 10th, 15th and nth term of each of the following APs.

(a) $8, 11, 14, 17, \ldots$ (b) $1000, 996, 992, 988, \ldots$

2 The fifth term of an AP is 8.6 and the eighth term is 6.2. Find its first, second and 30th term.

3 Find the number of terms in the following APs:

(a) $1, 2, 3, \ldots, 18$ (b) $51, 52, 53, \ldots, 68$

4 Find the sums of the following arithmetic progressions:

(a) a (first term) $= 2$, d (difference) $= 6$, n (number of terms) $= 10$

(b) $a = 4$, $d = -2$, $n = 5$

(c) $a = 2$, l (last term) $= 29$, $n = 10$

(d)   a series whose first term is 6, fourth term is 18, and which has 10 terms in all

5 Find the sums

(a) $1 + 2 + 3 + \cdots + 20$ (b) $1 + 1.6 + 2.2 + \cdots + 22.6$

6 Find the sum of the first 20 terms of the following APs:

(a) $5, 4.8, 4.6, \ldots$ (b) $2, 2\frac{1}{3}, 2\frac{2}{3}, \ldots$

7   The 8th term of an AP is three times the second term and the first term is 6. Find the 20th term, the sum of the first 10 terms and the sum of terms from the 11th to the 20th.

8   For the sum of the first n natural numbers

$$S_n = 1 + 2 + 3 + \cdots + n$$

   (a)   Prove that $S_n = \dfrac{1}{2}n(n+1)$

   (b)   Find the lowest value of n for which
         (i) $S_n > 100$ (ii) $S_n > 1000$.

9   The sum of the first n terms of a sequence is $S = n(5n-4)$. Show that the sequence is an AP; write down the first four terms and check out the formula for S in this case.

10* How many terms of the arithmetic series $1 + 4 + 7 + \cdots$ is it necessary to take to obtain the sum 1520?

11* In an AP the 9th term is half the second term and the 4th term is 24. Find the first term, the common difference and the sum of the first 10 terms.

12   As part of a productivity arrangement a patent office begins business by screening 5000 applicants per month, increasing that number by 100 each month thereafter. How many applications are processed in the first year?

Geometric progressions

A sequence in which each term is obtained from its predecessor by multiplying by a fixed number each time is called a **geometric progression**, written GP. As an example consider the sequence

$$3, \; 6, \; 12, \; 24, \; 48.$$

The first term, a, is equal to 3, the last term, l, is equal to 48 and each term is obtained from its predecessor by multiplying by the **common ratio** 2, denoted by r. The terms in the sequence can be written as a, ar, ar^2, ar^3, ar^4. The number of terms $n = 5$ and the last term $l = ar^{n-1}$.

Examples

1. In the GP

$$1, \frac{1}{2}, \frac{1}{4}, \frac{1}{8}, \frac{1}{16}, \frac{1}{32}, \frac{1}{64}$$

$$a = 1, r = \frac{1}{2}, n = 7, l = ar^6$$

2. In the GP

$$1, -2, 4, -8, 16$$

$$a = 1, r = -2, r = 5, l = ar^4$$

3. In a GP, $a = 3$, $l = 729$ and $n = 6$. Hence

$$3r^5 = 729$$
$$r^5 = 243 \quad \text{and} \quad r = 3$$

4. In a GP, $a = 3$, $l = 2187$ and $n = 7$. Hence

$$3r^6 = 2187$$
$$r^6 = 729$$

Either $r = 3$ or $r = -3$ is possible ∎

Sum of a GP

The sum of a GP can be obtained fairly simply by noting that if we let

$$S = a + ar + ar^2 + \cdots + ar^{n-1}$$

then

$$rS = \quad ar + ar^2 + ar^3 + \cdots + ar^n$$

Upon subtraction and cancellation

$$S - rS = a \qquad - ar^n$$

so that

$$S(1 - r) = a - ar^n = a(1 - r^n)$$

Hence

$$S = \frac{a(1 - r^n)}{(1 - r)} \tag{8.3}$$

unless $r = 1$.

If $r = 1$ then $S = a + a + a + \cdots + a = na$. The sum of a GP is called a **geometric series**.

Examples

1. The GP

$$1, \ 0.5, \ 0.25, \ 0.125, \ 0.0625, \ 0.031\,25$$

has $a = 1$, $r = 0.5$ and $n = 6$. Its sum is therefore

$$S = \frac{1(1 - (0.5)^6)}{1 - 0.5} = \frac{1 - 0.15\,625}{0.5}$$
$$= 2 \times 0.984\,375 = 1.968\,75$$

as you can verify directly

2. The GP

$$3, \ -6, \ 12, \ -24, \ 48$$

has a sum given by

$$S = \frac{3(1 - (-2)^5)}{1 - (-2)} = \frac{3(1 + 32)}{3}$$
$$= 33$$

which you can verify directly. ∎

Sum to infinity of a GP

Consider the GP

$$\frac{1}{2}, \ \frac{1}{4}, \ \frac{1}{8}, \ \frac{1}{16}, \ \frac{1}{32}$$

which has $a = \dfrac{1}{2}$, $r = \dfrac{1}{2}$ and $n = 5$. Its sum is

$$S = \frac{\dfrac{1}{2}\left(1 - \left(\dfrac{1}{2}\right)^5\right)}{1 - \dfrac{1}{2}} = \frac{\dfrac{1}{2}}{\dfrac{1}{2}}\left(1 - \dfrac{1}{32}\right) = \frac{31}{32}$$

Suppose we allow the GP to extend to 10 terms and then find its sum. We obtain

$$S = \frac{\dfrac{1}{2}\left(1 - \left(\dfrac{1}{2}\right)^{10}\right)}{\dfrac{1}{2}} = 1 - \frac{1}{1024} = \frac{1023}{1024}$$

If the GP has 20 terms its sum is

$$S = 1 - \left(\frac{1}{2}\right)^{20} = 0.999\,999\,046 \qquad \text{(9 d.p.)}$$

Suppose we allowed the GP to continue indefinitely. The sum of the first n terms is

$$S = 1 - \left(\frac{1}{2}\right)^n$$

and as n increases indefinitely $\left(\dfrac{1}{2}\right)^n$ approaches the value zero (never actually getting there). The value of S then approaches 1. We call this the **sum to infinity** of the GP.

Consider a square of size 1 m made from paper. Suppose I cut it in half and hand one of the two parts to you. I then cut the remaining piece of paper in half and hand one part to you. Assuming that I can continue to cut my piece of paper in half repeatedly, you will accumulate the areas

$$\frac{1}{2}, \frac{1}{4}, \frac{1}{8}, \frac{1}{16}, \ldots$$

and the sum of these areas, namely

$$\frac{1}{2} + \frac{1}{4} + \frac{1}{8} + \frac{1}{16} + \cdots$$

will get closer to the value of the original area of paper, namely 1.

In general we say that, *provided* $|r| < 1$, i.e. the terms of the GP are getting smaller in magnitude, the sum to infinity of a GP is given by

$$S = \frac{a}{1-r} \tag{8.4}$$

This follows because when $|r| < 1$, i.e. $-1 < r < 1$, then r^n gets progressively smaller as n increases and the term $(1 - r^n)$ in equation (8.3) approaches the value 1.

Examples

1. Find the infinite sum

$$\frac{3}{7} + \frac{3}{49} + \frac{3}{343} + \cdots$$

Since $a = \frac{3}{7}, r = \frac{1}{7}$ then

$$S = \frac{\dfrac{3}{7}}{1 - \dfrac{1}{7}} = \frac{\dfrac{3}{7}}{\dfrac{6}{7}} = \frac{1}{2}$$

You can check out the plausibility of this by calculating the first few terms of the sequence.

2. Express as a fraction the decimal $0.\dot{9}\dot{0} = 0.909\,090\ldots.$
   Here, the decimal is the infinite sum

$$\frac{9}{10} + \frac{9}{1000} + \frac{9}{100\,000} + \cdots$$

which is the sum of a GP with $a = \dfrac{9}{10}$ and $r = \dfrac{1}{100}$.
   The sum is given by

$$S = \frac{\dfrac{9}{10}}{1 - \dfrac{1}{100}} = \frac{\dfrac{90}{100}}{\dfrac{99}{100}} = \frac{90}{99} = \frac{10}{11}$$

3. The third term of a GP is half the difference between the first and the second terms. Write down the first four terms of the two possible sequences and find the sum to infinity of the one for which this sum exists.

With the usual notation,

$$ar^2 = \frac{1}{2}(a - ar)$$

so that $2r^2 = 1 - r$, i.e. $2r^2 + r - 1 = 0$. Factorising gives

$$(2r - 1)(r + 1) \quad \text{and} \quad r = \frac{1}{2} \quad \text{or} \quad -1$$

The sequences are

$$a, \ \frac{a}{2}, \ \frac{a}{4}, \ \frac{a}{8}$$

and

$$a, \ -a, \ a, \ -a$$

We require $|r| < 1$ for the sum to infinity to exist; the second series will have a sum which alternates between a and 0. For the first series

$$S = \frac{a}{1 - \frac{1}{2}} = 2a \qquad \blacksquare$$

Exercise 8.2

1   Find the common ratio and the term indicated for the following GPs:

(a) $5, 10, 20, \ldots$ (10th)   (b) $2, -10, 50, \ldots$ (7th)

(c) $3.2, 1.28, 0.512, \ldots$ (5th)

2   Find the sum of the following GPs:

(a) $4, 8, 16, 32, 64, 128, 256$ (b) $4, -8, 16, -32, 64, -128, 256$

(c)   the first 20 terms of $1, \frac{1}{3}, \frac{1}{9}, \ldots$

(d)   first term 2, common ratio -1; 101 terms

3   Find the sums of the following geometric series:

(a) a (first term) $= 3$, r (common ratio) $= 2$, $n = 5$

(b) $a = 5$, $r = \dfrac{1}{6}$, $n = 10$ (c) $a = 3$, $r = -\dfrac{1}{2}$, $n = 5$

4   A GP has a first term of 9 and a common ratio of 0.8. Find the sum of

(a)   the first 20 terms   (b)   the first 10 terms

(c)   terms 11 to 20

5   Write down the nth term of the following GPs:

(a)   first term 2, common ratio 3

(b)   first term $\dfrac{1}{4}$, common ratio $\dfrac{1}{2}$

(c)   first term $-\dfrac{1}{4}$, common ratio $\dfrac{1}{2}$

(d)   first term $-\dfrac{1}{4}$, common ratio $-\dfrac{1}{2}$

6   Write down the sum to n terms of a geometric series when

(a) $r = 1$ (b) $r = -1$

7   Write down the sum to infinity of a geometric series when $|r| < 1$. Deduce that
$$1 - x + x^2 - x^3 + x^4 - \cdots = \frac{1}{1+x}, \ |x| < 1.$$
What is the sum $1 + x^2 + x^4 + x^6 + \cdots$ for $|x| < 1$?

8   Find the common ratio of a GP whose first term is 4 and whose 9th term is $\dfrac{1}{64}$. What is the sum of the first 9 terms?

9   The sum to infinity of a geometric series is three times the first term. What is the common ratio?

10 Express the following decimals as simple fractions:

(a) $0.\dot{4}$ (b) $0.\dot{4}\dot{5}$ (c) $0.\dot{2}8\dot{8} = 0.288\,288\,288\ldots$ (d) $3.\dot{2}8\dot{8}$

11* A sum of £1000 is invested at an annual rate of 5% per annum, with the interest being compounded annually, i.e. in the nth year the amount A is given by the formula

$$A = P\left(1 + \frac{r}{100}\right)^n$$

where r is the annual percentage rate, P is the sum invested at the start of the first year. What is the total sum invested at the end of 4 years, 10 years? What is the effect of compounding interest every six months at a rate of 2% per half-year?

12 The dividend paid in an investment is agreed to be £1000 in the first year, rising by 7% per annum thereafter in the ensuing years.

(a)   Calculate the dividends paid in the second and third years.

(b)   Calculate the total amount which is paid out over the first 10 years.

13 A radioactive substance decays at a rate such that the amount of substance left halves every 100 years. How long will it be before the amount left is 0.01% of the original amount?

14 An object is dropped from a height of 5 m above ground level. It rebounds many times and on each bounce it reaches a height of 0.7 times its previous height. What height does it reach after 3 bounces, 5 bounces, n bounces? What is the total distance it travels after 3 bounces, 5 bounces? What is the distance travelled before it comes to rest (in theory)?

15 Note the following definition for two numbers a and b:

$$\text{Arithmetic mean} \quad \frac{1}{2}(a + b)$$

$$\text{Geometric mean} \quad \sqrt{ab}$$

$$\text{Harmonic mean} \quad \frac{2}{\dfrac{1}{a} + \dfrac{1}{b}}$$

Note that the harmonic mean of two numbers is the reciprocal of the arithmetic mean of their reciprocals.

(a)   Find these three quantities when $a = 4$, $b = 9$.

(b)   Show that, in general, the arithmetic mean is greater than or equal to the geometric mean. When is equality achieved?

(c)   If a, b and c are successive terms of an AP show that b is the arithmetic mean of a and c.

(d)   If a, b and c are successive terms of a GP show that b is the geometric mean of a and c.

The sigma notation

Consider the sequence

$$\frac{1}{2}, \frac{1}{4}, \frac{1}{8}, \frac{1}{16}, \frac{1}{32}, \frac{1}{64}$$

We can denote the terms as

$$t_1 = \frac{1}{2^1}, \; t_2 = \frac{1}{2^2}, \; t_3 = \frac{1}{2^3}, \; t_4 = \frac{1}{2^4}, \; t_5 = \frac{1}{2^5}, \; t_6 = \frac{1}{2^6}.$$

A typical term is $t_r = \dfrac{1}{2^r}$. The sum of the series

$$t_1 + t_2 + t_3 + t_4 + t_5 + t_6$$

can be written in a compact notation as

$$\sum_{r=1}^{6} t_r$$

The capital sigma $\sum$ means 'sum of'; the notation means the first value taken is t_1 and the 6 on top of the $\sum$ means that the last value taken is t_6. Similarly,

$$\sum_{r=2}^{5} t_r = t_2 + t_3 + t_4 + t_5$$

We could also write this sum as

$$\sum_{r=2}^{5} \frac{1}{2^r}$$

Example   Write the following series using the sigma notation:

(a) $\dfrac{1}{3} + \dfrac{1}{9} + \dfrac{1}{27} + \cdots + \dfrac{1}{243}$

(b) $2 + 4 + 6 + \cdots + 68$

(c) $2 + 6 + 18 + 54 + 162$

(a) The rth term is $\dfrac{1}{3^r}$, as you can check. Hence the sum is given by

$$S = \frac{1}{3^1} + \frac{1}{3^2} + \frac{1}{3^3} + \cdots + \frac{1}{3^6} = \sum_{r=1}^{6} \frac{1}{3^r}$$

(b) The rth term is $2r$ and there are 34 terms, so

$$S = \sum_{r=1}^{34} 2r \quad \left(= 2 \times \sum_{r=1}^{34} r \quad \text{as you can verify} \right)$$

(c) The first term is 2, the second is 2×3, the third is 2×3^2 and the rth is $2 \times 3^{r-1}$, so

Hence $S = \displaystyle\sum_{r=1}^{5} 2 \times 3^{r-1}$

$$\left(\text{Note that } 5 = 2 \times \sum_{r=1}^{5} 3^{r-1} = \frac{2}{3} \sum_{r=1}^{5} 3^r; \text{ verify this.} \right) \qquad \blacksquare$$

The sum of the infinite series

$$\frac{1}{2} + \frac{1}{4} + \frac{1}{8} + \frac{1}{16} + \cdots$$

can be written as

$$\sum_{r=1}^{\infty} \left(\frac{1}{2} \right)^r$$

The use of the symbol ∞ on top of the $\sum$ sign implies that the number of terms of the series is infinitely large. $\qquad \blacksquare$

Exercise 8.3

1   Write the following series in sigma notation:

(a) $1 + 4 + 9 + 16 + 25$

(b) $3 + 6 + 9 + 12 + \cdots + 30$

(c) $\dfrac{1}{2} + \dfrac{1}{3} + \dfrac{1}{4} + \dfrac{1}{5} + \cdots + \dfrac{1}{100}$

(d) $1 + \dfrac{1}{3} + \dfrac{1}{9} + \dfrac{1}{27}$

(e) $7 + 2 - 3 - 8 - 13$

(f) $5 + 1 + \dfrac{1}{5} + \dfrac{1}{25} + \cdots$

(g) $1 + x^2 + x^4 + x^6$

(h) $1 - x + x^2 - x^3 + \cdots + x^{10}$

2 Write out the sum of the first four terms and the number of terms in each of the following series:

(a) $\displaystyle\sum_{r=1}^{16} 3^r$

(b) $\displaystyle\sum_{r=1}^{8} (r + 3)$

(c) $\displaystyle\sum_{r=1}^{10} (12r - 5)$

(d) $\displaystyle\sum_{r=1}^{20} \dfrac{1}{4^4}$

(e) $\displaystyle\sum_{r=4}^{6} r^2$

(f) $\displaystyle\sum_{r=5}^{12} (r + 4)$

(g) $\displaystyle\sum_{r=0}^{11} 2^r$

(h) $\displaystyle\sum_{r=0}^{14} r^3$

(i) $\displaystyle\sum_{r=-2}^{2} 2^r$

(j) $\displaystyle\sum_{r=1}^{\infty} (r^2 + r)$

8.2 POLYNOMIALS

In Chapter 3 we dealt with linear expressions of the form $mx + c$. In Chapter 4 we met quadratic expressions $ax^2 + bx + c$ and cubic expressions $ax^3 + bx^2 + cx + d$. These are all examples of polynomials in x.

In general, the expression $p(x) \equiv a_n x^n + a_{n-1} x^{n-1} + \cdots + a_0$ where $a_n, a_{n-1}, \ldots, a_0$ are constants and n is a positive integer is a **polynomial** in the variable x.

The coefficient, a_n is called the **leading coefficient**. If $a_n \neq 0$ then the polynomial has **degree** n.

Examples of polynomials of degree 4 are

$$x^4, \quad -2x^4, \quad \frac{1}{2}x^4 + x^2 - 6, \quad x^4 + x^3 - 2x + 7$$

The important point is that the highest power of x which appears is the fourth power. Any, or all, of the lower powers of x and the constant term may be absent. Only positive integer powers are allowed.

A polynomial of degree 3 is a cubic (polynomial), a polynomial of degree 2 is a quadratic (polynomial) and a polynomial of degree 1 is a linear polynomial. To be complete we call a polynomial of degree zero a constant (polynomial).

The polynomial

$$a_n x^n + a_{n-1} x^{n-1} + \cdots + a_1 x + a_0$$

is in **descending order** whereas the polynomial

$$a_0 + a_1 + x + \cdots + a_{n-1} x^{n-1} + a_n x^n$$

is in **ascending order**.

Evaluation by nested multiplication

To find the value of a polynomial expression for a given value of x, we can substitute the value of x directly. Hence the value of $5x^3 - 2x^2 + 3x + 4$ when $x = 2$ is

$$5 \times (2)^3 - 2 \times (2)^2 + 3 \times 2 + 4 = 40 - 8 + 6 + 4 = 42$$

This was awkward enough but if we wanted to find the value when $x = -1.732$ that would be much more tedious. The method of **nested multiplication** reduces the number of multiplications needed and therefore leads to less round-off error, in addition to being quicker.

Example

Consider the polynomial

$$5x^3 - 2x^2 + 3x + 4$$

We write this first as $[5x^2 - 2x + 3]x + 4$ then as $[(5x - 2)x + 3]x + 4$.

Notice how we first collect all the terms in x and factorise out x; then we take the other factor and factorise out x from these terms which contain it. This process can be continued until we obtain a linear expression, in this case $5x - 2$. Hence

$$5x^3 - 2x^2 + 3x + 4 \equiv [(5x - 2)x + 3]x + 4$$

You can verify that the two expressions are equivalent. To evaluate the expression on the right-hand side when $x = 2$ we proceed as follows:

(i) Multiply 5 by x and subtract 2 $5 \times 2 - 2 = 8$

(ii) Multiply the result by x and add 3 $8 \times 2 + 3 = 19$

(iii) Multiply the result by x and add 4 $19 \times 2 + 4 = 42$ ■

Graphs of $y = x^n$

We saw in Chapter 4 that the graphs of $y = x^2$ and $y = -x^2$ were essentially of the same shape as the general curves $y = ax^2 + bx + c$ where $a > 0$ and $a < 0$ respectively. In the same way, the graphs of $y = x$ and $y = -x$ are typical of the general straight line $y = ax + b$ where $a > 0$ and $a < 0$ respectively. However, we saw that the curves $y = x^3$ and $y = -x^3$ are *not typical* of the curves $y = ax^3 + bx^2 + cx + d$.

In general, for $n > 2$ $y = x^n$ and $y = -x^n$ are *not typical* of the polynomials of degree n. Figure 8.1 shows the graphs of $y = x^4$ and $y = -x^4$. Notice that the graph of $y = x^4$ looks,

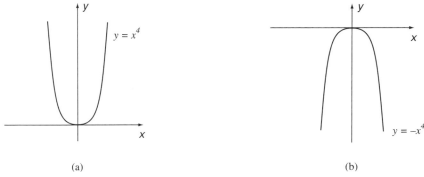

(a)   (b)

Figure 8.1 Graphs of (a) $y = x^4$ and (b) $y = -x^4$

at first sight, similar to the graph of $y = x^2$. However, $y = x^4$ is a more 'flat-bottomed' curve than $y = x^2$ and has steeper sides. This is because for $|x| < 1$, $x^4 < x^2$,

e.g. $\left(\dfrac{1}{2}\right)^4 = \dfrac{1}{16} < \dfrac{1}{4} = \left(\dfrac{1}{2}\right)^2$, whereas for $|x| > 1$, $x^4 > x^2$, e.g. $3^4 = 81 > 9 = 3^2$.

The graph of a more typical polynomial of degree 4 is shown in Figure 8.2. It has two local minima between which is sandwiched a local maximum. This is usually the case when the coefficient of x^4 is positive.

The precise location of the local minima and maxima with respect to the axes depends on the coefficients of the polynomial. Some other options are shown in Figure 8.3.

One feature of interest concerns the values of x where x-axis is crossed or touched. In the case of the polynomial $y = x^4 - 3x^3 + 2x^2$, shown in Figure 8.3(a), the axis is crossed at $x = 1$ and $x = 2$ and touched at $x = 0$. We say that the polynomial has four **zeros**: $x = 1$ and $x = 2$ and a **double zero** $x = 0$.

Equivalently, we say that the polynomial equation $x^4 - 3x^3 + 2x^2 = 0$ has four **roots**: $x = 1$, $x = 2$ and $x = 0$ (a double root).

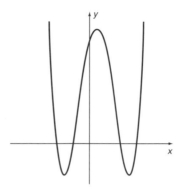

Figure 8.2 Graph of a typical polynomial of degree 4

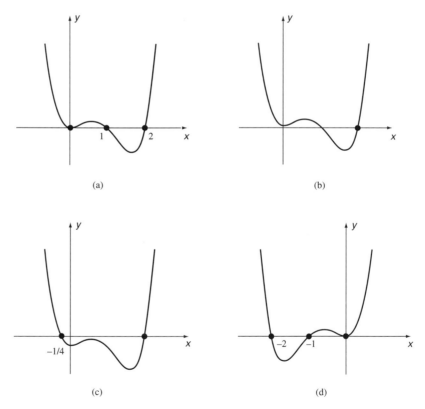

Figure 8.3 Graphs of four polynomials of degree 4 (a) $y = x^4 - 3x^3 + 2x^2$, (b) $y = x^4 - 3x^3 + 2x^2 + 0.625$, (c) $y = x^4 - 3x^3 + 2x^2 - 0.25$ and (d) $y = x^4 + 3x^3 + 2x^2$

In general, if $x = a$ is a **zero** of the polynomial $p(x)$ then $x = a$ is a **root** of the polynomial equation $p(x) = 0$. Furthermore, $(x - a)$ is a **factor** of the polynomial.

Example

The polynomial

$$p(x) = x^4 + 3x^2 + 2x^2 \equiv x^2(x + 1)(x + 2)$$

has factors x (twice), $(x + 1)$ and $(x + 2)$. The zeros of the polynomial are 0 (twice), -1 and -2. The roots of the equation $x^4 + 3x^3 + 2x^2 = 0$ are $x = 0$ (twice), $x = -1$ and $x = -2$. ∎

Notice that the zeros of the polynomial occur either side of the local maxima and minima. Looked at another way, between two different zeros a polynomial has either a local minimum or a local maximum.

However, something quite different happens with the polynomial x^4. The local minima and local maxima have been 'squeezed together' to coincide in a simple local minimum at $x = 0$. Other special cases can occur; for example, the polynomial

$$p(x) = x^4 - 4x^3$$

whose graph is shown in Figure 8.4(a) has its local maximum coinciding with one of the local minima. Note that $p(x) \equiv x^3(x - 4)$.

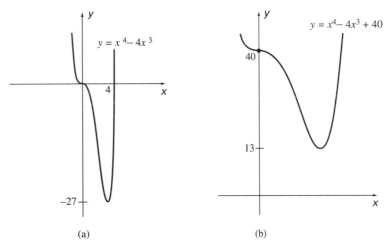

(a)   (b)

Figure 8.4 Graphs of two polynomials of degree 4

All the polynomials of degree 4 which we have considered have had four zeros or two zeros (including multiples). In Figure 8.4(b) the polynomial has no zeros.

One final point. If the leading coefficient is positive then a local minimum occurs first followed by a local maximum, then a local minimum, and so on. If the leading coefficient is negative then the sequence starts with a local maximum.

If n is even and a is positive then the graph of the polynomial 'starts' in the second quadrant and 'ends' in the first quadrant, as in Figure 8.4. If n is even and a is negative then the graph starts in the third quadrant and ends in the fourth. Here is a table of the possibilities for polynomials of degrees 1 to 6.

Degree of polynomial	Possible number of zeros
0 (constant)	0
1 (linear)	1
2 (quadratic)	2 or 0
3 (cubic)	3 or 1
4 (quartic)	4 or 2 or 0
5	5 or 3 or 1
6	6 or 4 or 2 or 0

Interpolation with polynomials

Figure 8.5(a) shows part of the graph of a function which passes through two given points (x_1, y_1) and (x_2, y_2). What is the y-coordinate of the point P'? It is not possible to say unless we know the equation of the curve.

However, there is only one straight line which passes through the given points and it may be used to approximate the given curve so that P in Figure 8.5(b) is reasonably close to the point P'. The use of the y-coordinate of P as an approximation for the y-coordinate of P' is known as **linear interpolation**.

Referring to Figure 8.5(b) we see that triangles QPM and QRN are similar, so

$$\frac{PM}{RN} = \frac{MQ}{NQ}, \quad \text{i.e.} \quad \frac{y - y_1}{y_2 - y_1} = \frac{x - x_1}{x_2 - x_1} \tag{8.5}$$

Therefore we can calculate the value of y.

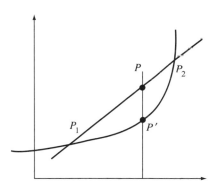

Figure 8.5 Linear interpolation

Example   A curve passes through the points (1, 3) and (4, 9). Estimate the y-coordinate of the point on the curve whose x-coordinate is $2\frac{1}{2}$.

Using equation (8.5) we obtain

$$\frac{y-3}{9-3} = \frac{2\frac{1}{2}-1}{4-1}$$

i.e. $$\frac{y-3}{6} = \frac{1\frac{1}{2}}{3} = \frac{1}{2}$$

so that $y - 3 = \dfrac{1}{2} \times 6 = 3$

i.e. $y = 6$

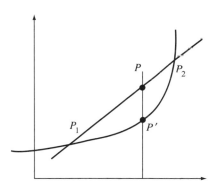

Figure 8.6 Linear interpolation

Refer to Figure 8.6 where $P_1 = (1, 3)$ and $P_2 = (4, 9)$.   ■

In the same way a quadratic curve can be used to approximate a curve that passes through three given distinct points which do not lie on a straight line. There is only one quadratic curve which passes through these three points.

Example   A curve passes through the points $(1, 9)$, $(3, 1)$ and $(6, 4)$. Use a quadratic approximation to estimate the y-coordinate of the point on the given curve whose x-coordinate is 4.2. Refer to Figure 8.7. A quadratic curve which passes through the three given points is $y = x^2 - 8x + 16$. You can verify this by showing that the coordinates of the three points satisfy this equation.

Since there is only one such quadratic, ours is unique.
When $x = 4.2$, $y = (4.2)^2 - 8 \times 4.2 + 16 = 0.04$.   ■

If *four* distinct data points are given then there is a unique cubic polynomial whose graph passes through them (assuming that the points do not lie on a straight line or on a quadratic curve) and the interpolation process can be carried out using this cubic curve.

One word of warning. If the point P is outside the range of given data points, e.g. outside the line segment QR in Figure 8.5(b) or the quadratic arc QRS in Figure 8.7, then the process is known as **extrapolation** and can give very poor results. Compare with Figure 8.6 and see why.

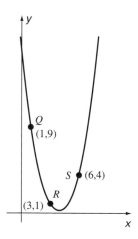

Figure 8.7   Quadratic interpolation

Exercise 8.4

1   On a graph with scale $-2 \leq x \leq 2$, $-10 \leq y \leq 10$, draw the graphs of $y = x^n$, $n = 1(1)5$. How do you think the graph would look like if n were very large? Consider the odd and even cases separately.

2   A quartic or fourth-order polynomial takes the form

$$y = a + bx + cx^2 + dx^3 + fx^4$$

Draw examples of sketches of the quartic when the following criteria are met:

(a) $a, c, f > 0$, $b = d = 0$ (note that $y(-x) = y(x)$)

(b) $a, b, c, d, f > 0$ (assume two roots only)

(c) $d < 0, f = 0$

(d) $a, b > 0$, $c = d = f = 0$

(e) $a = -f, f > 0$, $b = c = d = 0$

(f) $a = 0$, $b, c, d, f > 0$ (assume two roots only)

3   A polynomial of odd order has positive coefficients. Determine which of the following statements are true about it:

(a)   it has no roots   (b)   it has one positive root at least

(c)   it has one negative root at least

4   Draw the graphs of $x = y^4$ and $y = x^4$ for $-20 \leq x, y \leq 20$.

5   The equation

$$y = L(x) = \frac{(x - x_2)}{(x_1 - x_2)} y_1 + \frac{(x - x_1)}{(x_2 - x_1)} y_2$$

represents a straight line passing through (x_1, y_1) and (x_2, y_2).

(a)   By setting $x = x_1$ then $x = x_2$ verify that this is so, i.e. $y_1 = L(x_1)$ and $y_2 = L(x_2)$.

(b)   Rewrite the equation as $y = ax + b$, determining a and b.

(c)   What happens if $y_2 = y_1$?

6   Consider the expression

$$Q(x) = \frac{(x - x_2)(x - x_3)}{(x_1 - x_2)(x_1 - x_3)} y_1 + \frac{(x - x_3)(x - x_1)}{(x_2 - x_3)(x_2 - x_1)} y_2 + \frac{(x - x_1)(x - x_2)}{(x_3 - x_1)(x_3 - x_2)} y_3$$

It can be seen that $Q(x)$ is a quadratic in x, even though it is a complicated one.

(a)   Set $x = x_1, x_2, x_3$ in turn to show that $Q(x_1) = y_1$, $Q(x_2) = y_2$, $Q(x_3) = y_3$, i.e. $Q(x)$ passes through (x_1, y_1), (x_2, y_2), (x_3, y_3).

(b)   For the points $(0, 1)$, $(1, 0)$, $(2, 0)$ determine $Q(x)$.

(c)   What happens if (x_1, y_1), (x_2, y_2), (x_3, y_3) lie on a straight line?

7*   Prove that $Q(x)$ in Question 6 must be unique by assuming the result to be false and using *reductio ad absurdum*.

8*   In a manner similar to Question 6 it can be shown that a unique cubic passes through four points in a plane. Consider

$$C(x) = \frac{(x - x_2)(x - x_3)(x - x_4)}{(x_1 - x_2)(x_1 - x_3)(x_1 - x_4)} y_1 + \frac{(x - x_3)(x - x_4)(x - x_1)}{(x_2 - x_3)(x_2 - x_4)(x_2 - x_1)} y_2$$

$$+ \frac{(x - x_4)(x - x_1)(x - x_2)}{(x_3 - x_4)(x_3 - x_1)(x_3 - x_2)} y_3 + \frac{(x - x_1)(x - x_2)(x - x_3)}{(x_4 - x_1)(x_4 - x_2)(x_4 - x_3)} y_4$$

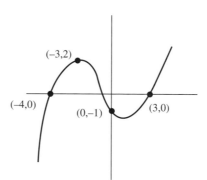

and show that $C(x)$ passes through (x_1, y_1), (x_2, y_2), (x_3, y_3) and (x_4, y_4). Find $C(x)$ for the points $(-4, 0)$, $(-3, 2)$, $(0, -1)$, $(3, 0)$, as shown.

9 Determine the polynomials formed by (a) adding, (b) subtracting and (c) multiplying the polynomials $2 - 3x + x^3$ and $6 - 5x^2 + 2x^5 - x^6 + x^7$.

10 Use nested multiplication to help plot the graph of $1 - x + 2x^2 - x^4 + x^5$ for $-2 \leq x \leq 2$.

11 Express the following as polynomials in the quantity shown in parentheses:

(a) $(\sqrt{x} - x\sqrt{x})(x^{1/4} - x^{3/4})$ $(x^{1/4})$

(b) $(9xy - 2\sqrt{xy})((xy)^{1/3} - (xy)^{4/3})$ $(xy)^{1/6}$

(c) $(1/\sqrt{x} - 1/x)(3/(x\sqrt{x}) - 4/x^2)$ $(1/\sqrt{x})$

12* By tabulating and identifying sign changes, estimate the values of x for which the following inequality holds:

$$x(x - 3)(x - 5) > (x + 7)(x + 1)(x - 2)(x - 4)$$

8.3 FACTOR AND REMAINDER THEOREMS

A theorem which we shall quote (but not prove) in a form applied to polynomials is the **intermediate value theorem**. In this form it states that if $p(x)$ is a polynomial in x and $p(a) \neq p(b)$ for two numbers a and b where $a < b$ then $p(x)$ takes every value between the values $p(a)$ and $p(b)$ at least once in the interval $a \leq x \leq b$.

In other words, for every number s between the values $p(a)$ and $p(b)$ there is at least one number c in the interval (a, b) for which $p(c) = s$. Refer to Figure 8.8(a).

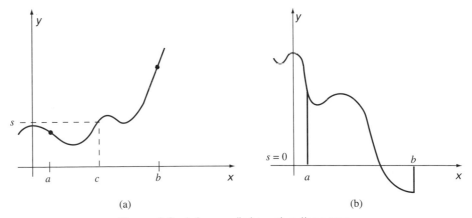

(a)   (b)

Figure 8.8 Intermediate value theorem

Corollary

An important corollary of the theorem applies when $p(a)$ and $p(b)$ are of the opposite sign. Then it follows that $p(x) = 0$ at least once in the interval $a \leq x \leq b$; see Figure 8.8(b).

The following examples illustrate the need for care when applying the intermediate value theorem and interpreting its result.

Example

Consider $p(x) = x^3 - 3x^2 + 4x - 12$.

It is straightforward to see that $p(0) = 12$ and relatively easy to check that $p(4) = 20$. Hence for at least one x in the interval $0 < x < 4$, $p(x) = 0$.

In fact there is only one such value of x and it is $x = 3$. ■

Examples

1. The polynomial $p(x) = x^2 + 4$ has no zeros; it is an **irreducible quadratic**. Since $p(-3) = 13$ and $p(4) = 20$ the intermediate value theorem indicates merely that the polynomial assumes all values between 13 and 20 at least once as x goes from -3 to 4. It *does not indicate* that the polynomial also takes values between 4 (its least value) and 13 as a sketch of the graph will show you.

   And the theorem does not tell us that, with the exception of the value 4 which occurs only when $x = 0$, the values from 4 to 13 are achieved twice.

2. The polynomial $q(x) = x^2 - 4$ has two zeros: $x = -2$ and $x = 2$. If we calculate $p(-3) = 5$ and $p(4) = 12$ and apply the intermediate value theorem, all we learn is that the polynomial achieves values between 5 and 12 at least once as x goes from -3 to 4. No mention is made of the values between -4 (the least value) and 5. ■

Every *polynomial of odd degree* has *at least one zero* but it may have more than one.

Example

Consider $p(x) = x^3 - 2x^2 - 5x + 6$.

Since $\quad p(-3) = -27 - 18 + 15 + 6 = -24 < 0 \quad$ and $\quad p(4) = 64 - 32 - 20 + 6 = 18 > 0$ then there is at least one zero of $p(x)$ in the interval $(0, 4)$. If we calculate $p(0) = 6$ then all we can conclude is that $p(x)$ takes every value between 6 and 18 at least once as x goes from 0 to 4.

In fact $p(x) \equiv (x + 2)(x - 1)(x - 3)$ as you can verify by multiplying out the right-hand side. Therefore there are three zeros: $x = -2$, $x = 1$ and $x = 3$. The intermediate value theorem does *not* help us to detect the latter two zeros.

However, had we checked that $p(2) = -4$ we should have noted that $p(x)$ changed sign in each of the intervals $(-3, 0)$, $(0, 2)$ and $(2, 4)$. Each of these intervals therefore contains a zero and, since a cubic polynomial has at most three zeros, we have located them all. ■

This approach is taken up again in *Mathematics in Engineering and Science* but for the moment we note the need to read the statement of a theorem carefully to check what it tells us *and* what it does not.

The following example illustrates the possibilities for a quartic polynomial.

Example

Consider the polynomials

$$p_1(x) = x^4 + 5x^2 + 4 \equiv (x^2 + 1)(x^2 + 4)$$
$$p_2(x) = x^4 + 3x^2 - 4 \equiv (x^2 - 1)(x^2 + 4) \equiv (x - 1)(x + 1)(x^2 + 4)$$
$$p_3(x) = x^4 - 5x^2 + 4 \equiv (x^2 - 1)(x^2 - 4) \equiv (x - 1)(x + 1)(x - 2)(x + 2)$$
$$p_4(x) = x^4 - 2x^3 + 5x^2 - 8x + 4 \equiv (x - 1)^2(x^2 + 4)$$
$$p_5(x) = x^4 - 4x^3 + 6x^2 - 4x + 1 \equiv (x - 1)^4$$

$p_1(x)$ has been factorised into the product of two irreducible quadratic factors. It has no roots.

$p_2(x)$ has been factorised into the product of two linear factors and one irreducible quadratic factor.

$p_3(x)$ has been factorised into the product of four linear factors.

$p_4(x)$ is a special case of $p_2(x)$ where the two linear factors are the same (indicating a double root of $p_4(x) = 0$).

$p_5(x)$ is a special case of $p_3(x)$ where all four linear factors are the same. ∎

In general a polynomial of degree n can be factorised into a product of linear factors and irreducible quadratic factors. If n is odd then there must be a least one linear factor.

The remainder theorem and the factor theorem

When we divide 18 by 6 the result is exactly 3; we say that the **quotient** is 3. When we divide 18 by 7 the result is $2\frac{4}{7}$; in terms of whole numbers we say that the quotient is 2 and the **remainder** is 4. This can be stated as

$$18 = \underset{\text{quotient}}{2} \times 7 + \underset{\text{remainder}}{4}$$

When the remainder is zero the quotient is a factor; hence 3 is a factor of 18, as we know. This result can be generalised to polynomials.

If $p(x)$ and $s(x)$ are polynomials and if $s(x) \not\equiv 0$ then we can find unique polynomials $q(x)$ and $r(x)$ such that

$$p(x) \equiv s(x)q(x) + r(x) \tag{8.6}$$

Note that if $r(x)$ is not identically zero then it is of lower degree than $s(x)$. The polynomial $q(x)$ is the **quotient** and $r(x)$ is the **remainder**.

Example

If $p(x) = x^3 - 3x^2 + x + 6$ and $s(x) = x - 2$ then you can verify that

$$x^3 - 3x^2 + x + 6 \equiv (x - 2)(x^2 - x - 1) + 4$$

Hence the quotient is $x^2 - x - 1$ and the remainder is 4. ∎

Later in this section we will see how to obtain the quotient. Suppose for the moment we just want to find the remainder.

In identity (8.6) suppose that $s(x)$ is a linear factor $(x - a)$. It follows that $r(x)$ must have degree less than 1; therefore it is a constant polynomial (i.e. a number) which we shall write as R. Then (8.6) becomes

$$p(x) \equiv (x - a)q(x) + R$$

If we now put $x = a$ we obtain the result

$$p(a) = 0 + R$$

so that $R = p(a)$.

This result can be stated as the **remainder theorem**:

> If a polynomial $p(x)$ is divided by the factor $(x - a)$ then the remainder is $p(a)$.

Example

We wish to divide the polynomial $p(x) = x^3 - 3x^2 + x + 6$ by the expression $(x - 2)$. Here $a = 2$ and the remainder is $p(2) = 8 - 12 + 2 + 6 = 4$ (as we found earlier). If we divide $p(x)$ by the expression $(x + 2)$ then $a = -2$ and the remainder is $p(-2) = -8 - 12 - 2 + 6 = -16$. This can be verified by direct division but we ask you now merely to verify that

$$x^3 - 3x^2 + x + 6 \equiv (x + 2)(x^2 - 5x + 11) - 16$$ ∎

An important result which partly follows from the remainder theorem is the **factor theorem**:

> A polynomial $p(x)$ has a factor $(x - a)$ if and only if $p(a) = 0$.

Proof

Using the remainder theorem, we obtain $p(x) = (x - a)q(x) + p(a)$ where $q(x)$ is the quotient. Note that there are two statements to be proved here:

(i) If $p(a) = 0$ then $p(x) \equiv (x - a)q(x)$ so that $(x - a)$ is a (linear) factor of $p(x)$.

(ii) Conversely, if $(x - a)$ is a factor of $p(x)$ then the remainder after dividing $p(x)$ by $(x - a)$ must be 0. Hence $p(a) = 0$, from the remainder theorem.

Example

Show that $(x + 2)$ is a factor of $p(x) = x^3 - 2x^2 - 5x + 6$ but that $(x - 2)$ is not.
    Since $p(-2) = -8 - 8 + 10 + 6 = 0$ then $(x + 2)$ is a factor. Since $p(2) = 8 - 8 - 10 + 6 = -4 \neq 0$ then $(x - 2)$ is not a factor. ∎

The statement that $(x - a)$ is a factor of $p(x)$ is equivalent to saying that $x = a$ is a root of the equation $p(x) = 0$. The factor theorem can therefore be used to establish roots of a polynomial equation.

It is worth remarking that whereas the equation $ax + b = 0$ has the simple solution $x = -b/a$ $(a \neq 0)$ and we have already used the formula for the solution of a quadratic equation (Chapter 3); this is about as far as it is reasonable to proceed by a formula approach. There *are* recipes (called **algorithms**) for finding the roots of cubic and of quartic equations but they are very tedious to apply. For polynomial equations of degree greater than 4 no such general recipes exist.

Finding quotients and factors

The remainder theorem identifies the remainder after division by linear expression, but it does not identify the quotient. The following examples develop a method for finding a quotient.

Examples

1. What is the quotient when $x^3 - 3x^2 + x + 6$ is divided by $(x - 2)$? The answer can be obtained progressively.

$$
\begin{aligned}
x^3 - 3x^2 + x + 6 &\equiv (x - 2)(\text{quadratic}) + && x \times x^2 \equiv x^3 \\
&\equiv (x - 2)(x^2) + && \text{gives } x^3 - 2x^2 \\
&\equiv (x - 2)(x^2 - x) + && \text{gives } x^3 - 2x^2 - x^2 + 2x \\
& && \equiv x^3 - 3x^2 + 2x \\
&\equiv (x - 2)(x^2 - x - 1) + && \text{gives } x^3 - 3x^2 + 2x - x + 2 \\
& && \equiv x^3 - 3x^2 + x + 2 \\
&\equiv (x - 2)(x^2 - x - 1) + 4
\end{aligned}
$$

Hence the quotient is $(x^2 - x - 1)$ and the remainder is 4.

2. Divide $2x^3 + 3x^2 - 5x + 9$ by $(x+3)$.

$$2x^3 + 3x^2 - 5x + 9 \equiv (x+3)(\text{quadratic}) + \qquad x \times 2x^2 \equiv 2x^3$$
$$\equiv (x+3)(2x^2) + \qquad \text{gives } 2x^3 + 6x^2$$
$$\equiv (x+3)(2x^2 - 3x) + \qquad \text{gives } 2x^3 + 6x^2 - 3x^2 - 9x$$
$$\equiv 2x^3 + 3x^2 - 9x$$
$$\equiv (x+3)(2x^2 - 3x + 4) + \qquad \text{gives } 2x^3 + 3x^2 - 9x + 4x + 12$$
$$\equiv (x+3)(2x^2 - 3x + 4) - 3$$

The quotient is $(2x^2 - 3x + 4)$ and the remainder is -3. ∎

The same approach can be used to factorise a polynomial completely.

Example   Verify that $(x - 1)$ is a factor of the polynomial $p(x) = x^3 + 4x^2 + x - 6$. Find the other factor(s). (Note that we do not yet know whether there are two other linear factors or simply an irreducible quadratic factor.) Since $p(1) = 1 + 4 + 1 - 6 = 0$ then $(x - 1)$ is a factor of $p(x)$.

$$x^3 + 4x^2 + x - 6 \equiv 0 = (x - 1)(\text{quadratic}) + \qquad x \times x^2 \equiv x^3$$
$$\equiv (x - 1)(x^2) + \qquad \text{gives } x^3 - x^2$$
$$\equiv (x - 1)(x^2 + 5x) + \qquad \text{gives } x^3 + 4x^2 - 5x$$
$$\equiv (x - 1)(x^2 + 5x + 6)$$

The quadratic expression factorises into the product $(x+2)(x+3)$ and therefore $p(x) \equiv (x - 1)(x + 2)(x + 3)$. ∎

Note that the same approach could have been used to find the other roots of $p(x) = 0$ given that $x = 1$ is a root. We should then have found the other roots to be $x = -2$ and $x = -3$.
If there is a multiple root this will not be detected immediately.

Example   Factorise $p(x) = x^4 - 2x^3 - 5x^2 - 8x + 4$, given that $p(1) = 0$. We know that $(x - 1)$ is a factor of $p(x)$, so

$$x^4 - 2x^3 + 5x^2 - 8x + 4 \equiv (x - 1)(\text{cubic}) \qquad x \times x^3 \equiv x^4$$
$$\equiv (x - 1)(x^3) + \qquad \text{gives } x^4 - x^3$$
$$\equiv (x - 1)(x^3 - x^2) + \qquad \text{gives } x^4 - 2x^3 + x^2$$
$$\equiv (x - 1)(x^3 - x^2 + 4x) + \qquad \text{gives } x^4 - 2x^3 + 5x^2 - 4x$$
$$\equiv (x - 1)(x^3 - x^2 + 4x - 4)$$

You may recall from an earlier example that $p(x) \equiv (x-1)^2(x^2+4)$. The fact that $(x-1)$ is a multiple factor (or $x=1$ is a multiple root of $p(x)=0$) has not emerged at this stage. ∎

Exercise 8.5

1 Determine the polynomials whose roots are as follows, and draw suitable sketches.

(a) -1, 1, 2

(b) -3 (twice), 0, 1

(c) 1, 2 (three times)

(d) -5, -2, 0 (twice), 1, 4

Assume in each case that the coefficent of the highest power of x is equal to 1, e.g. $(x+1)(x-1)(x-2)$ in (a).

2 Consider $f(x) = x^3 + x^2 - 6x + 21$.

(a) Prove that there are no roots of $f(x)=0$ for $x \geq 2$.

(b) Plot the graph of $f(x)$ for $x=-4$ (1) 3, using an $x{:}y$ scale of $1{:}10$. Estimate the only root to two significant figures.

(c) By considering $f(x) - 21$, show that $f(x)$ crosses the line $y=21$ in three points.

3 (a) For $y=f(x)$ in Question 2, what is the effect of scaling the x-axis by a factor a and the y-axis by a factor b?

(b) Write down the new equation for $a=2$ and $b=5$. What has happened to the root(s)?

4 When a polynomial $P(x)$ is divided by $(x-a)$ we can write $P(x) = (x-a)Q(x) + R$.

(a) By putting $x=a$, prove that $R=P(a)$.

(b) What is the conclusion if $R=0$?

(c) Given that

$$x^3 - 7x^2 + 2x + 11 \equiv (x-1)Q(x) + R$$

write down R then find $Q(x)$.

(d) If $Q(x)$ is as in part (c) and $R=0$, what is the cubic polynomial?

5 (a) Determine a and b if $x^3 - 2x^2 + ax + b$ has roots $x = 1, 2$.

   (b) If instead $x = -7$ is a repeated root and $x = 12$ the other root of the cubic in part (a), what are a and b?

6 The equation $cx^3 + x^2 + dx + 10 = 0$ has roots $x = -2, \frac{1}{4}, 1\frac{2}{3}$. Find c and d.

7 Determine the unknown constants in the following identities:

   (a) $Ax(x - 1) + Bx(x - 2) + C(x - 1)(x - 2) \equiv 2$

   (b) $3x^2 - 4x + 3 \equiv a(x - 1)^2 + b(x - 1) + c$

   (c) $Ax(x - 1)(x - 2) + Bx(x - 2)(x - 3) + Cx(x - 1)(x - 3)$
        $+ D(x - 1)(x - 2)(x - 3) \equiv 12$

   (d) $(x - 3)^3 - (x - 4)^3 \equiv 3x^2 + cx + d$

   In each case recognise that the identity holds for *every value of x*, so select values of x in the first instance to eliminate or simplify terms (e.g. set $x = 0$, 1 or 2 in (a)). Systematic multiplying out should be undertaken when it is the only option left.

8 For the expression

$$a^3(b - c) + b^3(c - a) + c^3(a - b)$$

   (a) Show that $a - b$, $b - c$, $c - a$ are factors.

   (b) For the remaining factor note that symmetry, i.e. complete interchangeability between a, b, c, is needed. Identify this factor and write the expression as a product of four factors.

9* (a) Identify three factors of

$$a^n(b - c) + b^n(c - a) + c^n(a - b)$$

   where n is an integer, negative or positive.

   (b) What happens if $n = 0$?

   (c) Prove that

$$\frac{b - c}{a} + \frac{c - a}{b} + \frac{a - b}{c} \equiv -\frac{(a - b)(b - c)(c - a)}{abc}$$

10 Multiply out

$$(x + y + 1)(x - y + c) \quad \text{and choose } c \text{ to factorise}$$

   (a) $x^2 - y^2 + x - y$ (b) $x^2 - y^2 + 2x + 1$ (c) $x^2 - y^2 + 5x + 3y + 4$

11 Write down the square of $(x - y)$ and express $x^2 - 2xy + 2y^2 + 4y + 4$ as the sum of two squares. Deduce that the expression is zero for only one pair of values of x and y.

12* (a) The expression

$$2x^2 - xy - y^2 - 3x - 3y + c$$

is known to be the product of two linear factors. Accepting that the coefficients of x in the factors must be 1 and 2 respectively and the coefficients of y must be 1 or -1, trial and error establishes that $(2x + y + a)$ and $(x - y + b)$ must be the factors, so the coefficient of xy in the product is -1. Multiply the two factors together and find c.

(b) Factorise the following polynomial as a product of two linear factors with a suitable value of c:

$$6x^2 + xy - y^2 - x + 2y + c.$$

13 Determine the quotient $Q(x)$ and the remainder R when the polynomials $P(x)$ are divided by the linear factors given

(a) $x^4 - 3x^3 + 2x + 1$ by $(x - 1)$
(b) $3x^3 + 2x^2 - x - 1$ by $(x + 2)$
(c) $4x^3 - 2x^2 - x + 30$ by $(x - 5)$
(d) $11 - 5x - x^3$ by $(3 - 2x)$

14 In the following cases the factors are exact; identify the unknowns.

	$P(x)$	Factor(s)
(a)	$x^3 - 9x^2 + ax + 15$	$x - 3$
(b)	$5 + 6x - x^2 + ax^3$	$x + 1$
(c)	$(x + 1)(x^2 + ax + 1)$	$x + 1$
(d)	$x^2 + px + q$	$x + c$
(e)	$60 + ax + bx^2 + x^3$	$x + 5, x + 3$
(f)	$x^3 + ax^2 + bx + c$	$x - 1, x - 2, x - 5$

15 Determine the quotient and remainder in the following cases:

(a) $x^3 - 5x^2 - x + 6$ $\div x^2 - x + 12$
(b) $2 - 3x + 4x^2 - 5x^3$ $\div x^2 - 2x + 4$

(c) $x^4 - x^3 + 8x^2 + 9x + 7 \ \div x^2 + x + 1$

(d) $x^4 + 5x^3 + 6x^2 + 4x + 3 \div x^3 + 2x^2 + 3x + 1$

(e) $11x^4 - 6$ $\qquad\qquad\qquad \div 11x^2 - x - 5$

(f) $9 - 4x^2 - x^3 - x^4 \qquad \div 9 - x^3$

16 Expand out

$$x(x+2) + 1 = \frac{12}{x(x+2)}$$

as a quartic equation and verify that there are only two real roots.

17* Solve the equations

(a) $3x(x+3) - 19 = \dfrac{-28}{x(x+3)}$

(b) $x(x+2) + \dfrac{3}{x(x+2)} = 4$

(c)   Expand out the sixth-order polynomial

$$(x(x+1)(x+2))^2 + 4x(x+1)(x+2) + 17$$

and prove that it has no real roots.

18* (a)   Solve the equation

$$\sqrt{x+4} + \sqrt{x+7} = \sqrt{15 + 2x}$$

(b)   Solve the simultaneous equations

$$x + y = 5$$
$$x^2y^2 + 3xy = 28$$

Start by putting $z = xy$ to solve the second equation for xy.

(c)   Solve the simultaneous equations

$$\frac{7}{x+y} + \frac{1}{x-y} = 2$$
$$x^2 - y^2 = 7$$

(d)   Determine the two values of x which satisfy

$$\sqrt{x+2} + \sqrt{11+2x} = 2\sqrt{2(x+3)}$$

8.4 ELEMENTARY RATIONAL FUNCTIONS

We have seen that when two polynomials are added together the resulting expression is also a polynomial. The degree of this polynomial is the higher of the degrees of the two original polynomials. For example

$$(2x^3 + x - 7) + (x^4 - x^3 + x^2) \equiv x^4 + x^3 + x^2 + x - 7$$
$$(2x^3 + x - 7) + (x^3 - x^2 + 2x) \equiv 3x^3 - x^2 + 3x - 7$$

When one polynomial is subtracted from another the resulting expression is also a polynomial. The degree of this polynomial is the larger of the degrees of the two original polynomials with one exception. For example

$$(2x^3 + x - 7) - (2x^3 - x^2 + 2x) \equiv x^2 - x - 7$$

In the special case where the leading terms of the two polynomials are equal then the resulting polynomial is of lower degree.

When two polynomials are multiplied together, the result is also a polynomial and its degree is the sum of the degrees of the two original polynomials. For example

$$(2x^3 - 7) \times (x^4 + x^2) \equiv 2x^7 + 2x^5 - 7x^4 - 7x^2$$

However, when one polynomial is **divided** by another it is the exception that the result is also a polynomial. For example

$$\frac{x^3 - x^2 + 4x - 4}{x^2 + 4} = x + 1$$

because $(x^2 + 4)$ is a factor of $x^3 - x^2 + 4x - 4$.

However, the expression

$$\frac{x^3 - x^2 + 4x - 4}{x^2 + 5}$$

cannot be simplified further. A **rational function** has the form

$$\frac{p(x)}{q(x)}$$

where $p(x)$ and $q(x)$ are polynomials.

When the degree of $p(x)$ is less than the degree of $q(x)$, the fraction is known as a **proper** rational function. If the degree of $p(x)$ is greater than or equal to the degree of $q(x)$, the fraction is an **improper** rational function.

Graphs of $y = \dfrac{a}{x^n}$

The case where $q(x)$ is a constant really cannot be considered a true rational function. Ignoring this case, the simplest rational function is one for which $p(x)$ is a constant. In this section we consider some examples. We have taken $a = 1$ for simplicity; the only effect of taking positive values of a different from 1 is to change the vertical scale of the graph. Question 2(e) of Exercise 8.6 asks you to do this.

Example

Figure 8.9 shows the graph of $y = \dfrac{1}{x}$. Note the following features:

(i) The graph appears in two sections, one in the first quadrant and one in the third quadrant. This is because when $x > 0$, $y > 0$ and when $x < 0$, $y < 0$.

(ii) If we were to rotate the part of the graph in the first quadrant by $180°$, using the origin as the point of rotation, it would coincide *exactly* with that part of the graph in the third quadrant. (As an example note that when $x = 2$, $y = 0.5$ and when $x = -2$, $y = -0.5$.)

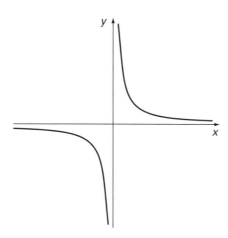

Figure 8.9   Graph of $y = \dfrac{1}{x}$

(iii) The graph is symmetrical about the line $y = x$.

(iv) The graph is symmetrical about the line $y = -x$.

(v) As $|x|$ increases, $|y|$ decreases.

In other words, if x becomes larger (more positive) then y decreases but stays positive. If x becomes more negative then y increases but stays negative. Geometrically, as we move to the extreme right or to the extreme left on the graph the curve gets closer to the x axis but does not cross it or even touch it. We say that the curve approaches the x-axis asymptotically or, equivalently, that the x-axis is an **asymptote** to the curve.

When $x = 0$ we cannot assign a value to y. As x approaches 0 through positive values (from the right) the values of y increases ever more rapidly and the curve gets ever steeper, becoming almost vertical. Similarly, as x approaches 0 through negative values (from the left) the values of y decrease ever more rapidly and the curve gets ever steeper, becoming almost vertical. As x approaches 0, therefore, the curve approaches the y-axis asymptotically; equivalently, the y-axis is an asymptote to the curve. We could also say that the curve possesses both horizontal and vertical asymptotes. ∎

Example

Figure 8.10 shows the graph of $y = \dfrac{1}{x^2}$. There are clear differences from the graph of $y = \dfrac{1}{x}$. Note the following features:

(i) The graph appears in two sections, one in the first quadrant and one in the second quadrant. This is because when $x > 0$, $y > 0$ and when $x < 0$, $y > 0$.

(ii) The part of the graph in the second quadrant is a reflection in the y-axis of the part in the first quadrant. (For example, note that when $x = 2$, $y = 0.25$ and when $x = -2$, $y = 0.25$ again.) Therefore the graph is symmetrical about the y-axis.

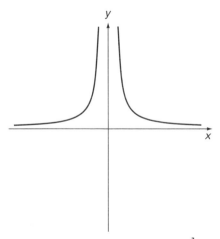

Figure 8.10 Graph of $y = \dfrac{1}{x^2}$

(iii) As |x| increases, |y| decreases. In other words, for large positive or large negative values of x, y gets closer to zero but stays positive. Hence the x-axis is an asymptote to the curve.

(iv) When $x=0$ we cannot assign a value to y. However, as x approaches zero (from either direction) the values of y increase ever more rapidly and the curve gets ever steeper, becoming almost vertical. The y-axis is an asymptote to the curve. The curve possesses both horizontal and vertical asymptotes. ∎

When a is negative the picture is altogether different.

Example   Figure 8.11(a) shows the graph of $y = -\dfrac{1}{x}$ and Figure 8.11(b) shows the graph of $y = -\dfrac{1}{x^2}$. At first sight it looks as though the graphs of $y = -\dfrac{1}{x}$ and $y = -\dfrac{1}{x^2}$ have been turned upside down. More precisely, they have been reflected in the x-axis. The general features are similar to those of the graphs in the previous two examples with obvious modifications to allow for the reflection. ∎

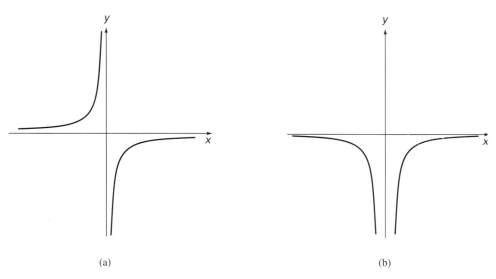

(a)   (b)

Figure 8.11 Graphs of (a) $y = -\dfrac{1}{x}$ and (b) $y = -\dfrac{1}{x^2}$

We now consider briefly the case where $q(x)$ is of the form $(x-a)$ or $(x+a)$.

Example   Figure 8.12(a) shows the graph of $y = \dfrac{1}{x-2}$. It looks as though the graph of $y = \dfrac{1}{x}$ has been moved to the right by 2 units, and this is just what *has* occurred.

The trouble spot is $x=2$ where the value of y cannot be defined. As x approaches the value 2 from above, the y-values get ever more positive; as x approaches the value 2 from below, the y-values get ever more negative.

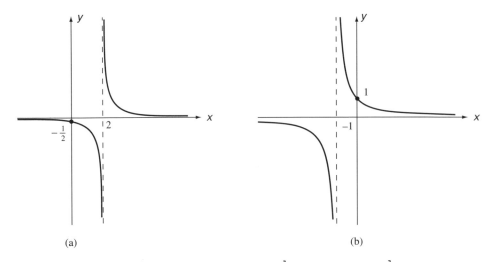

Figure 8.12 Graphs of (a) $y = \dfrac{1}{x-2}$ and (b) $y = \dfrac{1}{x+1}$

The asymptotes are $x=2$ (shown dashed) and the x-axis (or $y=0$). Note that the asymptotes are not part of the graph.

Figure 8.12(b) shows the graph of $y = \dfrac{1}{x+1}$. This time the graph of $y = \dfrac{1}{x}$ has been moved to the left by 1 unit. The asymptotes are $x=-1$ and the x-axis. If we write $y = \dfrac{1}{x-(-1)}$ then the analogy with Figure 8.12(a) should be clearer. In both cases the y-axis is crossed. The curve of $y = \dfrac{1}{x-2}$ cuts it at $y = -\dfrac{1}{2}$ $(x=0)$ and the curve $y = \dfrac{1}{x+1}$ cuts it at $y=1$. ∎

Ratio of linear polynomials

We consider the case where both $p(x)$ and $q(x)$ are linear polynomials. We examine one specific example and make some general remarks.

Example

$$y = \frac{2x-5}{x-3}$$

Note that $\dfrac{2x-5}{x-3} \equiv \dfrac{2x-6+1}{x-3} \equiv 2 + \dfrac{1}{x-3}$, hence we consider $y = 2 + \dfrac{1}{x-3}$.

Notice that as x approaches the value 3 the values of $\dfrac{1}{x-3}$ gets larger in size.

The add-on constant 2 is really unimportant; for example, when $x=3.0001$, $y = 2 + 10\,000 \simeq 10\,000$ and when $x=2.9999$, $y = 2 - 10\,000 \simeq -10\,000$.

However, when x gets increasingly large and positive, y approaches the value 2 from above; for example, when $x = 10\,003$, $y = 2.0001$.

When x gets increasingly large and negative, y approaches the value 2 from below; for example, when $x = -9997$, $y = 1.9999$.

The graph of $y = \dfrac{2x - 5}{x - 3}$ is shown in Figure 8.13. The lines $x = 3$ and $y = 2$ are asymptotes.

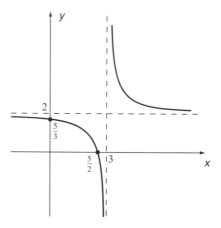

Figure 8.13 Graph of $y = \dfrac{2x - 5}{x - 3}$

The graph of $y = \dfrac{1}{x}$ has been moved to the right by 3 units then upwards by 2 units, i.e. to be centred at the point (3, 2). The curve cuts the y-axis at $y = \dfrac{5}{3}$ ($x = 0$) and cuts the x-axis at $x = \dfrac{5}{2}$ ($y = 0$). ∎

Finally we remark that the general expression

$$\frac{ax + b}{cx + d} \equiv \frac{a}{c} \left(\frac{cx + \dfrac{bc}{a}}{cx + d} \right) \equiv \frac{a}{c} \left(\frac{cx + d + \dfrac{bc}{a} - d}{cx + d} \right)$$

$$\equiv \frac{a}{c} \left(1 + \frac{\dfrac{bc}{a} - d}{cx + d} \right)$$

$$\equiv \frac{a}{c} + \frac{b - \dfrac{ad}{c}}{cx + d}$$

$$\equiv \frac{a}{c} + \frac{bc - ad}{c(cx + d)}$$

If $bc \neq ad$ then we can apply the method in the last example to study the behaviour of

$$y = \frac{ax + b}{cx + d}$$

What happens when $bc = ad$ and none of the four constants a, b, c and d is zero is the subject of Question 4 in Exercise 8.6.

Exercise 8.6

1 Sketch graphs of

(a) $y = \dfrac{1}{x}$

(b) $y = \dfrac{1}{|x|}$

(c) $x = \dfrac{1}{|y|}$

(d) $|y| = \dfrac{1}{|x|}$

2 Sketch the graphs of

(a) $y = -\dfrac{1}{x}$

(b) $y = \dfrac{2x + 3}{x + 1}$

(c) $y = \dfrac{2x + 1}{x + 1}$

(d) $y = \dfrac{3x - 1}{x + 2}$

(e) $y = \dfrac{1}{ax}, a = \dfrac{1}{2}, 1, 2, -\dfrac{1}{2}, -1, -2$

3 Find the vertical and horizontal asymptotes of the following:

(a) $y = \dfrac{5x + 4}{3x - 2}$

(b) $y = \dfrac{2 - 3x}{2x - 3}$

(c) $y = \dfrac{11 - x}{19 - x}$

(d) $y = 6 + \dfrac{2 - x}{2 + x}$

4 Given $y = \dfrac{ax + b}{cx + d}$ what happens if $ad = bc \neq 0$ or if $ad = bc = 0$?

5 Where are the roots $y = 0$ of the functions in Question 3?

6 On the same axes sketch the graphs of $y = \dfrac{1}{x}$ and $y = 3x + 2$. Determine the coordinates of the two points of intersection.

7* Determine necessary and sufficient conditions on a and b in the following three cases:

(a) $y = ax + b$ intersects $y = \dfrac{1}{x}$

(b) $y = ax + b$ touches $y = \dfrac{1}{x}$

(c) $y = ax + b$ neither intersects nor touches $y = \dfrac{1}{x}$

For part (b) write down the coordinates of the common point in terms of b. What is the sign of a?

8 Determine the coordinates of the points where the curve $y = \dfrac{x + 3}{x - 2}$

(a) crosses the coordinate axes

(b) intersects the horizontal line $y = 4$

(c) intersects the vertical line $x = -2$

(d) meets the straight line $y = \dfrac{x}{3} + 1$

9 Sketch the graphs of

(a) $y = x + 3 + \dfrac{1}{x + 2}$

(b) $y = \dfrac{x^2 + x + 6}{x - 1}$

Identify the points where the x and y axes are crossed.

10 Plot the graphs of $y = \dfrac{1}{x}$, $y = \dfrac{1}{x^2}$, $y = \dfrac{1}{x^3}$. Then plot $x = \dfrac{1}{y^2}$.

11 Using graphs as necessary, determine the coordinates of the intersection points of the following curves:

(a) $y = x$, $y = \dfrac{1}{x^3}$ \hspace{2em} (2 points)

(b) $y = \dfrac{1}{x}$, $y = -\dfrac{5}{x^2}$ \hspace{2em} (1 point)

(c) $y = \dfrac{1}{x^2}$, $y = \dfrac{1}{(x - 2)^2}$ \hspace{2em} (1 point)

(d) $y = \dfrac{1}{x^2}$, $y = \dfrac{1}{6}(x + 5)$ \hspace{2em} (3 points)

(e) $y = \dfrac{1}{x}$, $y = \dfrac{1}{6}(x^2 - 6x + 11)$ \hspace{2em} (3 points)

Note that in cases (d) and (e) a cubic equation in x needs to be solved. Use the remainder theorem to eliminate a root, noting that a solution exists in both cases when x is a small integer. Sketch the graphs in case (e).

12 Find the questions of the tangent and normal to

(a) $y = \dfrac{1}{x}$ at $\left(2, \dfrac{1}{2}\right)$ $\left(m = -\dfrac{1}{2}\right)$

(b) $y = \dfrac{1}{x+3}$ at $(-2, 1)$ $(m = -1)$

13 On the same axes draw graphs of $y = \dfrac{1}{x-1}$ and $y = \dfrac{1}{x-2}$. From the graphs determine the values of x for which $\dfrac{1}{x-1} > \dfrac{1}{x-2}$.

14* Draw the graph of $y = 2 + \dfrac{1}{x}$ and dot in the graph of $\left|2 + \dfrac{1}{x}\right|$ where this differs. For which values of x is $\left|2 + \dfrac{1}{x}\right| < \dfrac{3}{2}$?

15* Show that $\dfrac{3x-4}{x-3} = 3 + \dfrac{5}{x-3}$ and draw the graph of $y = \dfrac{3x-4}{x-3}$. On the same axes, draw the graph of $y = |x+2|$ and observe that the two curves meet in three points. Determine the values of x which satisfy

$$|x+2| < \frac{3x-4}{x-3}$$

and notice that one of the four related equalities is spurious.

16* With reference to Question 15, determine the values of x which satisfy

$$|x+2| < \left|\frac{3x-4}{x-3}\right|.$$

SUMMARY

- **Arithmetic progression (AP):** A sequence $a, a + d, a + 2d, \ldots$, in which each term is obtained from its predecessor by adding a constant amount, the common difference

- **Sum of an AP**

$$S = \frac{n}{2}(a + l)$$

  where a is the first term, l is the last term and n is the number of terms.

- **Geometric progression (GP):** A sequence $a, ar, ar^2, \ldots$, in which each term is obtained from its predecessor by multiplying it by a constant amount, the common ratio

- **Sum of a GP**

$$S = \frac{a(1 - r^n)}{1 - r}, \quad r \neq 1$$

  where a is the first term, r is the common ratio and n is the number of terms. If $-1 < r < 1$ the sum to infinity is

$$S = \frac{a}{1 - r}$$

- **Arithmetic mean** of two numbers a and b is $\frac{1}{2}(a + b)$

- **Geometric mean** of two numbers a and b is $\sqrt{ab}$

- **Sigma notation**

$$\sum_{r=1}^{n} t_r = t_1 + t_2 + \cdots + t_n$$

  If the sequence is infinite then

$$\sum_{r=1}^{\infty} t_r = t_1 + t_2 + \cdots$$

- **A polynomial of degree n** is written $p(x) \equiv a_0 + a_1 x + a_2 x^2 + \cdots + a_n x^n$.

- **Nested multiplication:** used to evaluate a polynomial.

- **Remainder:** when a polynomial $p(x)$ is divided by $(x - a)$ the remainder is $R = p(a)$. If $R = 0$ then $(x - a)$ is a linear factor of the polynomial.

- **Rational function:** the ratio of two polynomials, $\dfrac{p(x)}{q(x)}$. Where $p(x) = 0$ the function has a zero; where $q(x) = 0$ the function has a vertical asymptote.

Answers

Exercise 8.1

1 (a) $d=3$; 35, 50, $8+3(n-1)=5+3n$
  (b) $d=-4$; 964, 944, $1000+4(n-1)=1004-4n$

2 $3d=-2.4$ so that $d=-0.8$; 11.8, 11.0, -11.4

3 (a) 18   (b) 18

4 (a) 290   (b) 0   (c) 155   (d) 240

5 (a) 210   (b) 436.6

6 (a) 62   (b) $\dfrac{310}{3}=103\dfrac{1}{3}$

7 63, 195, $690-195=495$

8 (b) (i) 14 (ii) 45

9 $S=\dfrac{n}{2}(2+(n-1)10)$; 1, 11, 21, 31; $S=64$

10 32

11 $a=30$, $d=-2$, $S=210$

12 66 000

Exercise 8.2

1 (a) $r=2$; 2560   (b) $r=-5$; 31 250   (c) $r=0.4$; 0.081 92

2 (a) 508   (b) 172
  (c) $\dfrac{3\,486\,784\,400}{2\,324\,522\,934}\approx1.5$ (d) 2

3 (a) 93   (b) $6\left(1-\dfrac{1}{6^{10}}\right)=\dfrac{6^{10}-1}{6^{9}}\simeq6$
  (c) $\dfrac{33}{16}$

4 (a) $45[1 - (0.8)^{20}]$ (b) $45[1 - (0.8)^{10}]$

(c) $45[(0.8)^{10} - (0.8)^{20}] = 45 \times (0.8)^{10}[1 - (0.8)^{10}]$

5 (a) $2 \times 3^{n-1}$ (b) $\frac{1}{4} \times \left(\frac{1}{2}\right)^{n-1} = \frac{1}{2^{n+1}}$

(c) $-\frac{1}{2^{n+1}}$ (d) $\left(-\frac{1}{2}\right)^{n+1}$

6 (a) na (b)   0 for n even, a for n odd

7 $\frac{1}{1 - x^2}$

8 $r = \frac{1}{2}, \frac{127}{16}$

9 $r = \frac{2}{3}$

10 (a) $\frac{4}{9}$,   (b) $\frac{45}{99} = \frac{5}{11}$,   (c) $\frac{288}{999} = \frac{32}{111}$,   (d) $\frac{365}{111}$

11 £1215.51 approx., £1628.89 approx., £1218.40 approx., £1638.62 approx.

12 (a)   £1070, £1144.9   (b)   £13816 approx.

13 1328 years approx.

14 1.715 m, 0.840 35 m, $5 \times (0.7)^n$ m

$5 \times 0.7 + 5 \times (0.7)^2 + 5 \times (0.7)^3$ to 3 bounces

$= 1.7 \times [5 + 5 \times (0.7) + 5 \times (0.7)^2] = 1.7 \times \dfrac{5(1 - (0.7)^3)}{1 - 0.7} = 18.615$ m

$1.7 \times \dfrac{5[1 - (0.7)^5]}{1 - 0.7} = 23.571$ m (5 bounces)

$1.7 \times \dfrac{5}{1 - 0.7} = 28.\dot{3}$ m.

15 (a)   AM $= 6.5$, GM $= 6$, HM $= \dfrac{72}{13}$

(b)   Equality where $a = b$.

Exercise 8.3

1 (a) $\displaystyle\sum_{r=1}^{5} r^2$ (b) $\displaystyle\sum_{r=1}^{10} 3r$

(c) $\displaystyle\sum_{r=2}^{100} \frac{1}{r}$ (d) $\displaystyle\sum_{r=0}^{3} \frac{1}{3^r} = \sum_{r=1}^{4} \frac{1}{3^{r-1}}$

(e) $\displaystyle\sum_{r=1}^{5}(12 - 5r) = \sum_{r=0}^{4}(7 - 5r)$ (f) $\displaystyle\sum_{r=1}^{\infty} 25 \times \left(\frac{1}{5}\right)^r = 25 \sum_{r=1}^{\infty} \left(\frac{1}{5}\right)^r$

(g) $\displaystyle\sum_{r=0}^{3} x^{2r}$ (h) $\displaystyle\sum_{r=0}^{10}(-x)^r$

2 (a) $3 + 9 + 27 + 81$; 16   (b) $4 + 5 + 6 + 7$; 8

(c) $7 + 19 + 31 + 43$; 10   (d) $\dfrac{1}{4} + \dfrac{1}{16} + \dfrac{1}{64} + \dfrac{1}{256}$; 20

(e) $16 + 25 + 36$; 3   (f) $9 + 10 + 11 + 12$; 8

(g) $1 + 2 + 4 + 8$; 12   (h) $0 + 1 + 8 + 27$; 15

(i) $\dfrac{1}{4} + \dfrac{1}{2} + 1 + 2$; 5   (j) $2 + 6 + 12 + 20$; infinitely many

Exercise 8.4

1

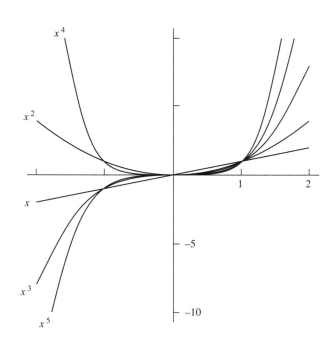

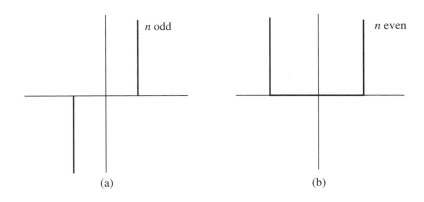

(a)   (b)

2   (a)   Here is an example   (b)   Here is an example

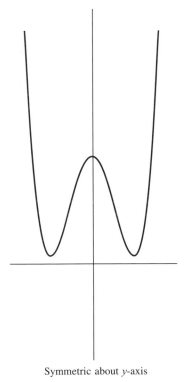

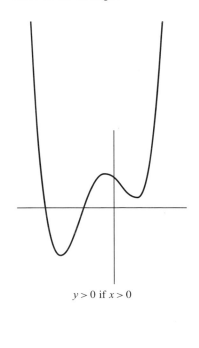

$y > 0$ if $x > 0$

Symmetric about y-axis

(c)   Here is an example

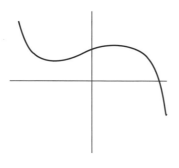

Cubic

(d)

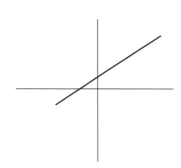

Straight line, positive slope, intercept to left
of origin

(e)

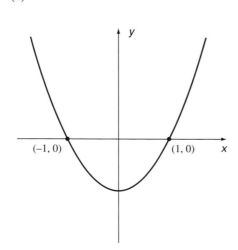

$$y = f(1 - x^2)(1 + x^2)$$

(f)

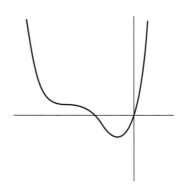

$y > 0$ if $x > 0$, $x = 0$ is a root

3   (a)   False

(b)   False ($y > 0$, if $x > 0$)

(c)   True ($y \ll 0$ if $x \ll 0$)

4

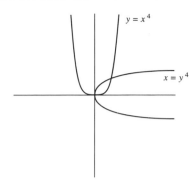

Intersections at $(0, 0)$, $(1, 1)$ only

5 (b) $a = \dfrac{y_1 - y_2}{x_1 - x_2}$ or $\dfrac{y_2 - y_1}{x_2 - x_1}$, $b = \dfrac{x_2 y_1 - x_1 y_2}{x_1 - x_2}$ or $\dfrac{x_1 y_2 - x_2 y_1}{x_2 - x_1}$

  (c) $L(x) = y_1$, i.e. the horizontal line $y = y_1$

6 (b) $Q(x) = \dfrac{1}{2} x^2 - \dfrac{3}{2} x + 1$

  (c) $Q(x)$ is that straight line; the coefficient of x^2 is zero

7 $q(x) = Q_1(x) - Q_2(x)$ is at most a quadratic which passes through $(x_1, 0)$, $(x_2, 0)$ and $(x_3, 0)$, i.e. it has three zeros; the only option is $q(x) \equiv 0$.

8 $C(x) = \dfrac{(x-3)(x+4)(5x+3)}{36} = \dfrac{5}{36} x^3 + \dfrac{2}{9} x^2 - \dfrac{19}{12} x - 1$

9 (a) $x^7 - x^6 + 2x^5 + x^3 - 5x^2 - 3x + 8$

  (b) $-x^7 + x^6 - 2x^5 + x^3 + 5x^2 - 3x - 4$

  (c) $x^{10} - x^9 - x^8 + 5x^7 - 8x^6 - x^5 + 21x^3 - 10x^2 - 18x + 12$

10

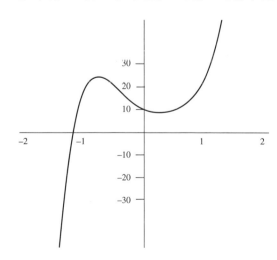

11 (a) $z^3(1 - z^4)(1 - z^2)$, $z = x^{1/4}$

(b) $z^5(9z^3 - 2)(1 - z^6)$, $z = (xy)^{1/6}$

(c) $z^4(1 - z)(3 - 4z)$, $z = 1/\sqrt{x}$

12 $-6 < x < -1$, $2 < x < 4$, approx.

Exercise 8.5

1 (a) $x^3 - 2x^2 - x + 2$

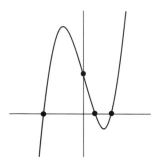

(b) $x^4 + 5x^3 + 3x^2 - 9x$

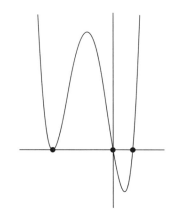

(c) $x^4 - 7x^2 + 18x^2 - 20x + 8$

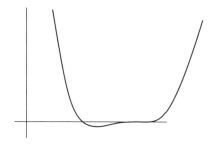

(d) $x^6 + 2x^5 - 21x^4 - 22x^3 + 40x^2$

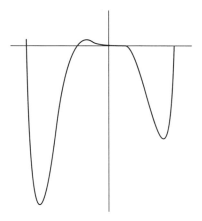

2 (a) $f(x) > 21$ if $x \geq 2$

(b)

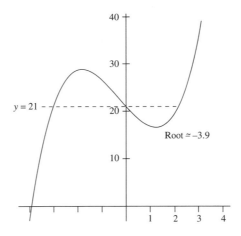

(c) $(-3, 21)$, $(0, 21)$, $(2, 21)$

3 (a) $x \mapsto x/a$, $y \mapsto y/b$

$$y = b\left(\left(\frac{x}{a}\right)^3 + \left(\frac{x}{a}\right)^2 - 6\left(\frac{x}{a}\right) + 21\right)$$

(b) if $a = 2$, $b = 5$ then $y = 5\left(\frac{x^3}{8} + \frac{x^4}{4} - 3x + 21\right)$

Only root doubled.

4 (b) $(x - a)$ is a factor   (c) $R = 7$, $Q(x) = x^2 - 6x - 4$

(d) $x^3 - 7x^2 + 2x + 4$

5 (a) $u = -1$, $b - 2$ (b) $a = 119$, $b = 588$

6 $c = 12$, $d = -41$

7 (a) $A = 1$, $B = -2$, $C = 1$ (b) $a = 3$, $b = 2$, $c = 2$

(c) $A = 2$, $B = 6$, $C = -6$, $D = -2$ (d) $c = -21$, $d = 37$

8 (b) $-(a - b)(b - c)(c - a)(a + b + c)$

9 (a) $a - b$, $b - c$, $c - a$ (b) Expression $\equiv 0$

10 (a) $(x+y+1)(x-y),$ $c=0$

     (b) $(x+y+1)(x-y+1),\ c=1$

     (c) $(x+y+1)(x-y+4),\ c=4$

11 $(x-y)^2 + (y+2)^2 \geq 0;\ x=y=-2$

12 (a) $(2x+y+1)(x-y-2),\ c=-2$

     (b) $(3x-y+1)(2x+y-1),\ c=-1$

13

	$Q(x)$	R
(a)	$x^3 - x^2 - 2x$	1
(b)	$3x^2 - 4x + 7$	-15
(c)	$4x^2 + 18x + 89$	475
(d)	$\dfrac{1}{2}x^2 + \dfrac{3}{4}x + 3\dfrac{5}{8}$	$\dfrac{1}{8}$

14 (a) $a=13$ (b) $a=-2$

     (c) $a=-1$ (d) $c^2 - pc + q = 0$

     (e) $a=47,\ b=12$ (f) $a=-8,\ b=17,\ c=-10$

15

	$Q(x)$	$R(x)$
(a)	$x-4$	$-17x + 54$
(b)	$-5x-6$	$5x+26$
(c)	$x^2 - 2x + 9$	$2x-2$
(d)	$x+3$	$-3x^2 - 6x$
(e)	$x^2 + \dfrac{x}{11} + \dfrac{56}{121}$	$\dfrac{111x - 446}{121}$
(f)	$x+1$	$4x^2 + 9x$

16 $x=-3,\ 1$

17 (a) $1, -4, \pm\dfrac{1}{6}\sqrt{165} - \dfrac{3}{2}$

     (b) $1, 3, \pm\sqrt{2} - 1$

     (c) $x^6 + 6x^5 + 13x^4 + 16x^3 + 16x^2 + 8x + 17$

     If $y = x(x+1)(x+2)$ then $(y+4)^2 + 1 = 0$, so no real y, hence no real x.

18 (a) $x = -3$

(b) $x = 1, 4, \dfrac{5}{2} \pm \dfrac{\sqrt{53}}{2}$

(c) $x = 4,\ y = 3$

(d) $x = -1,\ -\dfrac{33}{17}$

Exercise 8.6

1 (a)

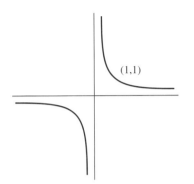

(b)

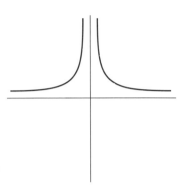

(c)

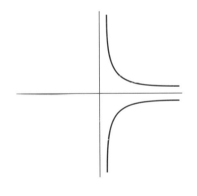

(d)

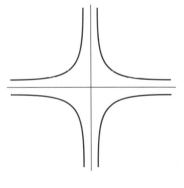

2 (a)

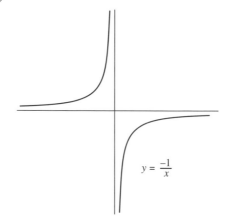

$y = \dfrac{-1}{x}$

(b)

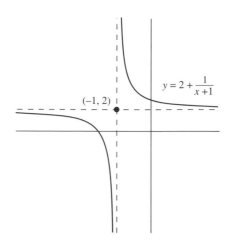

$y = 2 + \dfrac{1}{x+1}$

$(-1, 2)$

(c)

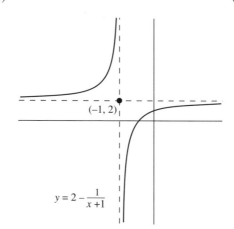

$(-1, 2)$

$y = 2 - \dfrac{1}{x+1}$

(d)

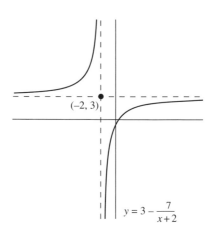

$(-2, 3)$

$y = 3 - \dfrac{7}{x+2}$

(e)

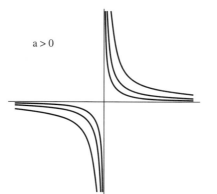

a > 0

(f)

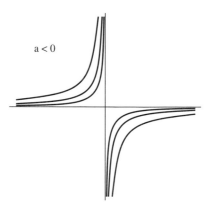

a < 0

The graphs move outward as a increases, e.g. for $a = -1$, the graph passes through $(1, -1)$, $(-1, 1)$ but for $a = -2$ it passes through $(\sqrt{2}, -\sqrt{2})$, $(-\sqrt{2}, \sqrt{2})$.

3 (a) $x = \dfrac{2}{3}, y = \dfrac{5}{3}$ (b) $x = \dfrac{3}{2}, y = -\dfrac{3}{2}$

  (c) $x = 19, y = 1$ (d) $x = -2, y = 5$

4 $y = b/d$ or a/c, i.e. constant

$$a = b = 0 \Rightarrow y = 0$$
$$c = d = 0, \; y = \infty, \; \text{undefined}$$
$$a = c = 0, \; y = \frac{b}{d}$$
$$b = d = 0, \; y = \frac{a}{c}, \; \text{as in the first case}$$

5 (a) $-\dfrac{4}{5}$ (b) $\dfrac{2}{3}$

  (c) 11 (d) $-\dfrac{14}{15}$

6

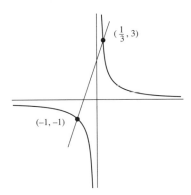

7 (a) $b^2 + 4a > 0$ (b) $b^2 + 4a = 0$

  (c) $b^2 + 4a < 0$; $\left(\dfrac{2}{b}, \dfrac{b}{2}\right)$, $a < 0$; here are two possibilities

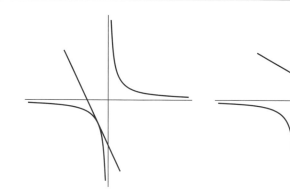

8   (a) $\left(0, -\dfrac{3}{2}\right), (-3, 0)$ (b) $\left(\dfrac{11}{3}, 4\right)$

(c) $\left(-2, \dfrac{1}{4}\right)$ (d) $(-3, 0), \left(5, \dfrac{8}{3}\right)$

9   (a)   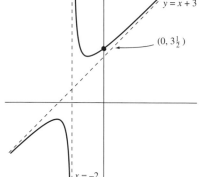   (b)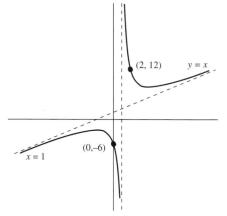

10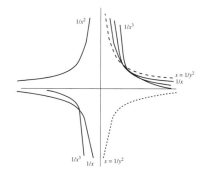

11 (a) $(1, 1), (-1, -1)$ (b) $\left(-5, -\dfrac{1}{5}\right)$ (c) $(1, 1)$

(d) $(1, 1), \left(-3+\sqrt{3}, \dfrac{1}{6}(2+\sqrt{3})\right), \left(-3-\sqrt{3}, \dfrac{1}{6}(2-\sqrt{3})\right)$

(e) $(1, 1), \left(2, \dfrac{1}{2}\right), \left(3, \dfrac{1}{3}\right)$

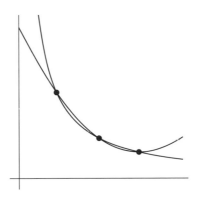

12 (a) $T: 2y + x = 3, \quad N: y = \dfrac{1}{2}(4x - 7)$

(b) $T: x + y + 1 = 0, \ N: y = x + 3.$

13

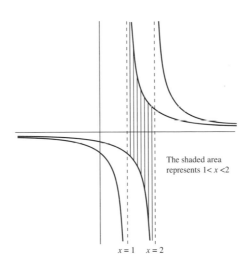

The shaded area represents $1 < x < 2$

$x = 1$ $x = 2$

14

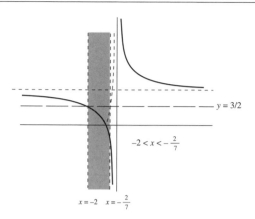

$$-2 < x < -\tfrac{2}{7}$$

15

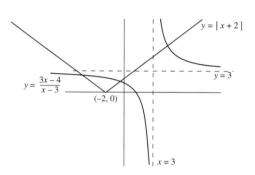

$$-1 - \sqrt{11} < x < 2 - \sqrt{6}, \text{ i.e. } -4.317 < x < -0.449,$$
$$3 < x < 2 + \sqrt{6}, \text{ i.e. } 3 < x < 4.449$$

16

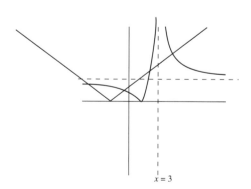

$$-4.317 < x < -0.449 \text{ (as Question 15) and } \sqrt{11} - 1 < x < 4.449, \text{ i.e. } 2.317 < x < 4.449.$$

9 COORDINATE GEOMETRY

INTRODUCTION

Coordinate geometry is a combination of geometry and algebra. Points, lines and geometric shapes are defined in algebraic terms using the position of points with respect to fixed lines, or axes. We are able to calculate distances, areas and volumes accurately, without recourse to scaled diagrams. In three dimensions, where diagrams would be difficult to produce, many of the formulae of coordinate geometry are simple extensions of corresponding formulae for two-dimensional cases. Robotics is a branch of engineering that makes extensive use of coordinate geometry.

OBJECTIVES

After working through this chapter you should be able to

- calculate the distance between two points with given coordinates
- find the position of a point which divides a line segment in a given ratio
- find the angle between two straight lines
- calculate the distance of a point from a given line
- calculate the area of a triangle knowing the coordinates of its vertices
- give simple examples of a locus
- recognise and interpret the equation of a circle
- define a parabola as a locus
- understand the concept of parametric representation of a curve
- use polar coordinates and convert to and from Cartesian coordinates
- interpret the equation of a straight line in three dimensions
- state and interpret the equation of a plane

9.1 DISTANCES AND AREAS IN TWO DIMENSIONS

In Chapter 3 we introduced the idea of the coordinates of a point. Strictly, they are known as **Cartesian coordinates**. In this section we shall use them to help us calculate distances and areas involving points which lie in a plane. In the heading we refer to two dimensions since to locate a point in the plane it is necessary to specify *two* coordinates.

Distances between two points

In Figure 9.1 P and Q are two points with coordinates (x_1, y_1) and (x_2, y_2) respectively. We wish to calculate the distance PQ.

We draw the right-angled triangle PQR where PR is parallel to the x-axis and RQ is parallel to the y-axis. The y-coordinate of R is therefore the same as the y-coordinate of P and the x-coordinate of R is therefore the same as the x-coordinate of Q. (PR lies on the line whose equation is $y = y_1$ and RQ lies on the line $x = x_2$.) Hence the coordinates of R are (x_2, y_1).

The lines MP and NQ are parallel to the y-axis and meet the x-axis at M and N respectively.

Then M is the point $(x_1, 0)$ and N is the point $(x_2, 0)$, hence

$$RP = NM = x_2 - x_1$$

In the same way

$$QR = y_2 - y_1$$

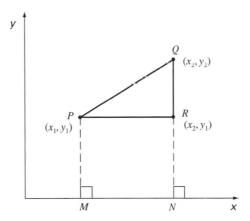

Figure 9.1 Distances between two points

Applying Pythagoras' theorem to the triangle PQR, we have

$$(PQ)^2 = (PR)^2 + (QR)^2 = (x_2 - x_1)^2 + (y_2 - y_1)^2$$

Taking the positive square root of both sides of this equation gives

$$PQ = \{(x_2 - x_1)^2 + (y_2 - y_1)^2\}^{1/2}$$

Notice that if the points P and Q are interchanged, the result is the same since $(x_2 - x_1)^2 \equiv (x_1 - x_2)^2$ and $(y_2 - y_1)^2 \equiv (y_1 - y_2)^2$.

Although we have drawn both P and Q in the first quadrant, the result applies quite generally.

The distance between the points $P(x_1, y_1)$ and $Q(x_2, y_2)$ is given by

$$PQ = \{(x_1 - x_2)^2 + (y_1 - y_2)^2\}^{1/2} \qquad (9.1)$$

Example

Find the distance between the points $(3, -8)$ and $(-2, 4)$.

Let $x_1 = 3, y_1 = -8, x_2 = -2, y_2 = 4$. Then the required distance is

$$\{(3 - (-2))^2 + (-8 - 4)^2\}^{1/2}$$

$$= \{(3 + 2)^2 + (-12)^2\}^{1/2}$$

$$= (25 + 144)^{1/2}$$

$$= 13$$

Dividing a line segment in a given ratio

In Figure 9.2 Q and R are the points (x_1, y_1) and (x_2, y_2) respectively. We want to find the coordinates of the point P which divides the line segment QR in the ratio $m : n$, i.e. $\dfrac{QP}{PR} = \dfrac{m}{n}$.

Let P have coordinates (x, y). The triangle QRT is drawn so that QT is parallel to the x-axis and TR is parallel to the y-axis. Therefore $Q\hat{T}R = 90°$.

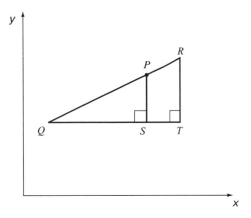

Figure 9.2 Dividing a line segment in a given ratio

Next, PS is drawn parallel to RT to meet QT at S. Hence $Q\hat{S}P = 90°$. In triangles QRT and QPS,

$$R\hat{Q}T = P\hat{Q}S \quad \text{(common angle)}$$

$$Q\hat{T}R = Q\hat{S}P \quad \text{(both 90°)}$$

$$Q\hat{R}T = Q\hat{P}S \quad \text{(angle sum in each triangle)}$$

Hence triangles QRT and QPS are similar. Therefore

$$\frac{QS}{QT} = \frac{PS}{RT} = \frac{QP}{QR}$$

But

$$\frac{QP}{QR} = \frac{QP}{QP + PR} = \frac{\dfrac{m}{n}PR}{\dfrac{m}{n}PR + PR} = \frac{\dfrac{m}{n}}{\dfrac{m}{n} + 1} = \frac{m}{m + n}$$

Therefore

$$\frac{QS}{QT} = \frac{SQ}{TQ} = \frac{m}{m + n}$$

Since $SQ = x - x_1$ and $TQ = x_2 - x_1$ we obtain the result

$$\frac{x - x_1}{x_2 - x_1} = \frac{m}{m + n}.$$

Cross-multiplying we have

$$(m + n)(x - x_1) = m(x_2 - x_1)$$

therefore $\qquad (m + n)x = mx_2 - mx_1 + (m + n)x_1 = mx_2 + nx_1$

Then $\qquad\qquad\qquad x = \dfrac{mx_2 + nx_1}{m + n}.$

Similarly, we find that

$$y = \frac{my_2 + ny_1}{m + n}.$$

The coordinates of the point which divides the line segment between (x_1, y_1) and (x_2, y_2) in the ratio $m : n$ are

$$\left(\frac{mx_2 + nx_1}{m + n}, \frac{my_2 + ny_1}{m + n} \right) \qquad\qquad (9.2)$$

There are several observations to make:

(i)   Each coordinate of P is a weighted average of the coordinates of Q and R, the weights (which sum to 1) being $\dfrac{n}{m + n}$ and $\dfrac{m}{m + n}$ respectively.

(ii) If $m = n$, i.e. P bisects QR, then coordinates of P are $\left(\dfrac{1}{2}(x_1 + x_2), \dfrac{1}{2}(y_1 + y_2) \right)$.

(iii) If $m > n$ the point P is closer to R than to Q and the weighting of the coordinates for R is greater than for Q.

Again, the diagram is for illustration only. The result is true wherever Q and R are located.

Example

Q is the point $(-4, -6)$ and R is the point $(2, 4)$. Find the coordinates of the point P which bisects QR and the point U which divides QR in the ratio 2:1.

Solution

The coordinates of P are $\left(\dfrac{1}{2}[(-4) + 2], \dfrac{1}{2}[(-6) + 4] \right)$, i.e. $(-1, -1)$.

The coordinates of U are given by

$$x = \frac{2 \times 2 + 1 \times (-4)}{2 + 1} = 0, \qquad y = \frac{2 \times 4 + 1 \times (-6)}{2 + 1} = \frac{2}{3}.$$

Hence U is the point $\left(0, \frac{2}{3}\right)$. ■

Since P lies between Q and R we should strictly say that P divides QR **internally** in the ratio $m : n$. If P were to divde QR **externally** in the ratio $m : n$ then $QP : PR = m : n$ but P lies outside QR; see Figure 9.3.

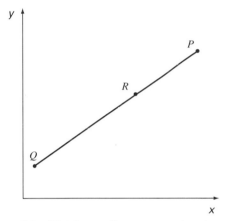

Figure 9.3 Dividing a line segment externally

The coordinates of P are given by

$$x = \frac{mx_2 - nx_1}{m - n}, \qquad y = \frac{my_2 - ny_1}{m - n}$$

Example Q is the point $(2, 4)$ and R is the point $(6, 8)$. Find the coordinates of the point P which divides QR externally in the ratio $2:1$ and U which divides QR externally in the ratio $1:2$.

Solution

The coordinates of P are given by

$$x = \frac{2 \times 6 - 1 \times 2}{2 - 1} = 10, \qquad y = \frac{2 \times 8 - 1 \times 4}{2 - 1} = 12$$

Hence P is the point $(10, 12)$.

The coordinates of U are given by

$$x = \frac{1 \times 6 - 2 \times 2}{1 - 2} = -2, \qquad y = \frac{1 \times 8 - 2 \times 4}{1 - 2} = 0.$$

Hence Q is the point $(-2, 0)$. ∎

Note that dividing QR externally in the ratio 1:2 is the same as dividing RQ externally in the ratio 2:1.

Angle between two straight lines

In Section 3.3 we stated without proof the conditions for two straight lines to be parallel or perpendicular. We generalise those conditions by stating, again without proof, the formulae for the angle between two straight lines.

The angle θ between the two straight lines $y = m_1 x + c_1$ and $y = m_2 x + c_2$ is given by

$$\tan \theta = \frac{m_1 - m_2}{1 + m_1 m_2}. \tag{9.3}$$

We make the following observations:

(i) If $m_1 = m_2$ then $\tan \theta = 0$, so $\theta = 0$, i.e. the lines are parallel.

(ii) If $m_1 m_2 = -1$ then $1 + m_1 m_2 = 0$, so $\tan \theta$ is infinitely large and $\theta = 90°$, i.e. the lines are perpendicular.

(iii) If $\tan \theta$ is calculated to be negative then θ is the obtuse angle between the lines whereas if $\tan \theta$ is positive then θ is the acute angle between the lines.

Examples

1. Find the acute angle between the lines $y = 2x - 5$ and $3x + y = 4$.

   The first line has a gradient $m_1 = 2$. The equation of the second line can be written $y = -3x - 4$, so the line has a gradient $m_2 = -3$. Then the angle θ is given by

   $$\tan \theta = \frac{2 - (-3)}{1 + 2 \times (-3)} = \frac{5}{-5} = -1$$

   so that $\theta = 135°$.
   The required *acute* angle is $180° - 135° = 45°$.

2. Find the equations of the lines which pass through the point $P(1, 2)$ and which make an angle of $45°$ with the line $y = \frac{1}{2}(x - 4)$; refer to Figure 9.4.

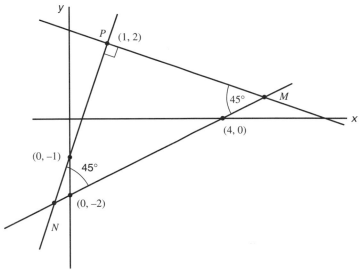

Figure 9.4 Diagram for the second example

The gradient of the given line is $m_2 = \dfrac{1}{2}$. Let the gradient of the line we seek be m_1. If the angle is $45°$ (line *NP*) then $\tan\theta = 1$ and

$$1 = \frac{m_1 - \dfrac{1}{2}}{1 + \dfrac{1}{2}m_1}, \text{ i.e. } m_1 = 3.$$

The equation of the line is

$$y - 2 = 3(x \quad 1)$$

i.e. $y = 3x - 1$.

If the angle is $135°$ (line *MP*) then $\tan\theta = -1$ and

$$-1 = \frac{m_1 - \dfrac{1}{2}}{1 + \dfrac{1}{2}m_1}, \text{ i.e. } m_1 = -\frac{1}{3}$$

The equation of the line is

$$y - 2 = -\frac{1}{3}(x - 1)$$

i.e. $\qquad y = -\frac{1}{3}x + \frac{7}{3}$

Note that the two lines are perpendicular since $3 \times \left(-\frac{1}{3}\right) = -1$. ∎

Equation (9.3) breaks down if one of the lines is parallel to the y-axis for then its gradient is infinitely large. Figure 9.5(a) shows that if the other line has a positive gradient $m_1 = \tan\theta$ then the required acute angle is $90° - \theta$. If $m_1 < 0$ then, as is shown in Figure 9.5(b), the required acute angle is $\theta - 90°$.

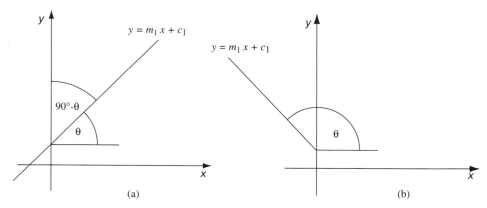

Figure 9.5 Angle between two straight lines

Distance of a point from a straight line

Let the line have equation $ax + by + c = 0$ and the point be $P(x_1, y_1)$. There are two cases to consider: where P is on the opposite side of the line from the origin, Figure 9.6(a), and where it is on the same side, Figure 9.6(b).

The line PN meets the given line perpendicularly at N and the required distance (the shortest distance from the point to the line) is d. PM is drawn parallel to the x-axis to meet the given line at M. Now

$$d = PN = PM \sin\theta$$

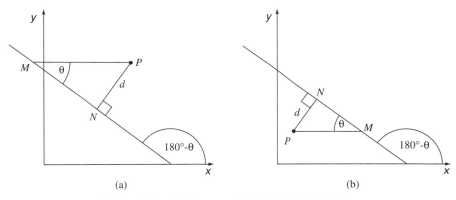

(a)   (b)

Figure 9.6 Distance of a point from a line

The y-coordinate of M is equal to that of P and is therefore y_1; since M lies on the line, its x-coordinate is given by $ax + by_1 + c = 0$, i.e.

$$x = -(by_1 + c)/a.$$

In Figure 9.6(a)

$$
\begin{aligned}
PM &= x_1 - (-(by_1 + c)/a) \\
&= x_1 + \left(\frac{by_1 + c}{a}\right) \\
&= \frac{ax_1 + by_1 + c}{a}.
\end{aligned}
$$

In Figure 9.6(b)

$$
\begin{aligned}
PM &= (-(by_1 + c)/a) - x_1 \\
&= -\frac{(ax_1 + by_1 + c)}{a}.
\end{aligned}
$$

Since we are concerned only with the length PM we may say that

$$PM = \frac{|ax_1 + by_1 + c|}{a}.$$

The equation of the line may be written

$$y = -\frac{ax}{b} - \frac{c}{b}$$

so that its gradient is $-(a/b)$.

In both diagrams the gradient of the line is $\tan(180° - \theta) = \tan \theta$. Hence $\tan \theta = a/b$ and $\sin \theta = \pm \dfrac{a}{(a^2 + b^2)^{1/2}}$. This result is now derived.

We know that $\sin^2 \theta + \cos^2 \theta \equiv 1$, so dividing by $\sin^2 \theta$ we get

$$1 + \frac{\cos^2 \theta}{\sin^2 \theta} \equiv \frac{1}{\sin^2 \theta}$$

i.e.

$$1 + \frac{1}{\tan^2 \theta} \equiv \frac{1}{\sin^2 \theta}$$

i.e.

$$\frac{1}{\sin^2 \theta} = 1 + \frac{b^2}{a^2} = \frac{a^2 + b^2}{a^2}.$$

Therefore

$$\sin^2 \theta = \frac{a^2}{a^2 + b^2}$$

and taking square roots,

$$\sin \theta = \pm \frac{a}{(a^2 + b^2)^{1/2}}.$$

Since it is the length of PN which is required and since $PN = PM \sin \theta$, we have the following result. The perpendicular distance from $P(x_1, y_1)$ to the line $ax + by + c = 0$ is given by

$$d = \frac{|ax_1 + by_1 + c|}{(a^2 + b^2)^{1/2}}. \tag{9.4}$$

Example

We wish to find the distances of the points $P(1, 2)$ and $Q(2, 1)$ from the line $2x + y + 1 = 0$ and to decide whether the points are on opposite sides of the line.

For P the required distance is

$$\frac{|2 \times 1 + 1 \times 2 + 1|}{(2^2 + 1^2)^{1/2}} = \frac{|5|}{\sqrt{5}} = \sqrt{5}$$

For Q the required distance is

$$\frac{|2 \times (-2) + 1 \times 1 + 1|}{(2^2 + 1^2)^{1/2}} = \frac{|-2|}{\sqrt{5}} = \frac{2}{\sqrt{5}}.$$

Since the signs of $ax_1 + by_1 + c$ are different for P and Q they lie on opposite sides of the line, as a scaled diagram will confirm. ∎

Area of a triangle

In Figure 9.7 the triangle ABC has vertices $A(x_1, y_1)$, $B(x_2, y_2)$ and $C(x_3, y_3)$. AL, CM and BN are drawn perpendicular to the x-axis.

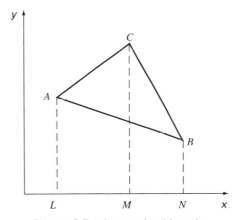

Figure 9.7 Area of a triangle

The area of the trapezium $ACML$ is

$$\frac{1}{2}(AL + CM) \times LM = \frac{1}{2}(y_1 + y_3)(x_3 - x_1).$$

Similarly the area of the trapezium $CBNM$ is

$$\frac{1}{2}(y_2 + y_3)(x_2 - x_3)$$

Write down the coordinates of the centroid of the triangle whose vertices are the points $(6, -2)$, $(-2, -1)$ and $(3, 5)$.

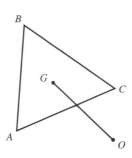

7 The line $y = 2x - 3$ is rotated about the point $(2, 1)$

(a) $45°$ clockwise (b) $\tan^{-1}(2)$ clockwise (c) $45°$ anticlockwise

Obtain the equation of the rotated lines in each case. Draw the graph in case (c).

8 The point $(1, 6)$ is 4 units distant from a line which lies to the right of it, and which intersects the x-axis at $45°$. What is the equation of the line?

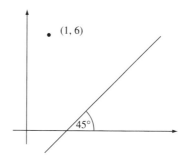

9* The point (x, y) is free to move so that it is the same distance from the two fixed points $(2, 5)$ and $(6, 1)$. Equate the squares of these distances and show that the coordinates x and y must satisfy the equation

$$x - y = 1$$

What does this mean?

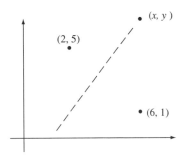

10* An enclosed garden is bounded by four walls and is therefore shaped in the form of a quadrilateral. A landscape gardener wishes to plant a tree in the *middle* so that the distance to the nearest wall is a maximum. How could the optimum position be found?

You are advised that there is no straightforward method to the optimum solution, even though this solution exists. Numerical methods are needed and these are beyond our scope. Try it first then look at two ideas in the answers.

9.2 LOCI AND SIMPLE CURVES

If a point moves to satisfy some condition then the path it traces out is called its **locus** (plural loci). As an example we may require it to be on the straight line joining the points (1, 1) and (2, 3); its coordinates then satisfy the equation $2x - y - 1 = 0$.

Example

A point P moves so that it is equidistant from the points $A(-1, 2)$ and $B(2, -1)$; describe its locus.

Figure 9.8 illustrates the situation. One position for P is the midpoint of the line segment AB. The condition on P is that $PA = PB$, so

$$(PA)^2 = PB)^2$$

i.e. $$(x - (-1))^2 + (y - 2)^2 = (x - 2)^2 + (y - (-1))^2$$

i.e. $$(x + 1)^2 + (y - 2)^2 = (x - 2)^2 + (y + 1)^2$$

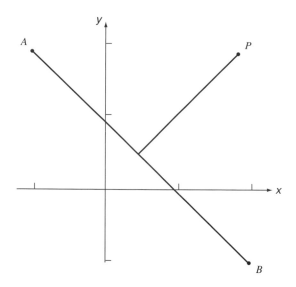

Figure 9.8 Locus of a point equidistant from two given points

Expanding and cancelling we obtain

$$x^2 + 2x + 1 + y^2 - 4y + 4 = x^2 - 4x + 4 + y^2 + 2y + 1$$

or
$$6x = 6y$$

i.e.
$$x = y$$

The locus of P is therefore a straight line passing through the origin. The standard equation of the locus is $y = x$. (It is the perpendicular bisector of AB.) ■

Intersection of two loci

Where two loci meet, the points of intersection can be found by solving simultaneously the equations of the loci.

Example

Find the points of intersection of the line $y = x + 2$ and the curve $y = 2x^2 - x + 2$.

Solution

Substituting the first equation into the second, we obtain

$$x + 2 = 2x^2 - x + 2$$

or $\qquad 2x^2 - 2x = 0$

i.e. $\qquad 2x(x - 2) = 0.$

Hence $x = 0$ or $x = 2$. When $x = 0$ either of the original equations gives $y = 2$ and when $x = 2, y = 4$. The points of intersection are therefore (0, 2) and (2, 4). ∎

A straight line which meets a curve in only one point is called a **tangent** to the curve at that point. The single point of intersection is known as the **point of contact**.

Example

Show that the line $y = mx + a/m$ is a tangent to the curve $y^2 = 4ax$ for all values of $m \neq 0$ and find the coordinates of the point of contact.

Substituting the equation of the line into the equation of the curve, we obtain

$$\left(mx + \frac{a}{m}\right)^2 = 4ax$$

i.e. $\qquad m^2x^2 + 2ax + \dfrac{a^2}{m^2} = 4ax$

i.e. $\qquad m^2x^2 - 2ax + \dfrac{a^2}{m^2} = 0$

or $\qquad \left(mx - \dfrac{a}{m}\right)^2 = 0.$

Hence $x = \dfrac{a}{m^2}$ is the only solution. Using the equation of the line we find that

$$y = \frac{a}{m} + \frac{a}{m} = \frac{2a}{m}.$$

The point of contact has coordinates $\left(\dfrac{a}{m^2}, \dfrac{2a}{m}\right)$. ∎

Some important loci

Circle

The locus of a point which moves so that its distance from a fixed point is constant is a **circle**. The point C is called the **centre** of the circle and the constant distance is the **radius** of the circle. Figure 9.9 shows the situation where the centre C is the point (x_0, y_0) and the radius is a.

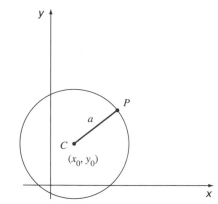

Figure 9.9 A circular locus

Let $P(x, y)$ be a point on the circumference of the circle. Then $PC = a$. Hence $(PC)^2 = a^2$,

i.e.
$$(x - x_0)^2 + (y - y_0)^2 = a^2 \tag{9.6a}$$

is the required equation. Note that in the special case where C is the origin the equation simplifies to

$$x^2 + y^2 = a^2. \tag{9.6b}$$

Example

Find the radius and the coordinates of the centre of the circle whose equation is $3x^2 + 3y^2 - 6x + 12y = 12$.

Solution

First we divide the equation by 3 to obtain

$$x^2 + y^2 - 2x + 4y = 4.$$

Then we collect the terms in x and y:

$$x^2 - 2x \equiv x^2 - 2x + 1 - 1 \equiv (x - 1)^2 - 1$$
and
$$y^2 + 4y \equiv y^2 + 4y + 4 - 4 \equiv (y + 2)^2 - 4$$

Hence the equation can be written

$$(x-1)^2 + (y+2)^2 - 5 = 4$$

or $\qquad (x-1)^2 + (y+2)^2 = 9 = 3^2.$

Comparing this with equation (9.6) we see that the radius of the circle is 3 and its centre is the point $(1, -2)$. ∎

Parabola

The locus of a point which moves so that its distance from a fixed point is equal to its distance from a given line is a **parabola**. The fixed point is the **focus** of the parabola and the given line is the **directrix**. Figure 9.10(a) shows the general case where S is the focus and $PS = PM$.

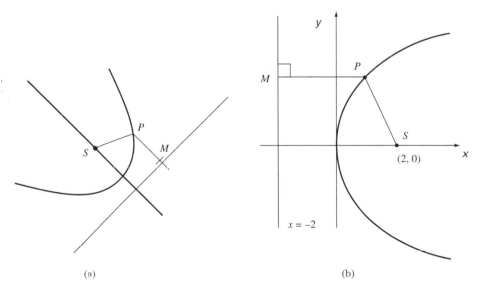

(a)   (b)

Figure 9.10 Parabola

Example

Find the equation of the locus of the point P which moves so that its distance from the line $x = 2$ is equal to its distance from the point $(2, 0)$.

From Figure 9.10(b) we know that $PS = PM$. Let P be the point (x, y). Then M is the point $(-2, y)$. Furthermore,

$$PM = x + 2$$

and

$$(PS)^2 = (x-2)^2 + (y-0)^2.$$

Since $(PS)^2 = (PM)^2$ it follows that

$$(x-2)^2 + (y-0)^2 = (x+2)^2$$

i.e. $$x^2 - 4x + 4 + y^2 = x^2 + 4x + 4$$

so that $$y^2 = 8x$$

which is the required equation. ∎

When the directrix is the line $x = a$ and the focus is the point $(a, 0)$ the equation becomes

$$y^2 = 4ax. \tag{9.7}$$

This way of writing the equation is called **standard form**.

Ellipse

The locus of a point which moves so that the ratio of its distance from a fixed point to its distance from a given line is constant and less than 1 is an **ellipse**. The ratio of the distances is the **eccentricity** of the ellipse, denoted e.

Figure 9.11(a) shows the general case and Figure 9.11(b) depicts the standard form of the ellipse. The directrix is $x = \dfrac{a}{e}$ and the focus is the point $(ae, 0)$. If we denote the

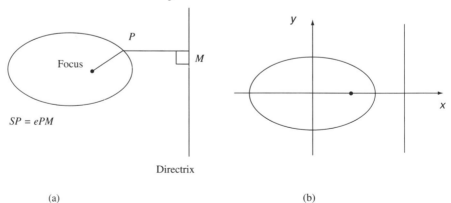

(a)   (b)

Figure 9.11 Ellipse

expression $a^2(e^2 - 1)$ by b^2 then the equation (in standard) form is

$$\frac{x^2}{a^2} + \frac{y^2}{b^2} = 1. \tag{9.8}$$

Hyperbola

The locus of a point which moves so that the ratio of its distance from a fixed point to its distance from a given line is constant and greater than 1 is a **hyperbola**. The ratio of the distances is the eccentricity of the hyperbola, denoted by e.

Figure 9.12(a) depicts the standard form of the hyperbola. The directrix is $x = \frac{a}{e}$ and the focus is the point $(ae, 0)$. If we denote the expression $a^2(e^2 - 1)$ by b^2 then the equation (in standard form) is

$$\frac{x^2}{a^2} - \frac{y^2}{b^2} = 1. \tag{9.9}$$

The dashed lines which the curve approaches as x approaches infinitely large positive or infinitely large negative values (i.e. on the far right or the far left) are the **asymptotes** of the hyperbola. They have the equations

$$y = \pm \frac{b}{a} x. \tag{9.10}$$

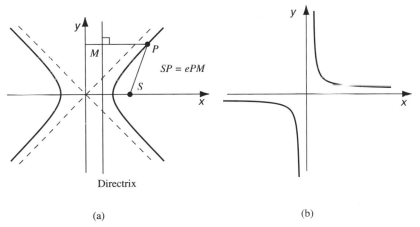

(a)   (b)

Figure 9.12 Hyperbola

When the asymptotes are perpendicular $b = a$ and the curve is called a **rectangular hyperbola**. In the example shown in in Figure 9.12(b) the asymptotes are the x and y axes. The standard form of the rectangular hypberbola is

$$xy = c^2. \tag{9.11}$$

Exercise 9.2

1   Determine the locus of a point which is equidistant from the origin and the point (3, 1). Draw a sketch to illustrate the answer.

2   Prove that the locus of a point equidistant from two fixed points (a, b) and (c, d) is a straight line. Determine its equation.

3   Write down the equations of the straight line which are equidistant from the following pairs of points.

(a)   (0, 5) and (6, 4)   (b)   (3, 4) and (2, −7).

4   Determine the two points which are exactly five units away from both of the points (−2, 0) and (6, 0).

5   Find the two straight line loci of a point equidistant from each of the coordinate axes.

6   Determine the equations of the two straight lines which are the loci of a point equidistant from both $3x + 4y = 5$ and $4x - 3y = 5$.

7   A point (x, y) moves such that its distance from the point (0, 5) is equal to the square of its distance from the origin. Prove that x and y satisfy the equation

$$x^4 + 2x^2y^2 + y^4 - x^2 - y^2 + 10y = 25.$$

8   Determine the Cartesian equation of each of the following circles, expanding it out in full.

(a)   centre (2, 1) and radius 3   (b)   centre (−5, 4) and radius 6.

9*   Determine the equation of the locus of a point which is equidistant from the point (0, 2) and the x-axis.

10* A point moves so that its distance from the line $y = 3$ is twice its distance from the origin. Find its locus.

9.3 CIRCLES

In the last section we found the equation for a circle. The standard form of the general equation is

$$x^2 + y^2 + 2gx + 2fy + c = 0 \qquad (9.12)$$

This can be rewritten as

$$(x + g)^2 + (y + f)^2 = g^2 + f^2 - c$$

which represents the circle whose centre is the point $(-g, -f)$ and whose radius is $(g^2 + f^2 - c)^{1/2}$.

Example   Find the radius and the coordinates of the centre of the circle whose equation is

$$x^2 + y^2 - 2x + 4y - 4 = 0$$

Here $g = -1, f = 2$ and $c = 4$. Hence the centre is at $(1, -2)$ and the radius is $[(-1) + 2^2 + 4]^{1/2} = \sqrt{9} = 3$. (This is an alternative to the approach on page 428.) ■

Equation of a circle on a given diameter

Suppose that AB is a diameter of the circle in Figure 9.13. Let $P(x, y)$ be any point on the circle. From Chapter 5 we know that the angle $A\hat{P}B = 90°$, so the line segments AP and PB are perpendicular.

The gradients of the lines AP and PB are respectively

$$\frac{y - y_1}{x - x_1} \quad \text{and} \quad \frac{y - y_2}{x - x_2}.$$

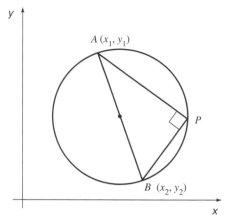

Figure 9.13   Circle on a given diameter

Hence

$$\left(\frac{y - y_1}{x - x_1}\right)\left(\frac{y - y_2}{x - x_2}\right) = -1$$

or $(x - x_1)(x - x_2) + (y - y_1)(y - y_2) = 0.$

Sometimes an alternative approach may be preferable as the following example shows.

Example   Find the equation of the circle on the diameter whose endpoints are $A(1, -4)$ and $B(5, 2)$.
The centre of the circle, C, is the midpoint of AB; its coordinates are therefore $(3, -1)$. The radius is the length of AC and

$$(AC)^2 = (3 - 1)^2 + (-1 - (-4))^2 = 4 + 9 = 13.$$

The radius is therefore $\sqrt{13}$ and the equation of the circle is

$$(x - 3)^2 + (y + 1)^2 = 13. \qquad\qquad\blacksquare$$

Equation of a circle through three given points

There is only one circle which passes through three given non-collinear points. In general it is necessary to find the solution of three simultaneous linear equations to determine the equation of the circle and this can be quite tricky. Sometimes, however, as the next example shows it can be straightforward.

Example   Find the equation of the circle which passes through the origin and the points (0, 2) and (−3, 3). Find the radius and centre of the circle.

The general equation for a circle is $x^2 + y^2 + 2gx + 2fy + c = 0$.

$$(0, 0) \text{ lies on the circle, so } c = 0$$

$$(0, 2) \text{ lies on the circle, so } 4 + 4f + c = 0$$

$$\therefore \quad f = -1$$

$$(1, 3) \text{ lies on the circle, so } 1 + 9 - 2g + 6f + c = 0$$

$$\therefore \quad g = 2.$$

The equation of the circle is therefore

$$x^2 + y^2 + 4x - 2y = 0.$$

This can be written

$$(x + 2)^2 + (y - 1)^2 = 5$$

so the centre of the circle is (−2, 1) and its radius is $\sqrt{5}$. ∎

Intersection of a straight line and a circle

The general case is complicated to handle, so for now we confine our attention to a circle whose centre is at the origin.

Let the circle be $x^2 + y^2 = a^2$ and the straight line be $y = mx + c$; these intersect where

$$x^2 + (mx + c)^2 = a^2$$

or $(1 + m^2)x^2 + 2mcx + c^2 - a^2 = 0.$

This quadratic equation has two real roots if its discriminant is positive, i.e.

ie   if $4m^2c^2 > 4(1 + m^2)(c^2 - a^2).$

This reduces to the condition

$$c^2 > (1 + m^2)a^2.$$

This is the condition for the line to cut the circle in two distinct points.
  Similarly we can argue that if

$$c^2 < (1 + m^2)a^2$$

then the line does not meet the circle at all.
  If there is only one point of intersection then the straight line is a **tangent** to the circle, so the line $y = mx + c$ is a tangent to the circle $x^2 + y^2 = a^2$ if

$$c^2 = (1 + m^2)a^2. \tag{9.13}$$

Examples

1. Show that the straight line $x + 3y = 10$ is a tangent to the circle $x^2 + y^2 = 10$ and find the coordinates of the point of contact. Find the equation of another tangent to the circle in the same direction, i.e. a parallel tangent.
   The equation of the line can be written $x = 10 - 3y$. This meets the circle where

$$(10 - 3y)^2 + y^2 = 10$$

i.e. $$10y^2 - 60y + 90 = 10$$

i.e. $$y^2 - 6y + 9 = 0$$

or $$(y - 3)^2 = 0.$$

There is only one solution, $y = 3$, and therefore $x = 10 - 9 = 1$. The point of contact is $(1, 3)$ and since it is unique the line is a tangent to the circle.
   Alternatively, since the equation of the line can be written as $y = -\dfrac{1}{3}x + \dfrac{10}{3}$, then

$c = \dfrac{10}{3}, m = -\dfrac{1}{3}$. Since $a^2 = 10$ then $(1 + m^2)a^2 = \dfrac{10}{9} \times 10 = \dfrac{100}{9} = c^2$. Hence,

from equation (9.13) the line is a tangent.
   From Figure 9.14 we see that the parallel tangent has $(-1, -3)$ as its point of contact. Its equation is $x + 3y = c$ and, since $(-1, -3)$ lies on the line, we can say that

$$-10 = c \quad \text{i.e. } c = -10.$$

The equation is therefore

$$x + 3y = -10.$$

2. A circle whose centre lies in the first quadrant touches both coordinate axes. The line $3y = 4x + 6$ is also a tangent to the circle. Find the equation of the circle and the point of contact with the given line.

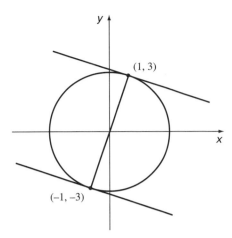

Figure 9.14 Tangents to a circle

Let the equation of the circle be

$$x^2 + y^2 + 2gx + 2fy + c = 0.$$

The circle meets the x-axis, i.e. $y = 0$, where $x^2 + 2gx + c = 0$.

Since there is only one point of contact, the left-hand side of this equation must be a perfect square, i.e. $(x^2 + 2gx + g^2)$, so $c = g^2$. Similarly the curve touches the y-axis, i.e. $x = 0$, if $f^2 = c$.

The equation of the circle can therefore be written

$$x^2 + y^2 + 2gx + 2gy + g^2 = 0.$$

The line $3y = 4x + 6$ meets the circle where

$$x^2 + \left(\frac{4x + 6}{3}\right)^2 + 2gx + 2g\left(\frac{4x + 6}{3}\right) + g^2 - 0$$

which reduces to

$$25x^2 + (48 + 42g)x + (36 + 36g + 9g^2) = 0. \qquad (9.14)$$

For the line to be a tangent this equation must have equal roots. Hence

$$(48 + 42g)^2 = 4 \times 25(36 + 36g + 9g^2).$$

This eventually reduces to the equation

$$2g^2 + g - 3 = 0$$

which has solutions $g = 1$ and $g = -\dfrac{3}{2}$.

The value $g = 1$ is rejected since this would give the centre of the circle as $(-1, -1)$, which is not in the first quadrant.

Hence $g = -\dfrac{3}{2}$, so the centre of the circle is $\left(\dfrac{3}{2}, \dfrac{3}{2}\right)$. The equation of the circle is

$$x^2 + y^2 - 3x - 3y + \frac{9}{4} = 0.$$

Equation (9.14) becomes

$$25x^2 - 15x + \frac{9}{4} = 0$$

i.e. $100x^2 - 60x + 9 = 0$

i.e. $(10x - 3)^2 = 0.$

Here the only solution is $x = \dfrac{3}{10}$ and, using the equation of the line, $y = \dfrac{12}{5}$.   ■

Length of the tangent from a given external point

A point $P(x_1, y_1)$ lies outside the circle $(x - x_0)^2 + (y - y_0)^2 = a^2$ if

$$(x_1 - x_0)^2 + (y_1 - y_0)^2 > a^2$$

In Figure 9.15 PT is one of the tangents from P to the circle $x^2 + y^2 + 2gx + 2fy + c = 0$. From Pythagoras' theorem we see that the length of the tangent PT from P to the circle is given by $(PT)^2 = (PC)^2 - (CT)^2$.

Since C is the point $(-g, -f)$ then

$$(PC)^2 = (x_0 + g)^2 + (y_0 + f)^2.$$

And CT is a radius of the circle, so

$$(CT)^2 = g^2 + f^2 - c.$$

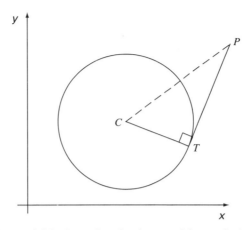

Figure 9.15   Length of a tangent to a circle

Therefore

$$(PT)^2 = x_0^2 + 2x_0 g + g^2 + y_0^2 + 2y_0 f + f^2 - g^2 - f^2 + c$$
$$= x_0^2 + y_0^2 + 2gx_0 + 2fy_0 + c.$$

This expression is the left-hand side of the equation of the circle with x replaced by x_0 and y replaced by y_0.

Parametric representation of a point on a circle

In Figure 9.16(a) the point P lies on the circle $x^2 + y^2 = a^2$. The line OP makes an angle θ with the positive x-axis. When P is at the position A this angle is zero, and θ increases as P moves anticlockwise round the circle from A. When $\theta = 360°$ the point P has returned to A.

Each point on its journey corresponds to a different value of θ and each value of θ in the range $0° \leq \theta < 360°$ corresponds to a different point on the circle. The angle θ is called a **parameter**.

From Figure 9.16(b) we can see that $OM = x = a\cos\theta$ and $PM = y = a\sin\theta$. The coordinates of P can be written $(a\cos\theta, a\sin\theta)$.

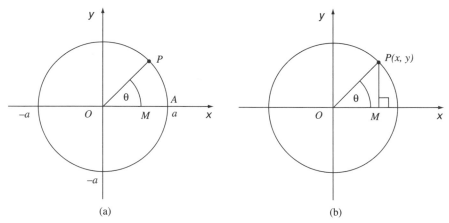

Figure 9.16 Parametric representation of a point on a circle

Note that

$$x^2 + y^2 = a^2 \cos^2 \theta + a^2 \sin^2 \theta$$
$$= a^2 (\cos^2 \theta + \sin^2 \theta)$$
$$= a^2 \times 1$$

which shows that the point whose coordinates are $(a \cos \theta, a \sin \theta)$ lies on the circle $x^2 + y^2 = a^2$.

Exercise 9.3

1 The line $y = x + c$ touches the circle $x^2 + y^2 = 9$ in the second quadrant. Determine c. Show that there is another choice for c in the fourth quadrant.

2* Of the two circles which are drawn in the diagram, one has radius 13 with centre (9, 12) and the other has radius 10 with centre $(-4, -6)$.

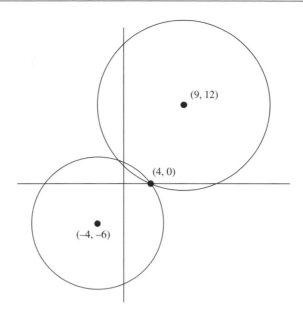

(a)   Write down the equations of the circles in expanded form.

(b)   Subtract one from the other to confirm that the coordinates of the points of intersection satisfy the equation $13x + 18y = 52$.

(c)   Substitute this relationship back into one of the equations to prove that the points of intersection are (4, 0) and (−0.8195, 3.4807).

(d)   Find the area common to both circles.

3 Determine the coordinates of the points of intersection of the straight line $y = \dfrac{1}{3}(x + 7)$ and the circle $3x^2 + 3y^2 - 18x - 20y + 57 = 0$.

4 Determine the equations of the circles which pass through the following sets of points:

(a)   (0, 0), (3, 1) and (3, 9)   (b)   (−2, 2), (2, 4) and (5, −5)

5 Show that the equation of the locus of a point which moves so that the sum of the squares of its distances from the points (1, 0) and (−1, 0) is constant and equal to 18 is $x^2 + y^2 = 8$.

6 The general equation of a circle is given by

$$x^2 + y^2 + 2gx + 2fy + c = 0$$

In terms of g, f and c find the radius and the coordinates of the centre of the circle.

7   Show that the equation for the locus of a point which moves such that its distance from the point $(-2, 0)$ is twice its distance from the point $(0, 1)$ is a circle. Where is the centre and what is the radius?

8   (a)   On the circle $x^2 + y^2 = 1$ mark the points

$$P_1\left(\cos\frac{\pi}{3}, \sin\frac{\pi}{3}\right), \ P_2(\cos\pi, \sin\pi), \ P_3\left(\cos\frac{5\pi}{2}, \sin\frac{5\pi}{2}\right)$$

(b)   Find (i) the Cartesian form and (ii) the parametric form for the circle of radius $\sqrt{2}$ centred at $(1, 1)$.

9*   By completing the square in x and y determine the coordinates of the centre and the radius of the following circles:

(a) $x^2 + y^2 + 4x - 6y + 5 = 0$ (b) $x^2 + y^2 - 5x + 9y - 3 = 0$

Interpret the meaning of the following 'circle equations':

(c) $x^2 + y^2 + 4x + 2y + 5 = 0$ (d) $x^2 + y^2 - 6x - 4y + 20 = 0$

10*   A circle has centre $(3, -2)$ and radius 5.

(a)   Determine the coordinates of the points where it crosses the x and y axes.

(b)   At what points does the line $y = x$ intersect the circle?

9.4 POLAR COORDINATES

In the Cartesian system of coordinates the position of a point lying on a plane is specified by its distance from two perpendicular lines in the plane, namely the x- and y-axes. In some situations an alternative system is more appropriate, especially when a key feature of the situation is the distance of an object from a fixed point.

In **polar coordinates** the fixed point is denoted by O, the **origin** of coordinates and the distance of a point P from O is denoted by r. Knowing that a point P, is a distance r from O does not fix its position; it merely confines P to a circle of radius r centred at the origin.

We specify a fixed ray (half-line) starting at the origin then measure the angle θ through which the ray must be rotated anticlockwise to lie along OP; see Figure 9.17(a). The numbers r and θ are the polar coordinates of P and P may be represented as (r, θ).

Figure 9.17(b) shows four points P_1 to P_4 and their polar coordinates. Note first that we have measured angles in degrees. We could have used radians instead; for example,

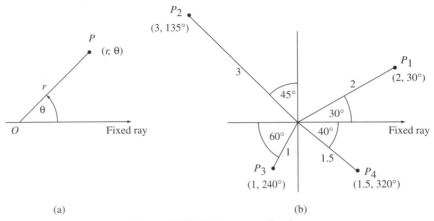

Figure 9.17   Polar coordinates

P_1 could be represented by the pair of coordinates $\left(2, \dfrac{\pi}{6}\right)$. Note also that if anticlockwise rotation is specified as positive then clockwise rotation is negative. As an example we could specify P_4 as $(1.5, -40°)$.

If the object rotates anticlockwise about O at a constant angular speed ω, measured in radians per second, then it will be at any particular position on several occasions. If the object is initially at P_0 then at any subsequent time t it will be at the point $P(a, \omega t)$; see Figure 9.18. If $\omega = \dfrac{\pi}{4}$ radians per second then the object is first at P_1 at $t = 1$ s. It returns to P_1 after 8 s, i.e. at $t = 9$ s; it is at P_1 again at $t = 17$ s, $t = 25$ s, and so on. According to the general formula at $t = 1$ s the position of P is $\left(a, \dfrac{\pi}{4}\right)$, at $t = 9$ s the position is $\left(a, \dfrac{9\pi}{4}\right)$ and at $t = 17$ s it is $\left(a, \dfrac{17\pi}{4}\right)$. All represent the same position, so the coordinate θ is not

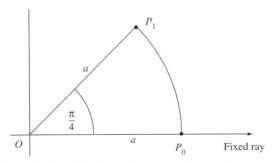

Figure 9.18   Rotation at constant angular speed

uniquely defined. Adding (or subtracting) a multiple of 2π radians (or $360°$) does not alter the position of a point.

To be specific, we limit θ to the range $-180° < \theta \le 180°$ (or $-\pi < \theta \le \pi$ in radians); sometimes it is limited to the range $0° \le \theta < 360°$ (or $0 \le \theta < 2\pi$).

The link between Cartesian and polar coordinates

By choosing the origin of polar coordinates to coincide with the origin of Cartesian coordinates, and choosing the fixed ray to be the positive part of the x-axis, we are able to relate the two coordinate systems to each other; see Figure 9.19.

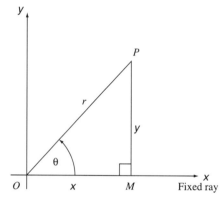

Figure 9.19 Polar and Cartesian coordinates

Using the fact that OPM is a right-angled triangle, it follows that

$$x = r\cos\theta, \qquad y = r\sin\theta. \tag{9.15}$$

Hence we can readily, and uniquely, convert from polar coordinates to Cartesian coordinates.

Example

P_1 is the point $(2, 30°)$, P_2 is the point $(3, 135°)$, P_3 is the point $(1, 240°)$. What are the Cartesian coordinates of these points?

Solution

For P_1 we have $r = 2$ and $\theta = 30°$, so

$$x = 2\cos 30° = 2 \times \frac{\sqrt{3}}{2} = \sqrt{3}, \qquad y = 2\sin 30° = 2 \times \frac{1}{2} = 1.$$

Hence P_1 is the point $(\sqrt{3}, 1)$ in Cartesian coordinates. The other two points can be obtained as follows:

$$P_2 \text{ is the point } (3\cos 135°, 3\sin 135°) = \left(3 \times \left(-\frac{1}{\sqrt{2}}\right), 3 \times \frac{1}{\sqrt{2}}\right)$$

$$= \left(-\frac{3\sqrt{2}}{2}, \frac{3\sqrt{2}}{2}\right)$$

$$P_3 \text{ is the point } (\cos 240°, \sin 240°) = \left(-\frac{1}{2}, \frac{\sqrt{3}}{2}\right). \qquad \blacksquare$$

Also from Figure 9.19 we see that

$$r^2 = x^2 + y^2, \qquad \tan\theta = y/x. \tag{9.16}$$

We can use these equations to convert from Cartesian coordinates to polar coordinates. There is a difficulty in identifying θ correctly as the following example shows.

Example   Find the polar coordinates of $P(2, 2)$ and $Q(-2, -2)$.

Solution

For $P, r^2 = 2^2 + 2^2 = 8 \qquad \therefore r = 2\sqrt{2}$

Also, $\tan\theta = \dfrac{2}{2} = 1 \qquad \therefore \theta = 45°.$

Hence P is $(2\sqrt{2}, 45°)$ in polar form.

For $Q, r^2 = (-2)^2 + (-2)^2 = 8 \qquad \therefore r = 2\sqrt{2}$

Also, $\tan\theta = \dfrac{-2}{-2} = 1.$

A calculator would provide the answer $\theta = 45°$, which cannot be so; P and Q are different points and must have different polar representations. Reference to Figure 9.20(a) helps to resolve the dilemma.

It is clear from the diagram that Q has the polar form $(2\sqrt{2}, 225°)$.

The difficulty arises since $\tan 225° = 1 = \tan 45°$ and using INV TAN on a calculator gives an answer in the range $-90° < \theta < 90°$. $\qquad \blacksquare$

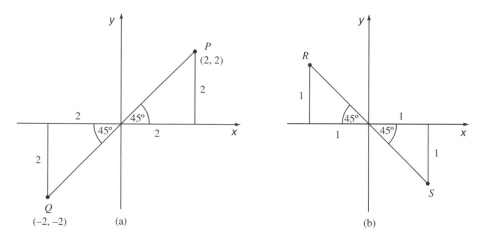

Figure 9.20 Converting from Cartesians to polars correctly

Figure 9.20(b) shows that in the Cartesian system $R = (-1, 1)$ and $S = (1, -1)$. Both R and S lead to

$$r^2 = (-1)^2 + (1)^2 = 2, \qquad \text{so } r = \sqrt{2}.$$

Calculators would give the result $\theta = -45°$, which identifies S. To get the correct angle for R we must add $180°$ to produce $135°$.

The lesson is clear. A sketch of where the point in question is positioned indicates whether the calculator value of θ is correct or whether we must add $180°$ to it.

Curves in polar form

We have already seen that the curve $r=$ constant is a circle centred at the origin. If we take the example $r = a$ then $r^2 = a^2$, and from (9.16) this becomes $x^2 + y^2 = a^2$, which is familiar to us from the previous section.

The form of a ray or half-line through the origin is $\theta =$ constant. In Figure 9.21 the solid half-line has equation $\theta = \dfrac{\pi}{4}$. It can also be specified as $y = x, x \geq 0$. The other ray, shown dashed, is $\theta = \dfrac{5\pi}{4}$ or $y = x, x \leq 0$.

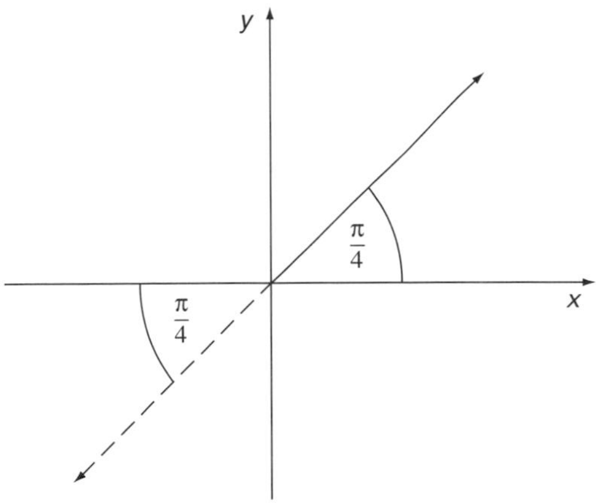

Figure 9.21 Two related rays

Examples 1. Write the equation $r = 4 \cos \theta$ in Cartesian form.
Multiplying both sides of the equation by r, we obtain

$$r^2 = 4r \cos \theta$$

and since $r^2 = x^2 + y^2$ we have

i.e. $\qquad x^2 + y^2 = 4x \quad \text{or} \quad x^2 - 4x + y^2 = 0$

We have to beware that this does not introduce points which were not present in the original equation. Multiplying by r allows the solution $r = 0$, i.e. $x = y = 0$, the origin. But when $\theta = 90°$ or $\dfrac{\pi}{2}$ radians $r = 0$, anyhow. No new points are therefore introduced.

Completing the square we see that

$$(x^2 - 4x + 4) + y^2 = 4$$
or $\qquad (x - 2)^2 + y^2 = 4$

which is a circle of radius 2 centred at the point (2, 0).

The circle is now easily drawn, but we could have sketched it directly from the polar form. Polar graph paper is designed to be radially symmetric as in Figure 9.22.

We measure the distance r along radial bearings. We choose a value of θ, work out r from the equation $r = 4 \cos \theta$ and plot a point distant r from the origin in the direction given by the angle θ. For example, $r = 4$ when $\theta = 0$.

Since $\cos(-\theta) = \cos \theta$ there is symmetry about the x-axis. Note too that $-\dfrac{\pi}{2} \leq \theta \leq \dfrac{\pi}{2}$ to keep $r \geq 0$.

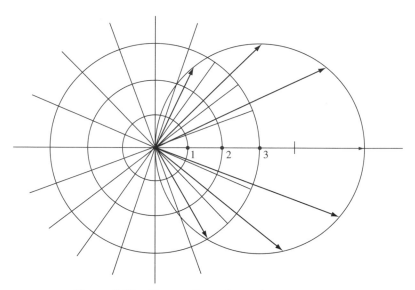

Figure 9.22   Generation of a polar curve

2. Plot the curve $r = a(1 + \cos\theta)$ where $a > 0$.

Taking $0 \leq \theta < 2\pi$ tabulate θ and $\dfrac{r}{a}$ for selection of values of θ. The polar graph is shown in Figure 9.23 and is known as a **cardioid** because it is heart-shaped.

θ (degree)	$\cos\theta$	r/a
0	1	2
30	0.87	1.87
45	0.71	1.71
60	0.5	1.5
90	0	1
120	−0.5	0.5
135	−0.71	0.29
150	−0.87	0.13
180	−1	0
210	−0.87	0.13
225	−0.71	0.29
240	−0.5	0.5
270	0	1
300	0.5	1.5
315	0.7	1.71
330	0.87	1.87
360	1	2

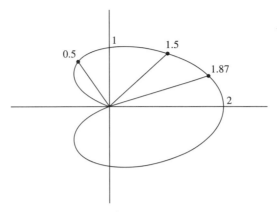

Figure 9.23 The cardioid $r = a(1 + \cos\theta)$ with $a > 0$: in this case the radial coordinate is r/a, which is equivalent to taking $a = 1$

Note that we took $0 \leq \theta < 2\pi$, starting on the positive x-axis and proceeding to the **cusp**; the cusp is the sharp corner on the curve when $\theta = \pi$. We then proceeded along the lower branch to $\theta = 2\pi$, where we started. Equally we could have taken $-\pi < \theta \leq \pi$ and plotted the curve anticlockwise from the cusp and back to the cusp. Note that the curve is symmetrical about the x-axis. This is always the case when θ appears as $\cos\theta$, because $\cos(-\theta) = \cos\theta$.

3. Plot the curve $r = 1 + 2\sin\theta$ for $0 \leq \theta < 2\pi$. First we tabulate values of θ and r; note that θ is in radians.

θ (degree)	$\sin\theta$	r
0	0	1
$\pi/6$	0.5	2
$\pi/4$	0.71	2.41
$\pi/3$	0.87	2.73
$\pi/2$	1	3
$2\pi/3$	0.87	2.73
$3\pi/4$	0.71	2.41
$5\pi/6$	0.5	2
π	0	1
$7\pi/6$	-0.5	0
$5\pi/4$	-0.71	-0.41
$4\pi/3$	-0.87	-0.73
$3\pi/2$	-1	-1
$5\pi/3$	-0.87	-0.73
$7\pi/4$	-0.71	-0.41
$11\pi/6$	-0.5	0
2π	0	1

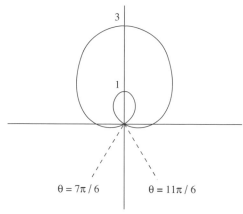

Figure 9.24 Polar graph for curve $r = 1 + 2\sin\theta, 0 \leq \theta < 2\pi$

We see that when $\theta = 7\pi/6$ the curve returns to $r = 0$ then r becomes negative. This is because the origin is crossed and negative values are essentially in the vertically opposite quadrant. Another change of sign takes place when $\theta = 11\pi/6$ with the curve returning to $r = 1$ on the x-axis when $\theta = 2\pi$. For larger values of θ, or negative θ, it merely repeats itself. Refer to Figure 9.24.

4. Plot the spiral $r = \theta/2\pi$, $\theta > 0$ as far as $\theta = 6\pi$. Here we see that the range of θ is not restricted and the spiral expands outwards indefinitely. Usually, however, restrictions do apply, especially when trigonometric, i.e. periodic, functions are involved; see Figure 9.25. ■

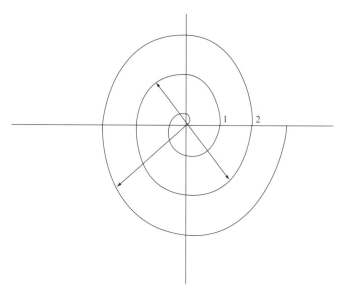

Figure 9.25 Polar graph for the spiral $r = \theta/2\pi, 0 < \theta \leq 6\pi$

Exercise 9.4

1 Determine the Cartesian equations of the following polar loci:

(a) $r = -4\cos\theta$ (θ second and third quadrants)

(b) $r = 2\sin\theta$

Sketch the curve in each case.

2 Determine the polar equation of the straight line $x = c$.

3 A straight line is a distance d from the origin at its closest point. Which of the following polar forms is correct?

(a) $d = r\cos(\alpha + \theta)$ (b) $d = r\cos(\alpha - \theta)$

(c) $d = r\sin(\alpha + \theta)$ (d) $d = r\sin(\alpha - \theta)$

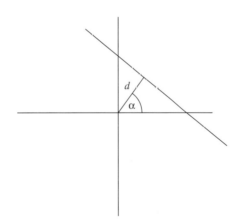

4 Determine polar forms for the following Cartesian equations:

(a) $x^2 + y^2 = 10y$ (b) $x^2 + y^2 = (x^2 + y^2 - y)^2$

(c) $3x = y(x^2 + y^2)^{1/2}$ (d) $x^2 + y^2 = (x - 1)^2 + (y - 1)^2$

In cases (a) and (d) interpret fully and draw sketches.

5

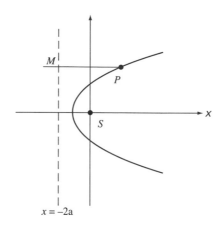

$x = -2a$

A parabola with focus at the origin has directrix $x = -2a$.

(a) Use the focus-directrix property, $SP = ePM$, to prove that the polar equation of the parabola is

$$\frac{2a}{r} = 1 - \cos\theta.$$

(b) Where is the vertex, in Cartesian coordinates?

(c) Prove that the Cartesian equation of the parabola is $y^2 = 4a(x + a)$.

6* Sketch the curve

$$r = 1 + c\cos\theta$$

where c is a constant. Consider the cases $c = 0.8$, $c = 1.2$. In what sense is $c = 1$ critical? What happens with $c = -1$?

7* Sketch the inward spiral

$$r = \frac{4\pi}{\theta + 2\pi}, \theta > 0.$$

How does this compare with the curve in Question 6?

8* Sketch the following curves:

(a) $r = \dfrac{10 \sin \theta}{\theta}, \theta > 0$; assume that $r = 10$ when $\theta = 0$

(b) $r = 6 \cos \theta + 8 \sin \theta$.

You may find that part (b) is sketched more easily by first putting it into Cartesian coordinates.

9.5 THREE-DIMENSIONAL GEOMETRY

In a plane (two dimensions) we need two independent coordinates to specify the position of a point; in three-dimensional space we need three such quantities. The rectangular Cartesian coordinate system can be extended for this purpose. Figure 9.26 shows a Cartesian coordinate system in three dimensions.

The point N is vertically below the point P. The x and y axes are, as usual, perpendicular to each other and intersect at the origin O. The third axis, the z-axis, is drawn to be perpendicular to the other two axes. The direction of positive z is chosen so that the whole system is 'right-handed'. A simple test for this is, if the first finger of your right hand points in the direction of positive x and the second finger points in the direction of positive y, then the thumb points in the direction of positive z.

The x and y axes lie in a plane called the x–y plane, the y and z axes lie in the y–z plane and the x and z axes lie in the x–z plane. These three planes divide three-dimensional space into eight parts known as **octants**. One such octant is where x, y and z are all ≥ 0.

The position of a point is specified by three distances, as shown in Figure 9.26. Each of the three coordinates can be positive or negative, giving eight possible combinations, one for each octant.

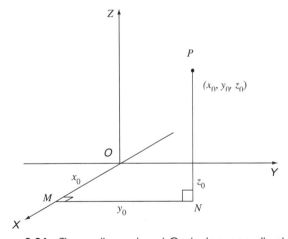

Figure 9.26 Three-dimensional Cartesian coordinates

Distance between two points

The distance between two points is found by applying Pythagoras' theorem twice in succession; refer to Figure 9.27. PM is parallel to the x-axis, MN is parallel to the y-axis and NQ is parallel to the z-axis.

Then

$$(PQ)^2 = (PN)^2 + (QN)^2$$
$$= (PM)^2 + (MN)^2 + (QN)^2$$

by applying Pythagoras' theorem first to the right-angled triangle PNQ and then to the right-angled triangle PMN.

Now M is the point (x_2, y_1, z_1) and N is the point (x_2, y_2, z_1). Hence the distance $PM = x_2 - x_1$, the distance $MN = y_2 - y_1$ and the distance $QN = z_2 - z_1$. We have the following result:

The distance d between the two points (x_1, y_1, z_1) and (x_2, y_2, z_2) is given by

$$d^2 = (x_2 - x_1)^2 + (y_2 - y_1)^2 + (z_2 - z_1)^2 \qquad (9.17)$$

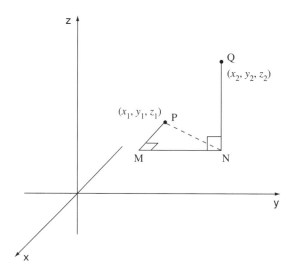

Figure 9.27 Distance between two points

Example   Which two of the points $P(2, 3, 4)$, $Q(0, -1, 0)$, $R(3, 3, -1)$, $S(4, 1, 2)$ are (i) furthest apart and (ii) closest together? (Note that $\therefore$ means 'therefore').

Solution

$(PQ)^2 = (0 - 2)^2 + (-1 - 3)^2 + (0 - 4)^2 = 4 + 16 + 16 = 36$ $\therefore PQ = 6$

$(PR)^2 = (3 - 2)^2 + (3 - 3)^2 + (-1 - 4)^2 = 1 + 0 + 25 = 26$ $\therefore PR = \sqrt{26}$

$(PS)^2 = (4 - 2)^2 + (1 - 3)^2 + (2 - 4)^2 = 4 + 4 + 4 = 12$ $\therefore PS = 2\sqrt{3}$

$(QR)^2 = (3 - 0)^2 + (3 - (-1))^2 + (-1 - 0)^2 = 9 + 16 + 1 = 26$ $\therefore QR = \sqrt{26}$

$(QS)^2 = (4 - 0)^2 + (1 - (-1))^2 + (2 - 0)^2 = 16 + 4 + 4 = 24$ $\therefore QS = 2\sqrt{6}$

$(RS)^2 = (4 - 3)^2 + (1 - 3)^2 + (2 - (-1)^2) = 1 + 4 + 9 = 14$ $\therefore RS = \sqrt{14}$

Hence P and Q are furthest apart, and P and S are closest together.   ■

Equation of a line in two dimensions (2D)

Suppose that we wish to find the equation of the straight line which passes through the points $Q(2, 1)$ and $R(4, 4)$. Referring to Figure 9.28, we can calculate the gradient of the line to be $\dfrac{RN}{NQ} = \dfrac{3}{2}$.

Let $P(x, y)$ be a point on this line, then the ratio $\dfrac{PM}{MQ} = \dfrac{3}{2}$ wherever P is placed. As we move from Q to P the distance travelled parallel to the y-axis is $\binom{3}{2}$ times the distance

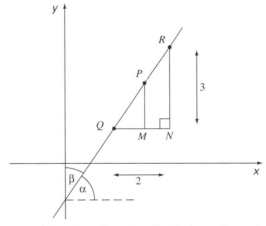

Figure 9.28 Equation of a line in two dimensions

travelled parallel to the x-axis. We could therefore write the coordinates of P as

$$x = 2 + u, \qquad y = 1 + \frac{3}{2}u$$

where u is a parameter.

Alternatively, to avoid fractions, we can write

$$x = 2 + 2t, \qquad y = 1 + 3t \tag{9.18}$$

where t is also a parameter, equal to $\frac{1}{2}u$. (We use Q as our starting-point but we could have used R instead.)

Each value of t corresponds to a unique point on the line. At Q we have $t = 0$ and at R we have $t = 1$. When $t = -\frac{1}{2}$ the point is $\left(1, -\frac{1}{2}\right)$ and when $t = 3$ the point is $(8, 10)$,

and so on. Points where t is negative lie to the 'south-west' of Q, points where $0 < t < 1$ lie between Q and R and points where $t > 1$ lie to the 'north-east' of R.

We can write equations (9.18) in the form

$$\frac{x - 2}{2} = \frac{y - 1}{3} = t. \tag{9.19}$$

Although this is not an elegant form, it will be useful for the case of lines in three dimensions. Starting with $\dfrac{x - 2}{2} = \dfrac{y - 1}{3}$, we multiply by 6 to obtain

$$3x - 6 = 2y - 2$$

or
$$y = \frac{3}{2}x - 2$$

which is a more familiar form of the equation of a straight line.

The ratio 2:3 formed from the denominators of the fractions in equation (9.19) is called the **direction ratio** for the line.

The angles α and β in Figure 9.28 are the angles made with the positive x-axis and the positive y-axis respectively. The values of $\cos \alpha$ and $\cos \beta$ are called the **direction cosines** of the line.

Since $R\hat{Q}N = \alpha$ then $\cos \alpha = \cos R\hat{Q}N = \dfrac{2}{\sqrt{5}}$ and since $\beta = 90° - \alpha$ then $\cos \beta = \sin \alpha = \sin R\hat{Q}N = \dfrac{3}{\sqrt{5}}$. Hence $\cos \alpha : \cos \beta = 2 : 3$, the direction of the line, and

$$\cos^2 \alpha + \cos^2 \beta = \cos^2 \alpha + \sin^2 \alpha = 1.$$

Equation of a line in three dimensions (3D)

Suppose we wish to find the equation of the straight line joining the points $Q(2, 1, 2)$ and $R(4, 4, 3)$ in Figure 9.29. OS is a straight line segment of unit length parallel to QR.

Moving from Q to R, the x-coordinate is increased by 2, the y-coordinate by 3 and the z-coordinate by 1. By analogy with the two-dimensional case the direction ratios of the line are 2:3:1. (Note we use the plural 'ratios'.)

If $P(x, y, z)$ is any point on the line then $x = 2 + 2t, y = 1 + 3t, z = 3 + t$. These equations can be rewritten as

$$\frac{x-2}{2} = \frac{y-1}{3} = \frac{z-3}{1} \tag{9.20}$$

since each fraction is equal to t.

Note that the direction ratios arc usually expressed in terms of integers. However, equations (9.20) can also be written as

$$\frac{x-2}{2} = \frac{y-1}{3d} = \frac{z-3}{d}$$

for any value $d \neq 0$.

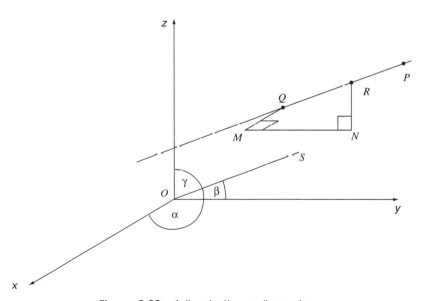

Figure 9.29 A line in three dimensions

The general result follows. The equations of the line which passes through the point (x_0, y_0, z_0) and which has direction ratios $l : m : n$ is

$$\frac{x - a}{l} = \frac{y - b}{m} = \frac{z - c}{n}.$$

(9.21)

Example

Find the equations of the straight line which passes through the points $(3, -1, 2)$ and $(-2, 3, 0)$.

Let Q be the point $(3, -1, 2)$ and R the point $(-2, 3, 0)$. Then in moving from Q to R the x-coordinate decreases by 5 (increases by -5), the y-coordinate increases by 4 and the z-coordinate increases by -2. The direction ratios are therefore $-5:4: -2$ and the equation of the line is

$$\frac{x - 3}{-5} = \frac{y - (-1)}{4} = \frac{z - 2}{-2}$$

i.e.
$$\frac{x - 3}{-5} = \frac{y + 1}{4} = \frac{z - 2}{-2}. \qquad \blacksquare$$

Note that if we use R as our 'base point' the equations become

$$\frac{x + 2}{-5} = \frac{y - 3}{4} = \frac{z}{-2}.$$

It is straightforward to show this version is equivalent to the equations we derived earlier.

In Figure 9.29 α, β, and γ are the respective angles made with the positive x, y and z axes by the line segment OS and therefore by the line through P and Q to which it is parallel.

The coordinates of S are $(\cos \alpha, \cos \beta, \cos \gamma)$. To see this, draw a perpendicular from S to each axis in turn and remember that $OS = 1$. Therefore the direction cosines of the straight line are $\cos \alpha : \cos \beta : \cos \gamma$ and the direction ratios are $l : m : n$, where $l : m : n = \cos \alpha : \cos \beta : \cos \gamma$.

By a double application of Pythagoras' theorem

$$(OS)^2 = \cos^2 \alpha + \cos^2 \beta + \cos^2 \gamma$$

We know that $OS = 1$, so the direction cosines satisfy the equation

$$\cos^2 \alpha + \cos^2 \beta + \cos^2 \gamma = 1$$

(9.22)

Example   Find the angles made with the positive coordinate axes by the line

$$\frac{x+1}{2} = \frac{y-1}{1} = \frac{z-2}{-2}.$$

The direction ratios of the line are $l : m : n = 2 : 1 : -2$.

Since $l^2 + m^2 + n^2 = 4 + 1 + 4 = 9 = 3^2$ we have the result that

$$\cos\alpha = \frac{2}{3}, \qquad \cos\beta = \frac{1}{3}, \qquad \cos\gamma = \frac{-2}{3}$$

so that $\cos\alpha : \cos\beta : \cos\gamma = l : m : n$ and equation (9.22) is satisfied.
Therefore, $\alpha = 48.2°$, $\beta = 70.5°$, $\gamma = 131.8°$, all results quoted to 3 s.f.   ■

In two dimensions it is the exception for two lines not to meet. In three dimensions the opposite is true. Two lines in a plane will not meet if they are parallel, and if no plane can be found which contains the two lines then the lines are **skew** and skew lines will not meet. Skew lines cannot be parallel.

As a simple example consider Figure 9.30. The line QR lies in the y–z plane. All points on it have z-coordinate 2 and x-coordinate 0. The line ST is the line $x + y = 1, z = 0$ and lies in the x–y plane. The lines are not parallel since there is no plane which contains them both and they clearly do not intersect. They are skew lines.

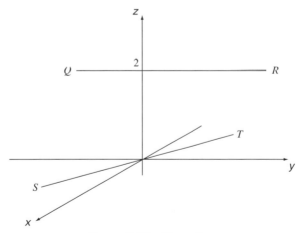

Figure 9.30 Skew lines

Example   Determine whether the lines

$$\frac{x-1}{2} = \frac{y+1}{1} = \frac{z}{-2} \quad \text{and} \quad \frac{x-1}{-3} = \frac{y-2}{0} = \frac{z-2}{4}$$

intersect and if so find the coordinates of the point of intersection. (The significance of the zero denominator for the second line is that $y = 2$ at all points on it. The line therefore lies on the plane $y = 2$, which is parallel to the x–z plane.)

Any point on the first line has coordinates given by

$$x = 1 + 2t, \qquad y = -1 + t, \qquad z = -2t \quad \text{for some } t.$$

Any point on the second line has coordinates given by

$$x = 1 - 3u, \qquad y = 2, \qquad z = 2 + 4u \quad \text{for some } u.$$

We try to find a value of t and a value of u which give the same point in each case.

Looking at the y-coordinates we must choose $t = 3$ for a match. Then for the x-coordinates $1 + 2t = 1 - 3u$, i.e. $7 = 1 - 3u$, so that $u = -2$. The real test is whether these values given the same z-coordinate. Now $-2t = -6$ and $2 + 4u = -6$. Hence the lines meet at the point $(7, 2, -6)$. ∎

The equation of a plane

We state without proof that the equation of a plane in three dimensions is

$$ax + by + cz = d \tag{9.23}$$

where a, b, c and d are constants. A geometrical interpretation of a, b and c is provided in Chapter 4 of *Mathematics in Engineering and Science*.

The origin lies on the plane if and only if $d = 0$. As special cases note

(i) $a \neq 0, b = c = 0$ gives $x = 0$, i.e. the y–z plane

(ii) $b \neq 0, a = c = 0$ gives $y = 0$, i.e. the x–z plane

(iii) $c \neq 0, a = b = 0$ gives $z = 0$, i.e. the x–y plane.

Note for instance that $x = $ constant is a plane parallel to the y–z plane. For example, the plane $x = 4$ cuts the x-axis at the point $(4, 0, 0)$; this is equation (9.23) with $a = 1, b = 0, c = 0, d = 4$.

Two planes which are not parallel meet in a line. A simple analogy is two consecutive pages of this book meeting in the line of the binding. As an example, the x–y plane meets the x–z plane at the y-axis. Since the x–z plane has equation $y = 0$ and the x–y plane has the equation $z = 0$, they meet where both y and z are zero.

The following examples illustrate some applications for the equation of a plane.

Examples

1. Find the equation of the plane which contains the points (2, 0, 0), (1, 1, 0) and (−2, 1, 1).

   If the equation of the plane at $ax + by + cz = d$ then

$$(2, 0, 0) \text{ lies on the plane so that } 2a = d \tag{1}$$
$$(1, 1, 0) \text{ lies on the plane so that } a + b = d \tag{2}$$
$$(-2, 2, 1) \text{ lies on the plane so that } -2a + 2b + c = d. \tag{3}$$

   From (1) and (2) $a = b - \dfrac{1}{2}d$ and from (3) $-d + d + c = d$, so $c = d$.

   Hence the equation of the plane is $\dfrac{1}{2}dx + \dfrac{1}{2}dy + dz = d$, or cancelling d, $x + y + 2z = 2$.

2. Find the equation of the plane which is parallel to the plane $x + 2y - 3z = 4$ and which passes through the point (4, −2, 1).

   The equation of the plane is of the form $x + 2y - 3z = d$. Since (4, −2, 1) lies on the plane

$$1 \times 4 + 2(-2) - 3(1) = d$$

   i.e.
$$d = -3.$$

   Hence the plane has equation

$$x + 2y - 3z = -3.$$

3. Where do the planes $x + y - 3z = -2$ and $2x - 3y - z = 6$ meet?

   The intersection of each plane with the plane $z = 0$ is a line; these lines meet where $x + y = 2$ and $2x - 3y = 6$, i.e. $x = 0, y = -2$, which is the point (0, −2, 0). Each plane meets the plane $z = 1$ in a line and these lines meet where

$$x + y - 3 = -2 \text{ i.e. } x + y = 1$$

   and
$$2x - 3y - 1 = 6 \text{ i.e. } 2 - 3y = 7.$$

   The point of intersection is where $x = 2, y = -1$ and $z = 1$, i.e. the point (2, −1, 1). The line of intersection of the two planes is the line joining (0, -2, 0) and (2, −1, 1). This has direction ratios (2, 1, 1) and hence equations

$$\frac{x}{2} = \frac{y + 2}{1} = \frac{z}{1}.$$

4. Find where the line $\dfrac{x+1}{1} = \dfrac{y}{1} = \dfrac{z-3}{1}$ meets the plane $3x - 4y - 2z = 15$.

Any point on the line has coordinates (x, y, z) where $x = -1 + t, y = t, z = 3 + t$ for some value of t. Substituting these expressions into the equation of the plane we obtain

$$3(-1 + t) - 4(t) - 2(3 + t) = 15$$

i.e.
$$-3t - 9 = 15$$

$$\therefore \quad t = -8.$$

Hence the coordinates of the point of intersection are

$$x = -1 - 8 = -9, \qquad y = -8, \qquad z = 3 - 8 = -5.$$

The point is $(-9, -8, -5)$. ∎

One final note. Many problems involving lines and planes in three dimensions are more easily handled by the use of vectors. See Chapter 4 of *Mathematics in Engineering and Science*.

Exercise 9.5

1 (a)   Write down an expression for the distance between the points whose coordinates   are (a, b, c) and (p, q, r).

(b)   How far is the point $(2, 1, 2)$ from the origin?

2 (a)   Depict the following points relative to a three-dimensional set of axes: $P_1(2, 3, 5)$, $P_2(-2, 1, 4)$, $P_3(3, -2, -4)$.

(b)   Determine the distances between each pair of them.

3 (a)   Depict a line in the direction $(-1, 1, -1)$ which passes through the point $P(1, 2, 1)$.

(b)   Write down the equation of the line.

(c)   Show that the line passes through the y-axis. Where does it intersect the plane $y = 0$?

4   What is the equation of the line passing through the points $P_1(1, 2, 3)$ and $P_2(-1, 1, 5)$? What are the coordinates of the point where the line intersects the horizontal plane $z = 1$?

5   The plane

$$\frac{x}{a}+\frac{y}{b}+\frac{z}{c}=1$$

intersects the coordinate axes at the points $(a, 0, 0)$, $(0, b, 0)$ and $(0, 0, c)$ respectively. Write down the intersection points of the following planes with the coordinate axes:

(a) $\dfrac{x}{5}+\dfrac{y}{6}+\dfrac{z}{7}=1$

(b) $3x + 5y + 9z = 45$

(c) $2x - 3y + z + 6 = 0$

(d) $3x - 7y = 42.$

Explain the result in case (d).

6   Solve the equations for the following pair of planes to determine the equation of their common line

$$x + 2y + 3z = 6$$
$$x - y + 2z = 2.$$

Where does this line intersect the plane $y = 0$?

7   A curve in space has the equation $y = x^2$, $z = 0$. What kind of curve is it? Draw a sketch.

8*   A triangle is formed by the origin and the points $P_1(x_1, y_1, z_1)$ and $P_2(x_2, y_2, z_2)$.

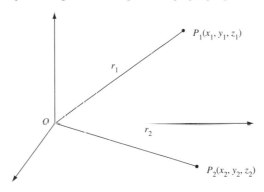

(a)   Write down the expressions for the distances r_1 and r_2.

(b)   Obtain an expression for the distance P_1P_2.

(c)   Use the cosine formula to prove that

$$\cos P_1\hat{O}P_2 = \frac{x_1x_2 + y_1y_2 + z_1z_2}{r_1r_2}$$

(d)   Find the cosines of the angles $P_1\hat{O}P_2$ in the cases

(i) $P_1 = (3, 1, 2), P_2 = (2, 2, 1)$

(ii) $P_1 = (4, -3, 0), P_2 = (3, -3, 4)$

(e)   What is the area of each of the triangles P_1OP_2?

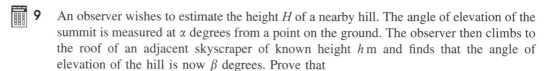

9   An observer wishes to estimate the height H of a nearby hill. The angle of elevation of the summit is measured at α degrees from a point on the ground. The observer then climbs to the roof of an adjacent skyscraper of known height h m and finds that the angle of elevation of the hill is now β degrees. Prove that

$$H = \frac{h \sin \alpha \cos \beta}{\sin \alpha \cos \beta - \cos \alpha \sin \beta}.$$

If $h = 100$ m, $\alpha = 10°$ and $\beta = 9°$ estimate H. How far away is the bottom of the hill? Give your answer in kilometres.

10* A large building of height h is observed across a lake from A and B, two points on the opposite shore at the same horizontal level, a distance d apart. Given the angles shown in the diagram, prove that

$$h = \frac{d \sin \alpha \sin \beta}{\sin(\beta + \gamma)}.$$

If $\alpha = 7°$, $\beta = \gamma = 80°$ and $d = 400$ m, determine h. How far away is the building from A?

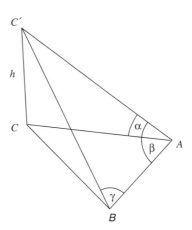

11 Determine the equation of the plane which passes through the points $(0, 0, 1)$, $(11, -2, 3)$ and $(5, -6, 7)$.

 12 Draw the section of the plane $x + y + z = 1$ for $x, y, z \geq 0$.

    (a)   Determine the angle of minimum slope.

    (b)   Determine the equation of the line lying in the plane which passes through the z-axis and the plane $z = 0$ in the direction of maximum slope.

13 Determine the volume of the tetrahedron formed by points $(0, 0, 0)$, $(-3, 0, 0)$, $(0, 6, 0)$ and $(-2, 4, 3)$.

14 A plane containing the point $(1, 2, 1)$ has a normal in the direction $(2, -1, 2)$. Determine its equation.

 15* A rectangular building with a symmetric pitched roof has coordinates as illustrated, where the numbers represent distances in metres. The line AB forms the ridge of the roof and the building is symmetrical. Determine the maximum angle of slope for the sides of the roof and the maximum angle of slope for the ends of the roof.

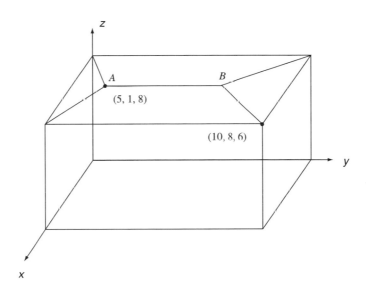

SUMMARY

- **The distance between the points** (x_1, y_1) and (x_2, y_2) is

$$\sqrt{(x_1 - x_2)^2 + (y_1 - y_2)^2}$$

- **To divide the line segment joining** (x_1, y_1) and (x_2, y_2) in the ratio $m : n$ find the point

$$\left(\frac{mx_2 + nx_1}{m + n}, \frac{my_2 + ny_1}{m + n} \right)$$

- **The angle** θ **between the lines** with gradients m_1 and m_2 is given by

$$\tan \theta = \frac{m_2 - m_1}{1 + m_2 m_1}$$

- **The distance of the point** (x_1, y_1) **from the line** $ax + by + c = 0$ is

$$d = \frac{|ax_1 + by_1 + c|}{\sqrt{a^2 + b^2}}$$

- **The area of the triangle** with vertices (x_1, y_1), (x_2, y_2) and (x_3, y_3) is

$$\frac{1}{2} \{ x_1(y_2 - y_3) + x_2(y_3 - y_1) + x_3(y_1 - y_2) \}$$

- **A locus** is a set of points which satisfy a given condition. For example, the set of points that are equidistant from points P and Q form the perpendicular bisector of the line segment PQ.

- **The equation of a circle** centre $(-g, -f)$ and radius $(g^2 + f^2 - c)^{1/2}$ is

$$x^2 + y^2 + 2gx + 2fy + c = 0$$

- **Parabola**: the locus of points which are equidistant from a fixed point (the focus) and a given line (the directrix)

- **Parametric representation**: a curve in the x–y plane specified by a pair of equations expressing x and y in terms of a third variable, the parameter.

- **Polar coordinates**: a point can be represented as (r, θ) where $x = r \cos \theta, y = r \sin \theta$

- **Equation of a straight line**: the equation of a straight line in three dimensions which passes through the point (x_1, y_1, z_1) with direction ratios (a, b, c) is

$$\frac{x - x_1}{a} = \frac{y - y_1}{b} = \frac{z - z_1}{c}$$

- **Equation of a plane**: the general equation of a plane is

$$ax + by + cz = d$$

Answers

Exercise 9.1

1 $\{(a-c)^2 + (b-d)^2\}^{1/2}$

    (a) $\sqrt{74}$ (b) $\sqrt{50}$

    (c) $\sqrt{65}$; $(-5, -4)$ is furthest away

2 (a) $45°$ (b) $72.26°$ (c) $68.09°$

3 (a, b) $\dfrac{p}{a} = \dfrac{b}{(a^2 + b^2)^{1/2}}$

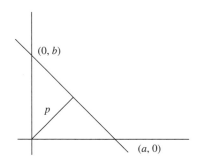

    (c) $p = \dfrac{|ab|}{(a^2 + b^2)^{1/2}}$, otherwise no change

    (d)   (i) $6/\sqrt{13}$ (ii) 0.48

4 (b) $y = ax/b$

5 (a) $\dfrac{n}{(l^2 + m^2)^{1/2}}$ (b) $\dfrac{3}{2\sqrt{10}}$

6 $(7/3, 2/3)$

7 (a) $y = \dfrac{1}{3}(x + 1)$ (b) $y = 1$

(c)

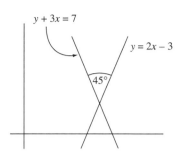

8 $y = x + 5 - 4\sqrt{2}$ (i.e. $y = x - 0.657$)

9 The locus is a straight line passing perpendicular to the line segment joining the two given points and passing through its midpoint.

10 An approximate solution can be obtained by (a) a classical approach or (b) a coordinate approach.

(a)

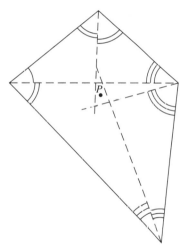

The bisectors of the angles are not concurrent as they are within a triangle but enclose a smaller quadrilateral. Choose P inside this smaller quadrilateral, and so on. The result applies to a *scalene* quadrilateral. Had we started with a square, the bisectors would have been concurrent at the centre.

(b)

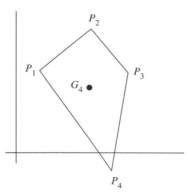

G_4 is the geometric centroid whose coordinates are

$$\left(\frac{(x_1 + x_2 + x_3 + x_4)}{4}, \frac{y_1 + y_2 + y_3 + y_4}{4}\right)$$

A cardboard cut-out of the quadrilateral would balance on G_4 and is usually very close to the true point. However, if the quadrilateral has a reflex angle, G_4 may lie outside it, whereas the true point is obviously inside, as shown in the diagram.

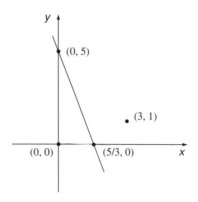

Exercise 9.2

1   $y = -3x + 5$

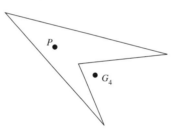

2 $(a-c)x + (b-d)y = \dfrac{1}{2}\{(a^2 - c^2) + (b^2 - d^2)\}$ Roles of a with c and b with d can be reversed.

3 (a) $12x - 2y - 27$ (b) $x + 11y + 14 = 0$

4 $(2, 3)$ and $(2, -3)$

5 Lines $y = x, y = -x$

6 $y = \dfrac{x}{7}, 7x + y = 10$

8 (a) $x^2 + y^2 - 4x - 2y - 5 = 0$ (b) $x^2 + y^2 + 10x - 8y + 5 = 0$

9 $y = 1 + \dfrac{x^2}{4}$

10 $4x^2 + 3y^2 + 6y = 9$

Exercise 9.3

1 $c = \pm 3\sqrt{2}$ (two values)

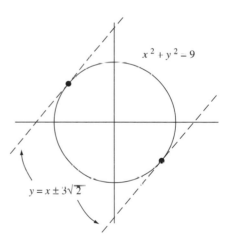

2 (a) (i) $x^2 + y^2 - 18x - 24y + 56 = 0$

       (ii) $x^2 + y^2 + 8x + 12y - 48 = 0$

   (d) 3.168 square units

3 $\left(4, \dfrac{11}{3}\right)$, $(2, 3)$

4 (a) $x^2 + y^2 - 10y = 0$ (b) $x^2 + y^2 - 4x + 2y - 20 = 0$

6 Centre $(-g, -f)$ radius $(g^2 + f^2 - c)^{1/2}$

7 Centre $\left(\dfrac{2}{3}, \dfrac{4}{3}\right)$ radius $= \dfrac{2}{3}\sqrt{5}$

8 (a)

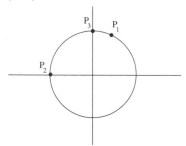

       (b)   (i) $x^2 + y^2 = 2(x + y)$

               (ii) $x = 1 + \sqrt{2}\cos t, y = 1 + \sqrt{2}\sin t$

9 (a)   Centre $(-2, 3)$, radius $2\sqrt{2}$

       (b)   Centre $\left(\dfrac{5}{2}, -\dfrac{9}{2}\right)$, radius $\dfrac{\sqrt{46}}{2}$

       (c)   Point $(-2, -1)$, hence $r = 0$

       (d) $(x - 3)^2 + (y - 2)^2 + 7 > 7$ for all x, y 'radius' negative, circle cannot exist

10 (a) $(3 + \sqrt{21}, 0), (3 - \sqrt{21}, 0); (0, 2), (0, -6)$

       (b) $(-2, -2), (3, 3)$

Exercise 9.4

1 (a) $x^2 + y^2 + 4x = 0$ (b) $x^2 + y^2 = 2y$

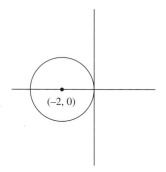

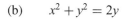

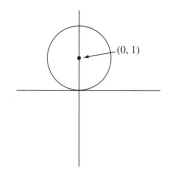

Both are circles, centres $(-2, 0)$ and $(0, 1)$ respectively

2 $r\cos\theta = c$

3 (b)

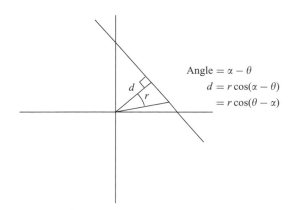

4 (a)

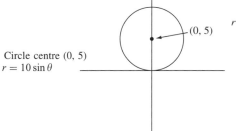 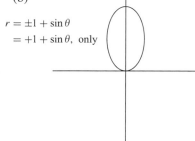

(a)   (b)

(c) $r = 3/\tan\theta$

(d) $r = \dfrac{1}{\cos\theta + \sin\theta} = \dfrac{1}{\sqrt{2}\cos(\theta - \pi/4)}$

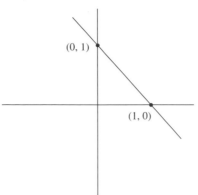

Line passing through $(0, 1)$ and $(1, 0)$

5 (b) $(-a, 0)$

6

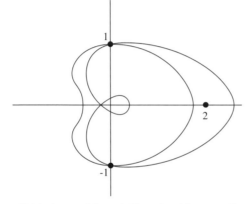

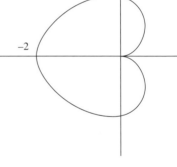

Origin is passed through if $c > 1$ and is a cusp if
$c = 1$

Cardioid reflected in y-axis

7

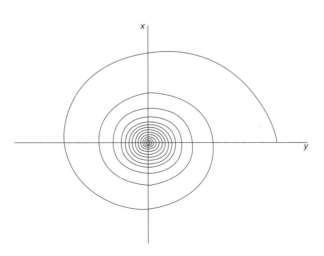

8 (a)

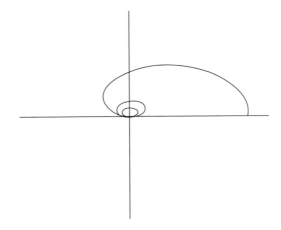

Inward spiral in quadrants I and II only. Note that $r < 0$ for $\pi < \theta < 2\pi$, then $r > 0$ again for $2\pi < \theta < 3\pi$, and $r < 0$ for $3\pi < \theta < 4\pi$, etc. The curve thus remains in quadrants I and II.

(b)

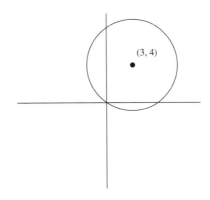

Circle centre (3, 4), radius 5, passing through the origin

Exercise 9.5

1 (a) $\{(a - p)^2 + (b - q)^2 + (c - r)^2\}^{1/2}$

(b) 3 units

2   (a)

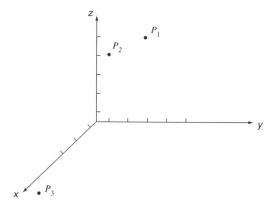

(b) $P_1P_2 = \sqrt{21}$ $P_2P_3 = \sqrt{98}$ $P_1P_3 = \sqrt{107}$

3   (a)

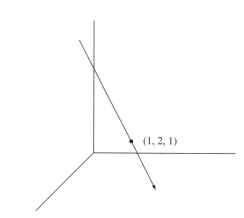

(b) $\dfrac{x-1}{-1} = \dfrac{y-2}{1} = \dfrac{z-1}{-1}$

(c)   Line crosses y-axis at $(0, 3, 0)$ and intersects plane $y = 0$ at $(3, 0, 3)$.

4 $\dfrac{x-1}{2} = \dfrac{y-2}{1} = \dfrac{z-3}{-2},$ $(3, 3, 1)$

5   (a) $(5, 0, 0), (0, 6, 0), (0, 0, 7)$ (b) $(15, 0, 0), (0, 9, 0), (0, 0, 5)$
      (c) $(-3, 0, 0), (0, 2, 0), (0, 0, -6)$ (d) $(14, 0, 0), (0, -6, 0)$ only

The plane in (d) is vertical.

6 $\dfrac{x+6}{7} = \dfrac{y}{1} = \dfrac{z-4}{-3};$ $(-6, 0, 4)$

7

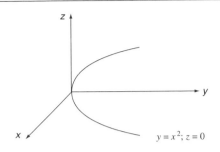

Parabola in plane $z = 0$

$y = x^2; z = 0$

8 (a) $r_1 = (x_1^2 + y_1^2 + z_1^2)^{1/2}, r_2 = (x_2^2 + y_2^2 + z_2^2)^{1/2}$

(b) $P_1 P_2 = \{x_1 - x_2)^2 + (y_1 - y_2)^2 + (z_1 - z_2)^2\}^{1/2}$

(d) (i) $\dfrac{10}{3\sqrt{14}}$ (iii) $\dfrac{21}{5\sqrt{34}}$

(e) (i) $\triangle = \dfrac{1}{2}\sqrt{26}$ (ii) $\triangle = \dfrac{1}{2}\sqrt{409}$

9 983 m 5.58 km

10 $h = 140.4$ m 1143 m from A

11 $y + z = 1$

12

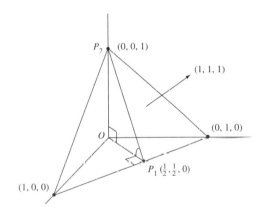

(a) $\tan O\hat{P}_1 P_2 = \sqrt{2}; O\hat{P}_1 P_2 = 54.74°$

(b) $1 - 2x = 1 - 2y = z$, i.e. $P_1 P_2$

13 9

14 $2x - y + 2z = 2$

15 Sides: $\tan^{-1}(2/5) = 21.8°$
Ends $\tan^{-1} 2 = 63.4°$

10 FUNCTIONS

INTRODUCTION

Functions are the verbs of mathematics. A function is a special kind of relationship between two variable quantities, which can be regarded as input and output variables. It is important to understand the operation of a function in its general sense so that specific functions can be handled properly. The exponential function can be used to model the spread of disease and the falling temperature of a substance as it cools down. Its inverse, the logarithmic function, can be applied to convert an exponential model into a straight line model.

OBJECTIVES

After working through this chapter you should be able to

- define a function
- understand the terms *domain* and *range*
- understand how a translation can alter a functional description
- understand how a reflection in the x axis or the y axis can alter a functional description
- understand how a scaling transformation can alter a functional description
- interpret the action of one function followed by another
- find the inverse of a given function both algebraically and graphically
- determine the domain and range of the inverse of a given function
- define and use the exponential function
- solve problems of exponential decay and exponential growth
- define and apply the logarithmic function
- state the domain and range of the inverse trigonometric functions

10.1 IDEAS, DESCRIPTIONS AND GRAPHS

Functions have been called 'the verbs of mathematics'. We have already referred to functions on several occasions and you are probably familiar with function keys on your calculator. In this section we put the ideas on a formal footing.

(a)   (b)

Figure 10.1   The function 'double'

A function is a rule by which an input is converted to a unique output. For example, the function 'double' converts any numerical input to a unique numerical output. The function can be represented in several ways. One way is shown in Figure 10.1(a), where the function is pictured as a 'black box'.

We often write the general input as x and the output as $2x$, as in Figure 10.1(b). We can regard the function as a mapping converting x to $2x$. In this approach we write

$$f : x \mapsto 2x$$

where the function has been named f. We read the statement as 'f maps x to $2x$'. For example, $f(2) = 4$; we say that 4 is the **image** of 2 under the mapping f. An alternative approach is to write

$$f(x) = 2x$$

which we read as 'f of x equals $2x$'; x is called the **argument** of the function.

It sometimes helps to present the function pictorially. For this purpose we write

$$y = 2x$$

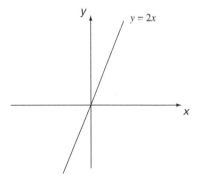

Figure 10.2   Graph of the function 'double'

then we can draw a graph of the relationship between y and x, as in Figure 10.2. In this context x is called the **independent variable** and y the **dependent variable**, where the nature of the dependency is provided by the function. To reinforce these ideas we examine two further examples.

Examples

1. The function 'cube' converts a number x uniquely to its cube. The number 3 is always converted to 27, the number -2 to -8, and so on. We can represent the function as the black box of Figure 10.3(a), as the mapping $f : x \mapsto x^3$ or as the relationship $f(x) = x^3$ or $y = x^3$.

Figure 10.3   Functions: (a) 'cube' and (b) 'double and add 3'

2. The function 'double and add 3' can be represented as $f : x \rightarrow 2x + 3$ or $f(x) = 2x + 3$ or $y = 2x + 3$. For example, $f(2) = 7$ and $f(-4) = -5$. Figure 10.4 shows the process in box form. ∎

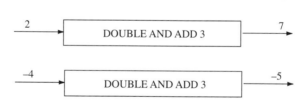

Figure 10.4   Applying the function 'double and add 3'

Note that x is merely a symbol to help illustrate the work of the function. In the last example we could write $f(u) = 2u + 3, f(a) = 2a + 3$, etc. It is usual at this level to use x as the independent variable.

In the three examples so far the function can receive any number as its input and every number can appear as an output. Consider now the function $f(x) = x^2$. It is clear that any number can be squared, but the output can never be a negative number. And the function $f(x) = \sqrt{x}$ (which corresponds to the $\sqrt{}$ key on your calculator) can only receive an input which is non-negative; the output also can never be negative.

A function can be thought of as a mapping between a set called the **domain** and a set called the **co-domain**. Each element of the domain is associated with *one* element of the co-domain.

If we consider the function $f : x \mapsto x^2$ as a mapping *from* the set of real numbers *to* the set of real numbers then we note that only *some* elements of the co-domain can result from the mapping, namely the non-negative elements. The subset of the co-domain which are the images under the mapping of elements in the domain is the **range** of the function.

In the case of the function $f(x) = 2x + 3$, which maps real numbers into real numbers, the range is the entire co-domain. Further examples follow.

Function	Domain	Range
x^3	All real numbers	All real numbers
x^2	All real numbers	All real numbers ≥ 0
$\sqrt{x}$	All real numbers ≥ 0	All real numbers ≥ 0
$\sin x$	All real numbers	All real numbers y satisfying $-1 \leq y \leq 1$
$\cos x$	All real numbers	All real numbers y satisfying $-1 \leq y \leq 1$
$\tan x$	All real numbers except odd multiples of $\dfrac{\pi}{2}$	All real numbers x
$\dfrac{1}{x}$	All real numbers except $x = 0$	All real numbers except 0
$\dfrac{1}{x^2}$	All real numbers except $x = 0$	All positive real numbers
$\lvert x \rvert$	All real numbers	All non-negative real numbers

The function $f(x) = \lvert x \rvert$ is the **absolute value** or **modulus** function defined by the statements

$$f(x) = \begin{cases} x & \text{if } x \geq 0 \\ -x & \text{if } x < 0 \end{cases}$$

Therefore, if $x = 3$ then $f(x) = 3$ and if $x = -2$ then $f(x) = 2$; in effect we ignore any minus sign. The graph of the function is shown in Figure 10.5; c is a positive value of x.

Note that some functions can be defined on different domains from those quoted in the table. For instance, if the function $f(n) = n^2$ is defined on the domain of integers then its co-domain is the non-negative integers, but only a subset of non-negative integers form the range, namely 0, 1, 4, 9, 16, etc.

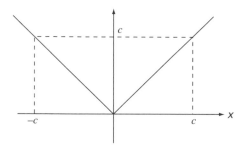

Figure 10.5   The function 'absolute value'

If we consider the function $f(x) = 2x$ then there is only one value of x for which $f(x) = 8$, namely $x = 4$. However, when we consider the function $f(x) = x^2$ there are *two* values of x for which $f(x) = 4$, namely $x = 2$ and $x = -2$. Indeed for any output $c > 0$ there are two values of x that are mapped into it; this causes difficulties when we come to 'undo' or invert a function, a process covered in Section 10.3.

Exercise 10.1

1   Using simple inequalities, e.g. $2 \leq x < 3$, write down the domain specified by the following:

(a)   All real numbers except $x = 0$ (b) $x \geq -1$ excluding $x = 0, \sqrt{2}$

(c) $|x| \geq 4$ (d) $\dfrac{1}{x + 5} > 5$.

2   Interpret the following functions (a) to (d) both in the form $f(x) = \ldots$ and as a mapping. In case (e) identify the quadratic function from the given inputs and outputs.

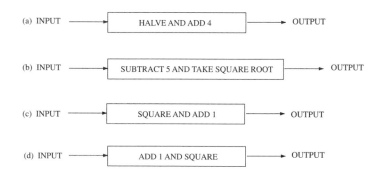

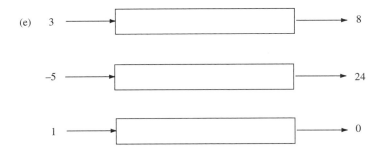

(e)   3 → [ ] → 8

  −5 → [ ] → 24

  1 → [ ] → 0

3   Identify the domain and range for each of the functions given in Question 2.

4   A function is defined as

$$f(x) = \begin{cases} x+2 & x \leq -2 \\ 0 & -2 < x < 1 \\ x-1 & x \geq 1 \end{cases}$$

Draw its graph. What are its domain and range?

5   Sketch the graph of $y = \dfrac{2+3x}{1+x}$.

Hence determine the domain and range of the function $f(x) = \dfrac{2+3x}{1+x}$.

6   (a)   On the same axes draw the graphs of $y = x$, $y = 3$ and $y - 10 \quad x$.

   (b)   The function $f(x) = \min(x, 3, 10 - x)$ is defined as the minimum of x, 3 and $10 - x$ for each x. For example, if $x = 4$ we look at x itself, which is 4, $10 - x = 10 - 4 = 6$, and 3. Out of 3, 4 and 6 the value 3 is the minimum so $f(x)$ takes the value 3, $f(4) = 3$. Highlight $f(x)$ on your graph.

   (c)   Given that the domain of x is the set of real numbers, what is the range?

   (d)   Rewrite $f(x)$ as

$$f(x) = \begin{cases} x & x \leq 3 \\ 3 & 3 < x < a \\ 10 - x & x \geq a \end{cases}$$

and determine a.

This is an example of a **piecewise function**, i.e. a function which has different definitions over different parts of its domain.

7 The function $f(x) = \max(1, 4 - x^2, x - 3)$ is defined over all real numbers to be the maximum of $1, 4 - x^2$ and $x - 3$.

(a)   Sketch the graph of $f(x)$.

(b)   Rewrite the function in the form of Question 6 part (d) by inserting the appropriate intervals in the expression below.

$$f(x) = \begin{cases} 1 \\ 4 - x^2 \\ x - 3 \end{cases}$$

8 From the sketches given below, identify the domain and range of the functions whose curves are drawn.

(a)

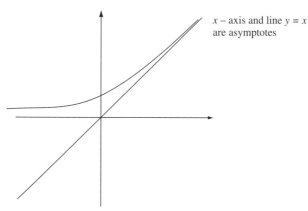

x – axis and line $y = x$ are asymptotes

(b)

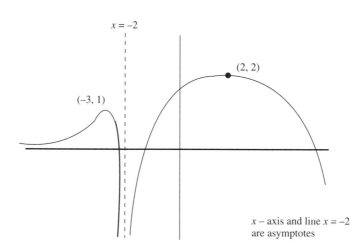

$x = -2$

$(2, 2)$

$(-3, 1)$

x – axis and line $x = -2$ are asymptotes

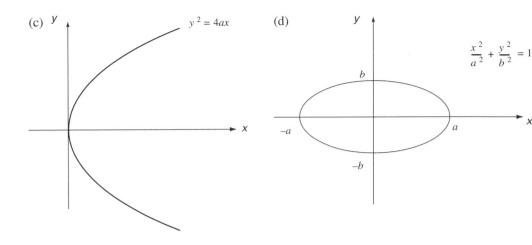

(c) $y^2 = 4ax$

(d) $\dfrac{x^2}{a^2} + \dfrac{y^2}{b^2} = 1$

9   Write down the domain and range of the following functions; write them in the form $f(x) = \cdots$.

(a) $f : x \longmapsto 2x^2$

(b) $f : x \longmapsto x^3$

(c) $f : x \longmapsto 1/x^3$

(d) $f : x \longmapsto 1/(x + 2)$

(e) $f : x \longmapsto \cos x,\ -\dfrac{\pi}{2} \le x \le \dfrac{\pi}{2}$

(f) $f : x \longmapsto \tan x,\ -\dfrac{\pi}{2} < x < \dfrac{\pi}{2}$

(g) $f : x \longmapsto |2x + 3|$

(h) $f : x \longmapsto \sin(1/x)$

10   In the following examples the ranges or domains are restricted; identify the corresponding domains or ranges for $f(x)$. In (f) consider that domain which lies between $-\pi/2$ and $\pi/2$ only.

	Function	Domain	Range
(a)	$x + 5$	$x \ge 0$	
(b)	$1/x$		$5 \le 1/x \le 20$
(c)	$x^2 + 1$	$1 < x < 1$	
(d)	$|2x + 3|$		$7 < |2x + 3| < 11$
(e)	$\sin x$	$\pi/2 \le x \le 3\pi/4$	
(f)	$\tan x$		$1/\sqrt{3} < \tan x < \sqrt{3}$

11   A **continuous function** is one whose graph has no breaks, i.e. no **discontinuities**.

(a)   Verify that the function below is continuous and sketch its graph.

$$f(x) = \begin{cases} x - 1 & x < 0 \\ x^2 - 1 & 0 \leq x \leq 2 \\ x + 1 & x > 2 \end{cases}$$

(b)   Where is the discontinuity for the function $f(x) = \dfrac{1}{x - 3}$?

(c)   Is $f : x \to |x|$ continuous?

(d)   Is the following function continuous? Sketch its graph.

$$\begin{aligned} f : x &\to x^2 + 1 & x < 0 \\ f : x &\to 0 & x = 0 \\ f : x &\to x^2 + 1 & x > 0 \end{aligned}$$

12* (a) Sketch the graph of $y = \sqrt{x - 1}, x \geq 1$.

(b) Now draw the graph of $y^2 = x - 1$. Does this define a function?

(c) From (b) express x as a function of y and write down its range and domain.

(d) Determine the domain and range of $f(x) = \sqrt{x - 1}$.

13* The domain of the function f is the set of all integers n where

$$f(n) = \frac{n + 1}{n^2 + 4n + 5}.$$

(a)   Write down $f(n)$ for $|n| \leq 4$.

(b)   Divide the top and bottom of the fraction by n to find out what happens when n is
(i) large and positive   (ii) large and negative

14   A train starts from rest at station A then increases its speed uniformly until a maximum speed is reached. It then cruises at constant speed until station B is approached, when the brakes are applied so that the speed is uniformly decreased and the train is brought to rest at B. Let t_0, t_1, t_2 and t_3 denote the departure time from A, the time when the maximum speed is reached, the time when the brakes are applied and the time of arrival at B. Then we can write the speed as a function

$$s(t) = \begin{cases} K_1(t - t_0) & t_0 \leq t < t_1 \\ K_2 & t_1 \leq t < t_2 \\ K_2 - K_3(t - t_2) & t_2 \leq t \leq t_3 \end{cases}$$

(a)   Show that if $s(t)$ has a continuous graph then $K_2 = K_1(t_1 - t_0) = K_3(t_3 - t_2)$.

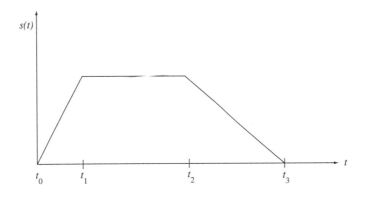

(b)   The train leaves A at 0910 and reaches its cruising speed of $180 \, \text{km hr}^{-1}$ or kph at 0913, which it maintains until 0945. It then decelerates to B, arriving at 0950.

(i)   Taking $t_0 = 0$ define s as a function of t.

(ii)   How far apart are A and B?

10.2 TRANSLATION, SCALING, FUNCTION OF A FUNCTION

Translation

We saw in Chapter 4 that it was possible to deduce the main features of the quadratic function $f(x) = ax^2 + bx + c$ by writing

$$ax^2 + bx + c \equiv a\left[\left(x + \frac{b}{2a}\right)^2 + \left(c - \frac{b^2}{4a}\right)\right]$$

then relating the graph of

$$y = a\left[\left(x + \frac{b}{2a}\right)^2 + \left(c - \frac{b^2}{4a}\right)\right]$$

to the graph of $y = x^2$.

The graph in diagram (a) of Figure 10.6 is the result of moving the graph of $y = x^2$ upwards (in the positive y-direction) by a distance 2 units. The graph in diagram (b) is the result of moving the graph of $y = x^2$ downwards (in the negative y-direction) by 1 unit.

In diagram (c) the graph of $y = x^2$ has been moved to the right (in the positive x-direction) by 1 unit; in diagram (d) the graph of $y = x^2$ has been moved to the left (in the negative x-direction) by 2 units.

To obtain the graph in diagram (e) we moved the graph of $y = x^2$ to the right by 1 unit and then upwards by 2 units. (We *could* have first moved the graph upwards by 2 units then to the right by 1 unit.)

These movements are known as **translations**. Note that we could achieve the result in diagram (e) by a single translation directly from the origin to the point (1, 2) but it would not have been easy to write down the new relationship between x and y.

Diagram (f) is the result of a **reflection** of the graph of $y = x^2$ in the x-axis followed by a translation of 3 units in the positive y-direction. The reflection gave the term in x^2 a negative coefficient.

Reflections and translations are examples of isometries. An **isometry** is a mathematical transformation which preserves distance.

Example

Sketch the graph of the function $f(x) = 4x - x^2$ by first writing the right-hand side in a form which indicates a reflection and two translations of the graph $y = x^2$.

First note that $4x - x^2 \equiv 4 - (x - 2)^2$. From x^2 we proceed to $(x - 2)^2$, which implies a translation to the right by 2 units; see Figure 10.7(a). Then we carry out a reflection in the x-axis to get $-(x - 2)^2$; see Figure 10.7(b). Finally we make a translation upwards by 4 units to obtain the required function in Figure 10.7(c). To carry out a reflection in the

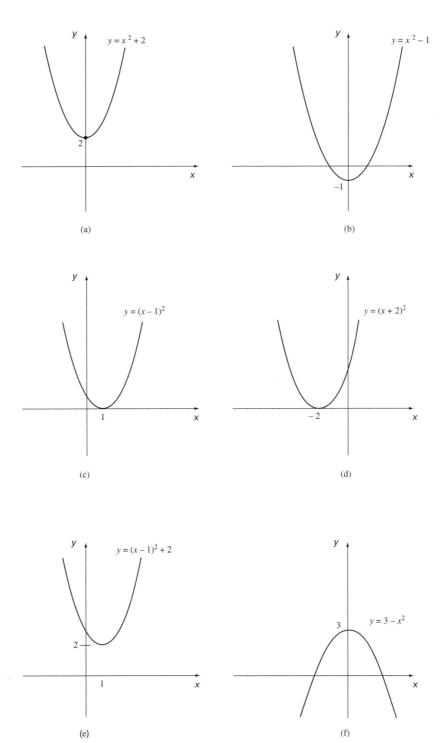

Figure 10.6 Translations of $y = x^2$

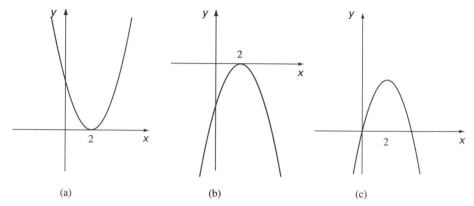

(a)   (b)   (c)

Figure 10.7 Creating a function using simple isometries: (a) begin with $y = (x - 2)^2$; (b) use a reflection to obtain $y = -(x - 2)^2$; (c) use a translation to obtain $y = -(x - 2)^2 + 4$, which simplifies to the required function, $y = 4x - x^2$

y-axis we change the sign of x as the first operation. For example, consider $f(x) = (x - 2)^2$; its graph is shown in Figure 10.8(a). If we replace x by $-x$ then $f(-x) = (-x - 2)^2 \equiv (x + 2)^2$, and its graph is shown in Figure 10.8(b). This second graph is the reflection in the y-axis of the first graph. ∎

We can generalise the results we have obtained so far by starting with the graph of an arbitrary function $y = f(x)$.

The graph of $y = f(x) + a$ is obtained from the graph of $y = f(x)$ by moving it a vertical distance a; if $a > 0$ then the distance is upwards, if $a < 0$ then the distance is downwards.

The graph of $y = f(x + a)$ is obtained from the graph of $y = f(x)$ by moving it a horizontal distance a; if $a > 0$ then the distance is to the left, if $a < 0$ then the distance is to the right.

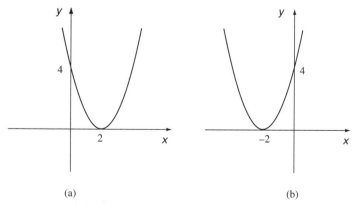

(a)   (b)

Figure 10.8 Reflection in the vertical axis

For a horizontal movement we modify x before applying the function f; for a vertical movement we apply f first and then modify the result. The vertical movement obeys the rule 'positive upwards' but the horizontal movement *contradicts* the rule 'positive to the right'.

Finally, the graph of $y = -f(x)$ is obtained from the graph of $y = f(x)$ by reflecting it in the x-axis; the graph of $y = f(-x)$ is obtained from the graph of $y = f(x)$ by reflecting it in the y-axis.

Scaling

Figure 10.9(a) shows the graph of the function defined by

$$f(x) = \begin{cases} 2, & -1 \le x \le 1 \\ 0, & \text{elsewhere} \end{cases}$$

Figure 10.9(b) shows the graph of $y = 1.5f(x)$ and Figure 10.9(c) shows the graph of $y = 0.5f(x)$. The scale on the vertical axis has, in effect, been changed.

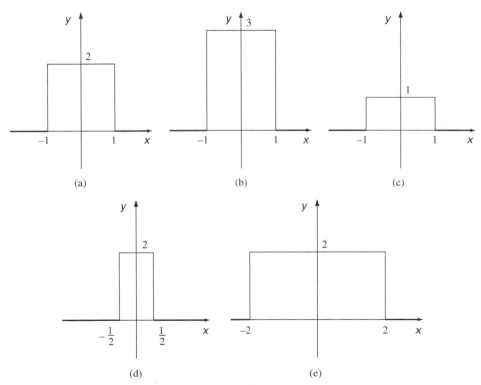

Figure 10.9 Scaling a function

Figure 10.9(d) shows the graph of $y = f(2x)$; to see this we replace x by $2x$ in the definition of f. Hence

$$f(2x) = \begin{cases} 2, & -1 \le 2x \le 1 \\ 0, & \text{elsewhere} \end{cases} \quad \text{i.e.} \quad -\frac{1}{2} \le x \le \frac{1}{2}.$$

Similarly, Figure 10.9(e) shows the graph of $y = f\left(\frac{1}{2}x\right)$.

In general, $y = af(x)$ indicates that the graph of $y = f(x)$ has undergone a change of vertical scale, stretching it by a factor $|a|$ if $|a| > 1$ and contracting it by a factor $|a|$ if $|a| < 1$. If $a < 0$ the graph is also reflected in the x-axis.

Similarly, $y = f(ax)$ indicates that the graph of $y = f(x)$ has undergone a change of horizontal scale, contracting it by a factor $|a|$ if $|a| > 1$ and stretching it by a factor $|a|$ if $|a| < 1$. If $a < 0$ the graph is reflected in the y-axis.

Note that to carry out a horizontal change we modify x before applying the function f; to carry out a vertical movement we apply f first and then modify the result.

The vertical effect is what we might expect, i.e. multiplying by 2 stretches the graph; the horizontal effect is the inverse of what we might expect, i.e. multiplying x by 2 *squashes* the graph.

Function of a function

In Section 10.1 we considered the function $f(x) = 2x + 3$ as a single operation on x. We could break this operation into two simpler ones, as in Figure 10.10(a).

Let f represent the mapping $f: x \mapsto 2x$ or $f(x) = 2x$ and g represent the mapping $g: x \mapsto x + 3$ or $g(x) = x + 3$. Then $g(f(x)) = g(2x) = 2x + 3$.

The order of applying the functions is important. Reversing the order gives

$$f(g(x)) = f(x + 3) = 2(x + 3)$$

which is depicted in Figure 10.10(b).

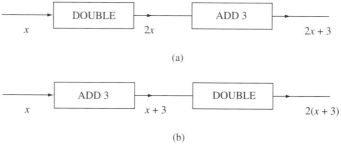

(a)

(b)

Figure 10.10 The composition of two functions depend on the order in which they are performed: (a) $2x + 3$ (b) $2(x + 3)$

Sometimes we write $g(f(x))$ as $g \circ f(x)$ and $f(g(x))$ as $f \circ g(x)$. The function $g \circ f$ is known as the **composition** of g with f. The term **function of a function** makes it more obvious what is being carried out.

We have to be careful when combining functions which have restrictions on their domain, and/or range,

Example   The functions $f(x) = \sqrt{x}$ and $g(x) = x - 2$ are given. Find $f(g(x))$ and $g(f(x))$ and state any restrictions on their domain and range.

f is the function 'take square root' and g is the function 'subtract 2'. Hence

$$f(g(x)) = \sqrt{(x - 2)} \quad \text{or } (x - 2)^{1/2}$$

$$g(f(x)) = \sqrt{x} - 2 \quad \text{or } x^{1/2} - 2.$$

Now the domain of f is the set of all real numbers ≥ 0 and its range is the set of all real numbers ≥ 0, whereas the set of all real numbers represents the domain and range of g.

In the case of $g(f(x))$, f requires an input ≥ 0 and g receives as input any real numbers which are ≥ -2. Hence the domain of $f(g(x))$ is all real numbers ≥ 0; its range is all real numbers ≥ -2.   ∎

Note that functions can be compounded more than once. For example, the function $f(x) = (x^2 + 1)^3$ can be regarded as the composition of the functions 'square', 'add 1' and 'cube'.

Exercise 10.2

1 A function $f(x)$ is defined for $x \geq 0$ as shown in the diagram. Dot in $f(x - 1)$ and $f(x + 1)$. How is the domain of x-values changed?

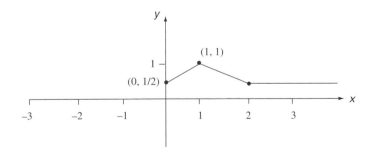

2   A function whose graph is the same after reflection in the y-axis is an **even** function; here is an example. Notice that $f(-x) = f(x)$ for every value of x, where the domain of x is $\mathbb{R}$.

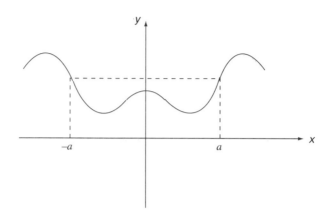

Likewise a function whose graph is an upside-down or inverse reflection in the y-axis, often with $f(0) = 0$, is an **odd** function; here is an example. Note $f(-x) = -f(x)$.

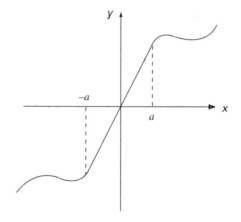

(a)   Determine whether the following are even or odd:
(i) x

(ii) x^2

(iii) x^3

(iv) x^n, n positive even integer

(v) x^n, n positive odd integer

(b)   Show that $\sin x$ is odd and $\cos x$ is even. What about $\tan x$?

(c)   Determine whether the following functions are even, odd or neither:

(i) $x^4 + x^2$ (ii) $|x|$ (iii) $x^2 + x$
(iv) $x \sin x$ (v) $|x + 3|$ (vi) $x \tan x$

3   If f is defined over $[-L, L]$ and g and h are defined to be

$$g(x) = \frac{1}{2}(f(x) + f(-x)), \qquad h(x) = \frac{1}{2}(f(x) - f(-x))$$

show that $g(x)$ is even, $h(x)$ is odd and that $h(0) = 0$.

4   Take the function $f(x)$ sketched in Question 1 and reflect it in the y-axis, i.e. extend the domain to include $x < 0$. Draw a new sketch.
      Now define

$$g(x) = \begin{cases} f(x) & x > 0 \\ f(0) = \dfrac{1}{2} & x = 0 \\ f(x) & x < 0 \end{cases}$$

(a)   What is the domain of g?

(b)   Sketch the graph of $g(x)$.

5   Take the function $f(x)$ from Question 1 and sketch

(a) $f(2x)$ (b) $f(x/2)$ (c) $f(-2x)$

6   By drawing sketches simplify the following as multiples of $\sin x$, $\cos x$ or $\tan x$:

(a) $\cos(x - \pi/2)$ (b) $\sin(x - \pi)$ (c) $\tan(x - \pi)$

(d) $\sin(x + 3\pi/2)$ (e) $\cos(x - 2\pi)$

7   On the same axes sketch the following functions:

(a) $\sin x$ (b) $\sin 2x$ (c) $\sin(x/3)$

(d) $\sin(x/3 - \pi/2)$

8*   Define $f(x)$ to be the function sketched in Question 1, i.e.

$$f(x) = \begin{cases} \dfrac{1}{2}(x + 1) & 0 \le x < \dfrac{1}{2} \\ 1 - \dfrac{x}{2} & \dfrac{1}{2} \le x < 1 \\ \dfrac{1}{2} & x \ge 1 \end{cases}$$

Identify the following graphs in the form $rf(px+q)+s$ in the best way possible

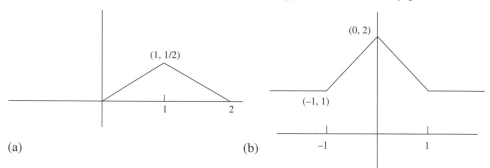

(a)

(b)

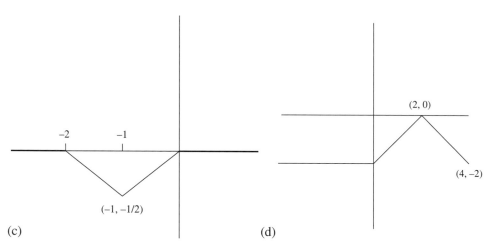

(c)

(d)

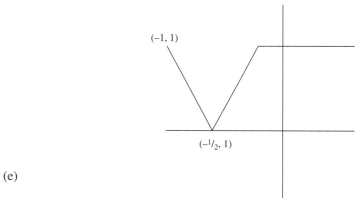

(e)

9 Draw the graph of $f(x) = \sin |x|$.

    (a)   Prove that $f(x)$ is even.   (b) Is $f(x)$ continuous for all $x \in \mathbb{R}$?

10* The function $f(x)$ has domain $\mathbb{R}$ and range $(0, M)$ where M is the maximum value of $f(x)$, and $f(x) \to 0$ as $x \to \pm\infty$.
Identify the domain and range of the following functions:

    (a) $f(2x)$ (b) $f(3x + 4)$ (c) $3f(x) + 4$ (d) $f(-x)$.

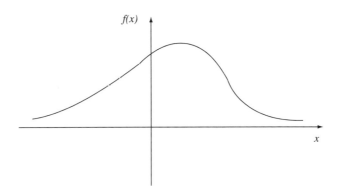

11 The function $f : x \mapsto 3x = 1$ maps the domain $[1, 2]$, i.e. $1 \le x \le 2$, into the range $[2, 5]$. Identify the domains and ranges of the following functions:

    (a) $f(2x + 1)$ (b) $2f(x) + 1$ (c) $3f(x - 5) + 2$

    (d) $f(x^2)$ (e) $f(2x^2 + 6x + 5)$ (f) $f(-x)$

    (g) $f(ax + b)$

    (h)   Were f to have domain $\mathbb{R}$, how would that alter the ranges in parts (a) to (f)?

12 For the function $f(x) = \dfrac{2}{x + 3}$ determine

    (a) $f(x^2)$ (b) $f(5x - 6)$

    (c) $f(3\sqrt{x}) + f(3/\sqrt{x})$ (d) $f(\sin x) + f(\cos x)$

In cases (c) and (d) place your answer over a compact common denominator.

13 Given that $3x^2 - 6x + 10 \equiv 3(x-3)^2 + 1$ and $f(x) = x^2$, identify which one of the following represents $3x^2 - 6x + 10$.

(a) $f(3(x-3)) + 1$ (b) $f(3(x-3)+1)$

(c) $3f(x-3) + 1$ (d) $3f(x^2) - 3f(x) + 1.$

14 Show that $ax^2 + bx + c \equiv a\left(x + \dfrac{b}{2a}\right)^2 + c - \dfrac{b^2}{4a}$. Given $f(x) = x^2$ show that any parabola

of the form $ax^2 + bx + c$ can be written as $rf(px + q) + s$ by identifying p, q, r and s in terms of a, b, c.

15 (a)   The general solution to the equation $\sin x = \dfrac{1}{2}$ is

$$x = (\pi/6) + 2n\pi, \qquad x = (5\pi/6) + 2n\pi, \qquad n \text{ integer.}$$

By setting $y = 2x - (\pi/2)$ determine the general solution to the equation

$$\sin(2x - \pi/2) = \dfrac{1}{2}.$$

(b)   Solve $\tan\left(3x + \dfrac{\pi}{4}\right) = 1$ given that the general solution to $\tan x = 1$ is

$$x = \left(\dfrac{\pi}{4}\right) + n\pi, \qquad n \text{ integer.}$$

10.3 INVERSE FUNCTIONS

The function $f(x) = 2x$ ('double') and $g(x) = \dfrac{1}{2}x$ ('halve') have a special relationship to each other:

$$f(g(x)) = 2 \times \dfrac{1}{2}x = x \quad \text{and} \quad g(f(x)) = \dfrac{1}{2} \times 2x = x.$$

In other words, the function 'undoes' the effect of the function g and vice versa. We say that f is the **inverse** of g and that g is the inverse of f.

Figure 10.11(a) represents the function f in box form. To undo the effect of f we can image pushing an input through the box from right to left; the box now becomes the inverse operation 'halve', as in Figure 10.11(b).

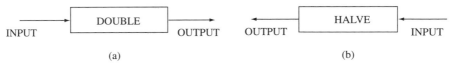

Figure 10.11 Inverting a function

Another way to obtain the formula for the inverse function is to take the relationship $y = 2x$ and rewrite it as $x = \frac{1}{2}y$, then interchange x and y to obtain $y = \frac{1}{2}x$.

Example Find the inverses of (a) $f(x) = x + 3$ and (b) $f(x) = x^3$.

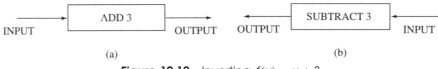

Figure 10.12 Inverting $f(x) = x + 3$

(a) If we write $y = x + 3$ then $x = y - 3$. Interchanging x and y we obtain $y = x - 3$. Hence the inverse function is $g(x) = x - 3$; see Figure 10.12.

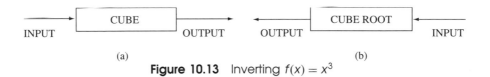

Figure 10.13 Inverting $f(x) = x^3$

(b) If $y = x^3$ then $x = y^{1/3}$. The inverse function is $g(x) = x^{1/3}$; see Figure 10.13. ■

The inverse function of $f(x)$ is written $f^{-1}(x)$. In the example above $f^{-1}(x) = x - 3$ in part (a) and $f^{-1}(x) = x^{1/3}$ in part (b).

There is a way of obtaining the graph of the inverse function if we already have the graph of the function itself. We simply reflect the graph of $y = f(x)$ in the line $y = x$. The reflection is the graph of $y = f^{-1}(x)$. Figure 10.14 shows the operation carried out for the functions of the example above.

Note that in diagram (a) the intercept of the inverse function on the x-axis is equal to the intercept of the original function on the y-axis. And the intercept of the inverse

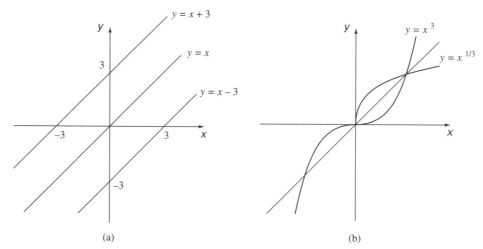

Figure 10.14 Graphical approach to inverting a function

function on the y-axis is equal to the intercept of the original function on the x-axis. In diagram (b) the graph of the function crosses the line $y = x$ and so does the graph of the inverse function, at the same points. These results are true in general.

You may have noticed that we have referred to *the* inverse function. This is because if an inverse function exists it is unique. But not all functions have an inverse.

Consider again the function $f(x) = x^3$, depicted in Figure 10.15(a). Suppose we know that $f(x) = 1.728$ and we want to know the value of x; there is a unique answer $x = 1.2$. What we have done is to apply $f^{-1}(x) = x^{1/3}$ so that $f^{-1}(1.728) = 1.2$.

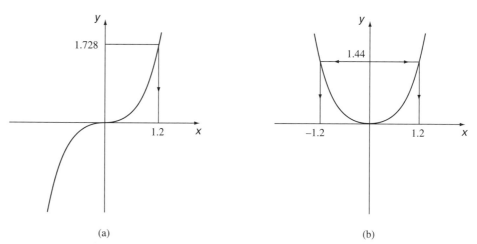

Figure 10.15 Not every inverse is a function: (a) $x = y^{1/3}$ is a function but (b) $x = \pm y^{1/2}$ is not

Figure 10.15(b) demonstrates the problem with $f(x) = x^2$. Suppose we know that $f(x) = 1.44$ there is no unique value of x. The point we made about a function at the beginning of the chapter is that a function maps each input into a *unique* output.

We could say that $x = \pm\sqrt{1.44}$, i.e. $x = \pm 1.2$ but this is not good enough. If we restrict the domain of $f(x) = x^2$ to the non-negative real numbers then we have effectively restricted the graph of $f(x)$ to the right-hand half in diagram (b). We are therefore able to obtain the positive square root only, which is what the $\sqrt{\ }$ button on your calculator will do. Figure 10.16 shows the effect of reflecting the graph of $y = x^2$ in the line $y = x$. In diagram (a) the domain of f is all real numbers, but in diagram (b) it is all non-negative real numbers.

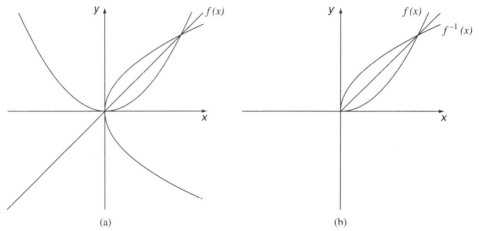

(a)   (b)

Figure 10.16   Inverting the function $f(x) = x^2$

The reflected graph in diagram (b) *is* a function but the rejected graph in diagram (a) is not, since all inputs $x > 0$ give *two* ouputs y. If the domain of $f(x) = x^2$ is $x \geq 0$ then the range is all $y \geq 0$. This is true of the inverse function $f^{-1}(x) = \sqrt{x}$.

Inverse of a function of a function

The inverse of a function of a function, if it exists, can be obtained rather like unwrapping a parcel with several layers of wrapping paper. The last layer of wrapping is taken off first and the first layer of wrapping is taken off last. The box representation of a function helps here.

Consider $f(x) = 2x + 3$ shown in Figure 10.17(a). If the input is put through the boxes right to left, each box now acts as the *inverse* of itself compared to when inputs went through left to right; see Figure 10.17(b). It suggests that the inverse function is $f^{-1}(x) = \dfrac{1}{2}(x - 3)$.

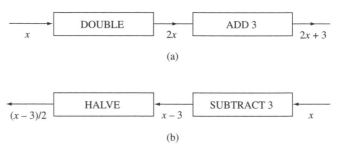

(a)

(b)

Figure 10.17 Inverting a function of a function

To check this write $y = 2x + 3$ then rearrange to get $x = \frac{1}{2}(y - 3)$. Interchanging x and y gives $y = \frac{1}{2}(x - 3)$. There are no restrictions on the domain and range of either function.

We can write the inverse of $f(g(x))$ as $g^{-1}(f^{-1}(x))$. In the alternative notation the inverse of $f \circ g$ is $g^{-1} \circ f^{-1}$. Care is needed when either f or g has a restriction on domain or range.

Example

Find the inverse function of $f(x) = (x + 1)^{1/2}$, restricting the domain of f as necessary. Rearranging $y = (x + 1)^{1/2}$ we get $x = y^2 - 1$ sign. Then $f^{-1}(x) = x^2 - 1$. This function has domain all real numbers and range real numbers ≥ -1. However, the function $f(x)$ has domain $x \geq -1$ and range $y \geq 0$, Figure 10.18(a). Figure 10.18(b) shows how the inverse function has domain $x \geq 0$ and range $y \geq -1$. ∎

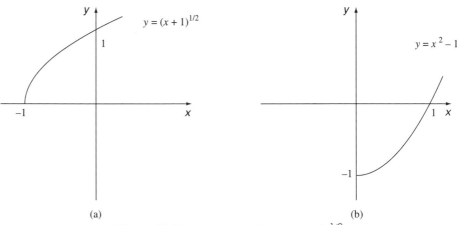

(a)   (b)

Figure 10.18 Inverse of $f(x) = (x + 1)^{1/2}$

Inverse trigonometric functions

Figure 10.19 shows the graph of $y = \sin x$ and its reflection in the line $y = x$. Notice that we must restrict the domain of $f(x) = \sin x$ if we are to produce an inverse function.

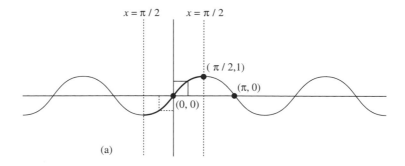

(a)

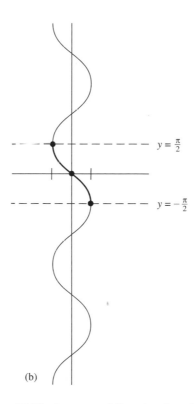

(b)

Figure 10.19 Inverse of the sine function

If we restrict the domain of $-\dfrac{\pi}{2} \leq x \leq \dfrac{\pi}{2}$ then $\sin x$ takes all values between -1 and 1 (its entire range) once only. The portion of the graph is highlighted in diagram (a). The corresponding portion of the reflected graph is highlighted in diagram (b).

Hence for $f(x) = \sin x$ the domain is $-\dfrac{\pi}{2} \leq x \leq \dfrac{\pi}{2}$ and the range is $-1 \leq y \leq 1$. For the inverse function the domain is $-1 \leq x \leq 1$ and the range is $-\dfrac{\pi}{2} \leq y \leq \dfrac{\pi}{2}$. This is consistent with pressing INV SIN on a calculator. The inverse function is written $\sin^{-1} x$ or $\arcsin x$. The value produced by the calculator is the **principal value** of $\sin^{-1} x$.

Exercise 10.3

1 By simply writing $y = f(x)$ and solving the equation for x in terms of y, determine inverse functions for the following:

(a) $f : x \mapsto x + 1$ (b) $f : x \mapsto x^5$ (c) $f : x \mapsto \dfrac{1}{1 - x}$

(d) $f : x \mapsto x^3 - 1$ (e) $f : x \mapsto \sqrt{x - 1}$

(f) $f : x \mapsto ax + b \; (a \neq 0)$ (g) $f : x \mapsto 1/x^3$

You may assume there are no additional restrictions on the domain for the above. Identify the domain and range in each case.

2 (a) What happens to part (f) in Question 1 if $a = 0$?

(b) For $f : x \mapsto \dfrac{ax + b}{cx + d}$ prove that f^{-1} exists provided that $ad \neq bc$. What happens if $ad = bc$?

3 Consider the function $f : x \mapsto x^2, x > 0$, i.e.

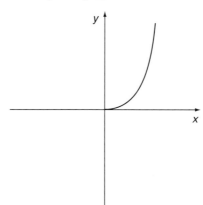

(a)   What is the range of f?

(b)   Determine f^{-1} by setting $y = x^2$ and solving for y in terms of x.

(c)   Define g: $x \mapsto x^2, x < 0$. Accepting that g^{-1} converts the range of g back into its domain, show that g^{-1}: $x \mapsto -\sqrt{x}$ inverts g. Put $y = -\sqrt{x}$ and proceed from there.

(d)   Given h: $x \mapsto x^2 + 2x + 2 = (x+1)^2 + 1$ show that h^{-1}: $x \mapsto \sqrt{x-1} - 1$ or $x \mapsto -\sqrt{x-1} - 1$, depending on the choices of domain for h. Identify the choices.

4* For the following quadratic functions solve the equation $y = f(x)$ for x in terms of y and use the information to define f^{-1} appropriately, giving domain and range.

(a) $f : x \mapsto x^2 + 2x + 5$ (b) $f : x \mapsto x^2 - 12x - 50$

(c) $f : x \mapsto 3x^2 + 2x - 11$.

Now try this for the general quadratic function

(d) $f : x \mapsto ax^2 + bx + c, (a \neq 0)$ (e)   Consider (d) for $a = 0$ when
                                                                    (i) $b \neq 0$ (ii) $b = 0$.

5* As we know, the graph of the inverse function is a reflection in the line $y = x$ of the graph of the original function.

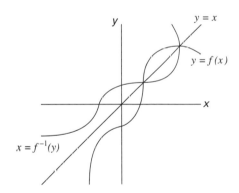

By laying a ruler perpendicular to $y = x$ draw graphs of the following functions, their inverses and the line $y = x$.

(a) $f : x \mapsto x^2 + 2x + 5$ (b) $f : x \mapsto \sin x$ (c) $f : x \mapsto \cos x$.

Note how the inverse functions on the graph represent the totality of $x = f^{-1}(y)$, giving the complete range of output x for input y, and note how complete consistency is preserved in that an input can have only one output. Alternatively if (x, y) is a point on the graph of $y = f(x)$, then (y, x) is a point on the graph of $x = f^{-1}(y)$. However, looking at the x-axis, notice there may be many values of y for a given x on the graph of f^{-1}. Perhaps this has highlighted the need for restrictions on ranges and domains where inverses are defined.

6 For the function $f : x \mapsto \tan^{-1} x$, written $y = \tan^{-1} x$, it is a small step to write $x = \tan y$. Use this fact to sketch the graph of f and identify a suitable domain for its principal value.

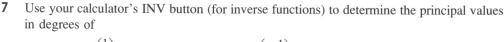 **7** Use your calculator's INV button (for inverse functions) to determine the principal values in degrees of

(a) $\quad \sin^{-1}\left(\dfrac{1}{2}\right)$ (b) $\quad \cos^{-1}\left(-\dfrac{1}{2}\right)$ (c) $\quad \tan^{-1}(100)$

(d) $\quad \sin^{-1}\left(\dfrac{\sqrt{2} - 1}{\sqrt{2} + 1}\right)$ (e) $\quad \tan^{-1}(2 + \sqrt{3})$.

8 An equation in two variables represents a function only when one of the variables is uniquely expressible in terms of the other, and the domain and range are specified. A non-functional equation is called an **implicit functional relationship**, e.g. $x^2 + y^2 = 1$, which represents a circle of radius 1.

(a) Show graphically that $x^2 + y^2 = 1$ is its own reflection in $y = x$.

(b) Write down functional equations for the arcs in the form $y = f(x)$. Define domain and range.

9* The equation of an ellipse is $\dfrac{x^2}{a^2} + \dfrac{y^2}{b^2} = 1$ where a and b are the lengths of the semi-major and semi-minor axes.

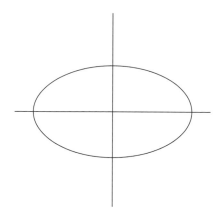

Notice there are two values of y for each value of x and vice versa, but only one of each value per quadrant. Define functions, one for each quadrant, which map x into y. In each case define the inverse function. Define the domain and ranges.

10* Repeat Question 9 for the hyperbola $\dfrac{x^2}{a^2} - \dfrac{y^2}{b^2} = 1, a > b$.

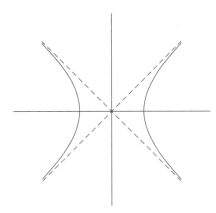

11 Let $f: x \mapsto 1/x, g: x \mapsto 1 - x, h: x \mapsto x^3$.

   (a)   With f, g and h chosen as above, write down or determine the following
       (i) $fg, gh, hf; fg$ or $f \circ g$ refers to $f(g(x))$, etc.
       (ii) f^{-1}, g^{-1}, h^{-1} and verify that $ff^{-1} = I$ (identify).

   (b)   Prove that $fg \neq gf$, etc. in general.

   (c)   Verify that $(fg)^{-1} = g^{-1}f^{-1}$, etc.

   (d)   What are $f(f(x)), g(g(x))$ and $h(h(x))$? Is there a pattern?

   (e)   Draw the graph of $f(x) + g(x)$ and from it demonstrate that $f + g$ is a well-defined function. What of $(f + g)^{-1}$?

   (f)   Write down $f(x)g(x), g(x)h(x)$ and $h(x)f(x)$, showing that all of them differ from the results in part (a).

12 A function satisfies $f(x + y) = f(x)f(y)$ for all real x, y.

   (a)   Verify that $f(2) = \{f(1)\}^2, f(3) = \{f(1)\}^3$.

   (b)   Suggest a form for $f(n)$ where n is a positive integer.

   (c)   Prove that $f(n) = 2^n$, where n is any integer, totally meets the requirements. Start by considering $n > 0$ then find $f(0)$ and finally prove for $n < 0$.

   (d)   By taking $x, y \in \mathbb{R}$ to be any number of your choice, on your calculator, verify that $f(x) = 2^x$ satisfies the provisions of f.

(e) Repeat the argument of (c) for $f(n) = a^n$ where a is another integer, e.g. $a = 3$. Will any integer do for a? What is $f(1)$?

(f) On your calculator experiment with generalising part (d) by finding a^x where a and x are any numbers, positive, negative or zero. What conclusions do you reach?

13* A function g satisfies the rule

$$g(xy) = g(x) + g(y).$$

Prove that

(a) $g(1) = 0$ (b) $g(\sqrt{x}) = \dfrac{1}{2}g(x)$ (assume $x > 0$)

(c) $g(1/x) = -g(x), x \neq 0$

(d) if $g(x_1) = y_1$ and $g(x_2) = y_2$ then $g(x_1 x_2) = y_1 + y_2$.

If f is the inverse function of g then $fg(x_1 x_2) = x_1 x_2 = f(y_1 + y_2)$. Prove that f must satisfy the condition in Question 12.

(e) Accepting that $f(x) = 2^x$ satisfies the provisions of f in Question 12, by setting $y = 2^x$ show that $x = \log_2 y$. Verify that $g : x \mapsto \log_2 x$ satisfies the provisions of g.

(f) Plot $y = 2^x, y = x$ and $y = \log_2 x$ on the same axes. (Note that $\log_2 1 = 0, \log_2 \left(\dfrac{1}{2}\right) = -1$, etc.).

14 The following functions are defined over the set $\mathbb{N}$ of positive integers. Where possible, determine the inverse function over $\mathbb{N}$ or state that no inverse exits.

(a) $f : n \mapsto n^2$

(b) $f : (2n - 1) \mapsto (2n - 1)^2, n$ odd; $f : (2n) \mapsto (2n)^3, n$ even

(c) $f : n \mapsto n, n$ not a perfect square; $f : n \mapsto \sqrt{n}, n$ a perfect square

(d) $f : (2n - 1) \mapsto 2n; f : (2n) \mapsto 2n - 1$

(e) $f : n \mapsto 1/n$ (f) $f : n \mapsto (n + 1)(n + 2)$

(g) $f : n \mapsto |2 - n|$.

15 By considering each section in turn, obtain an inverse function for the following continuous piecewise function:

$$f : x \mapsto 2x + 1 \qquad -\infty < x < 0$$
$$f : x \mapsto 1 + x^2 \qquad 0 \leq x \leq 3$$
$$f : x \mapsto 13 - x \qquad 3 < x < \infty$$

Draw the graph of $f^{-1}(x)$.

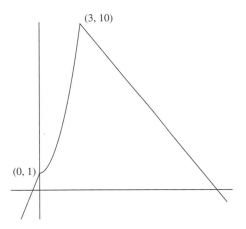

10.4 POWER LAWS, EXPONENTIAL AND LOGARITHMIC FUNCTIONS

Power laws

We know from Chapter 1 that the *numbers* $3^2, 3^3, 3^4$ are examples of powers of the number 3; so too are $3^0, 3^{-1}, 3^{-2}$.

The *function* $3^0, 3^{-1}, 3^{-2}$ is a **power function** with the number 3 as the base. The general form of a power function is $f(x) = a^x$. Figure 10.20 shows graphs of $y = f(x)$ for $a = 2, 3$ and $\left(\dfrac{1}{2}\right)$.

Can you see a connection between the cases $a = 2$ and $a = \dfrac{1}{2}$? When $a = \dfrac{1}{2}$, $f(x) = \left(\dfrac{1}{2}\right)^x = \dfrac{1}{2^x} = 2^{-x}$ so that if $f(x) = 2^x$, $\left(\dfrac{1}{2}\right)^x = f(-x)$ and the graph of

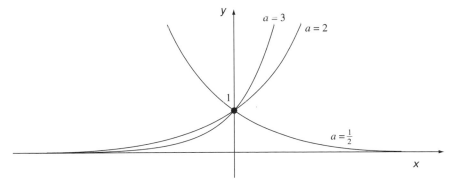

Figure 10.20   Power functions

$y = \left(\dfrac{1}{2}\right)^x$ is the reflection of the graph of $y = 2^x$ in the vertical axis. Note that all three curves pass through the point $(0, 1)$. This is true in general since $a^0 = 1$ (unless $a = 0$).

When $a > 1$ the values of $f(x)$ increase as x increases; this is an example of a **growth** function. When $0 < a < 1$ the values of $f(x)$ decrease as x increases; this is an example of a **decay** function. If $a = 1$ then $f(x) \equiv 1$.

Examples

1. The value of an antique increases by 5% each year. If its value in 1990 was £10 000 what was its value in 1995? What will be its value in 2000, 2010? If x denotes the number of years from 1990 then the value (in £) is given by

$$V = 10\,000 \times (1.05)^x$$

since a 5% increase means the value is multiplied by 1.05.

In 1995, $x = 5$ and $V = 10\,000 \times (1.05)^5 = 12\,763$ to the nearest integer, so the value to the nearest £ is £12 763.

In 2000, $x = 10$ and $V = 10\,000 \times (1.05)^{10}$. To the nearest £ the value is £16 289.

In 2010, $x = 20$ and the value is £26 533.

2. Each month an unused battery loses 10% of its charge. After how many months is the charge 1% of its original value?

Let the initial charge be C. Then after one month the charge is $0.9C$, after two months it is $(0.9)^2C$, after 3 months it is $(0.9)^3C$ and after x months it is $(0.9)^xC$ or $C(0.9)^x$.

When $(0.9)^xC = 0.01C$ then $(0.9)^x = 0.01$.

Taking logarithm to the base 10 we obtain

$$\log(0.9)^x = x\log(0.9) = \log(0.01)$$
$$x = \frac{\log(0.01)}{\log(0.9)} = 43.7.$$

After approximately 44 months the battery has retained only 1% of its original charge. ∎

The exponential function

An accurate plot of the graph of $y = 3^x$ was used to estimate the gradients of the tangents to the curve at values of $x = \dfrac{1}{2}, 1, 1\dfrac{1}{2}, 2$. Then the ratio of this gradient to the value of y was calculated in each case and the results displayed in a table. In fact, as we shall see in Chapter 11, the ratio can be found exactly by calculation. It is approximately 1.1.

x	Gradient of tangent	y	Ratio
$\dfrac{1}{2}$	1.9	1.732	1.1
1	3.3	3	1.1
$1\dfrac{1}{2}$	5.7	5.196	1.1
2	9.9	9	1.1

If we repeat the calculations for $y = 4^x$, we find the corresponding ratio is about 1.4 and for $y = 2^x$ the ratio is about 0.7. It seems reasonable to suppose there is a value of a between 2 and 3 for which the ratio is exactly 1. This value is about 2.718 and is denoted by e.

> The gradients of $y = e^x$ at any point is equal to the value of e^x at that point.

Your calculator should have the button e^x. In Figure 10.21 we have plotted the graph of $y = e^x$. We used it to calculate the gradient at $x = 2$ and found it to be about 7.4; the value of e^2 is 7.389 to 3 d.p.

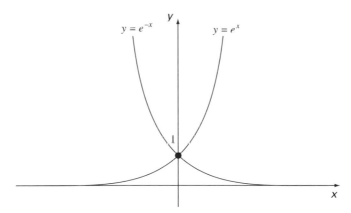

Figure 10.21 Graph of $y = e^x$ and $y = e^{-x}$

An equally important and related function is the **negative exponential function** $e^{-x}(= 1/e^x)$. Its graph, shown in Figure 10.21, can be obtained by reflecting the graph of $y = e^x$ in the vertical axis. It is used to model decay.

Examples

1. The temperature θ of a hot liquid cooling in an atmosphere of constant temperature θ_s is found from Newton's law of cooling as

$$\theta = \theta_s + (\theta_0 - \theta_s)e^{-kt}. \qquad (10.1)$$

where θ_0 is the temperature of the liquid at time $t = 0$ and k is a constant related to its thermal properties. It is instructive to build up the graph of θ. The multiplying factor $(\theta_0 - \theta_s)$ changes the scale on the vertical axis; see Figure 10.22(a). And adding θ_s effectively raises the graph a distance θ_s; see Figure 10.22(b).

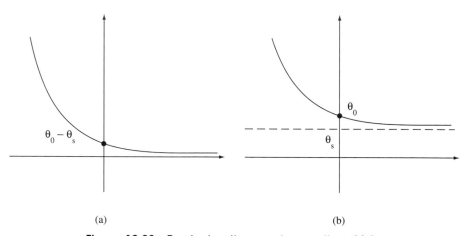

(a)   (b)

Figure 10.22 Producing the graph equation (10.1)

Suppose that $\theta_0 = 100\,°C, \theta_s = 15\,°C$ and $k = 0.1$ and suppose we wish to estimate the temperature of the liquid at $t = 5$ minutes, 10 minutes, 20 minutes. The formula (10.1) becomes

$$\theta = 15 + 85e^{-0.1t}.$$

Hence

when $t = 5$, $\theta = 65.6\,°C$
when $t = 10, \theta = 46.3\,°C$
when $t = 20, \theta = 26.5\,°C.$

2. An important model describes radioactive decay. It is known that the rate of decay of a radioactive substance is directly proportional to the amount of the substance remaining.

Mathematically it can be proved that the amount of the substance m left after time t is given by

$$m = m_0 e^{-kt} \tag{10.2}$$

where m_0 is the mass at time $t = 0$ and k is a constant of radioactive decay.

3. The voltage in a d.c. circuit containing a capacitance is known to build up to a maximum from zero when the circuit is switched on. The voltage v at time t is given by $v = V(1 - e^{-kt})$ where k is a constant. The graph of the function v is shown in Figure 10.23. We say that the graph approaches the steady-state maximum value V **asymptotically** or that the horizontal line $v = V$ is an **asymptote** to the curve, i.e. a tangent at infinity. ∎

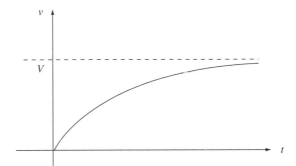

Figure 10.23   Voltage across the capacitor versus time

The graph has been obtained by reflecting the graph of e^{-kt} about the horizontal axis to obtain $-e^{kt}$. The constant 1 is added by lifting the curve through a distance 1; multiplication by V is achieved by scaling the vertical axis differently.

The logarithmic function

Suppose in the problem of the cooling liquid we wanted to find the time taken for the liquid to cool from $100\,^\circ$C to $20\,^\circ$C. We would need to solve the equation

$$20 = 15 + 85e^{-0.1t}$$

so $\qquad 5 = 85e^{-0.1t} = \dfrac{85}{e^{-0.1t}}$

hence $\quad e^{0.1t} = \dfrac{85}{5} = 17.$

Remember from Section 1.3 that, for some number $a > 0$, if $a^x = y$ then $x = \log_a y$. Hence if $e^x = y$ then $x = \log_e y$.

Logarithms to the base e are called **natural logarithms** (or sometimes Naperian logarithms). A notation which is used on calculators is to write $\log_e x$ as $\ln x$; the symbol 'ln' stands for 'logarithm natural'.

In the cooling problem we had $e^{0.1t} = 17$, so

$$0.1t = \log_e 17 = \ln 17$$

and $\qquad t = 10 \ln 17 = 10 \times 2.830 = 28.3.$

Natural logarithms obey all rules for logarithms stated in Section 1.3; here are some of them:

$$\ln(xy) = \ln x + \ln y$$
$$\ln(x/y) = \ln x - \ln y$$
$$\ln(x^n) = n \ln x$$

Example

We wish to express a^x in terms of natural logarithms. Let $a^x = e^y$ then

$$\ln(a^x) = x \ln a$$

But $\ln(e^y) = y$, so $a^x = e^{x \ln a}$. ∎

The **logarithmic function** is the inverse of the exponential function. Its graph can be obtained by reflecting the curve $y = e^x$ in the line $y = x$; see Figure 10.24.

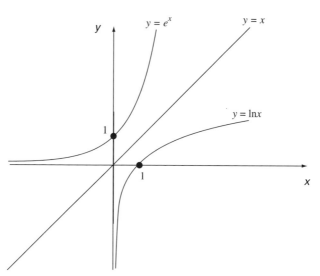

Figure 10.24 Graphs of the exponential and logarithmic functions

Note that $\ln x$ is not defined for $x < 0$. This can be seen as follows. We have $e^y > 0$ for all y; now consider $y = \ln x$ so that $x = e^y$, but $e^y > 0$, hence x must be greater than 0.

The gradient of $y = \ln x$ is always positive but decreasing. This means that the values of $\ln x$ are increasing with x but at a progressively slower rate. In fact we shall show in Chapter 11 that the gradient is given by $\frac{1}{x}$.

Example

We return to the problem of radioactive decay.

Consider the time elapsed when $m = 0.5 m_0$, i.e. half the original mass; we have

$$0.5 m_0 = m_0 e^{-kt}$$

so that $\quad 0.5 = \frac{1}{2} = e^{-kt} = \frac{1}{e^{kt}}$

and $\quad e^{kt} = 2.$

Hence $kt = \log_e 2 = \ln 2$ and $t = \frac{1}{k} \ln 2$.

This value is called the **half-life** of the substance and it is independent of the amount of mass present at the start of decay.

How long does it take the substance to decay from an amount 0.5 to 0.25? It is easily seen that the time is again the half-life. Hence if the half-life of a substance is 1 year, then after 1 year there will be half of the original substance remaining, after 2 years one quarter, after 3 years one eighth, and so on.

Exponential and logarithmic functions compared

	Exponential function	Logarithmic function
Domain	All real values	All real values > 0
Range	All real values > 0	All real values
Asymptote	Horizontal axis	Vertical axis
Gradient	e^x (increasing)	$1/x$ (decreasing)
	$e^0 = 1$	$\ln 1 = 0$
	$e^x \cdot e^y = e^{x+y}$	$\ln(xy) = \ln x + \ln y$
	$e^x \div e^y = e^{x-y}$	$\ln(x/y) = \ln x - \ln y$
	$(e^x)^n = e^{nx}$	$\ln x^n = n \ln x$
	$e^{\ln x} = x \; (x > 0)$	$\ln(e^x) = x \; (\text{all } x)$

Exercise 10.4

1 Show that $2^{10} = 1024$ and establish that $(2)^{10n} \sim (10)^{3n}$.

2 Any positive integer can be written uniquely in terms of powers of 2, or **binary form**:

$$58 = 1 \times 2^5 + 1 \times 2^4 + 1 \times 2^3 + 0 \times 2^2 + 1 \times 2 + 0 \text{ and is written } 111010$$

(a) Find the binary representation of the following integers:
(i) 7   (ii) 19   (iii) 41
(iv) 59   (v) 113   (vi) 520

(b) Roughly speaking, how many binary digits are there per decimal digit? A binary digit is called a **bit**.

3 When a sum of money, £S is invested at $p\%$ per annum its value increases by a factor of $\left(1 + \dfrac{p}{100}\right)$ for each year invested.

(a) Determine the value of the investment after
(i) 1 year   (ii) 2 years   (iii) n years

(b) A sum of £100 is invested at an annual interest rate of 5%.
(i) What is the investment after 5 years?

(ii) How long is it before the investment is doubled?

4 Without using tables, evaluate the following logarithms:

(a) $\log_2 16$

(b) $\log_3\left(\dfrac{1}{9}\right)$

(c) $\log_{10} 1\,000\,000$

(d) $\log_6 (36)^{-2/3}$

(e) $\log_2 8\sqrt{2}$

(f) $\log_7\left(\dfrac{49^x}{343^y}\right)$

(g) $\log_4 0.125$

(h) $\log_{100} \sqrt{10}$.

5 (a) If $x = \log_y z$ $y = \log_z x$ $z = \log_x y$,   express each in terms of powers and prove that $xyz = 1$.

(b) Where all the logarithms are taken to the same base prove that

(i) $\dfrac{\log 16 + \log 25 - \log 36}{\log 10 - \log 3} = 2$

(ii) $\dfrac{\log\sqrt{125}-\log 2+\log\sqrt{343}-\log\sqrt{2}}{\log 4900-\log 16}=\dfrac{3}{4}$.

6 By expressing in terms of powers, prove that $\log_a b=\dfrac{1}{\log_b a}$.

7 If $0<x<y$ then $\log x<\log y$, for any base. Use this fact to find the minimum value of n which satisfies the inequality $(0.9)^n<0.05$.

8 Solve the equations

(a) $(6.1)^x=3.515$ (b) $\log_{10}(x^2-x)=0.3010$

and the simultaneous equations

(c) $(3.5)^x(2.9)^{-y}=6.7$, $(4.5)^x(3.7)^{2y}=19.1$.

9 On the same axes sketch the graphs of $y=e^x$ and $y=e^{-x}$. Comment.

10 By taking natural logarithms, solve the following equation for x

$$3^x=e^p.$$

Prove that

(a) $a^x=e^{x\ln a}$, for $a>0$

(b) $\log_b c=\dfrac{\ln c}{\ln b}$, for $b,c>0$.

11 Given that if $f:x\mapsto e^x$ then $f^{-1}:x\mapsto \ln x$, draw on the same axes the graphs of $y=f(x),y=x$ and $y=f^{-1}(x)$. Comment.

12 Consider $f(x)=e^x, f^{-1}(x)=\ln x$, $g(x)=x^2$ and $g^{-1}(x)=\sqrt{x}$ with domains and ranges suitably defined.

(a)   Determine in the simplest (or lowest) form

(i) $f(g(2))$ (ii) $f^{-1}(g(3))$
(iii) $g^{-1}(f^{-1}(e^2))$ (iv) $f^{-1}\{g^{-1}(4)\}^2$

(b)   Which of the following statements are true?

(i) $f^{-1}(g(x))\equiv 2f^{-1}(x)$ (ii) $f^{-1}(g(f(x)))\equiv g(x)$
(iii) $f^{-1}(\sqrt{g(x)})\equiv f^{-1}(x)$

(c)   What is the domain of $\ln\ln x$? Hence explain why $\ln\ln\sin x$ and $\ln\ln\cos x$ cannot be defined.

13 A power law $x = C(1.4)^{-t}$ is used to predict the resale value of a desk-top computer when it is t years old.

(a) What is the meaning of C?

(b) Change the base of the exponent to e and express x in terms of C, e and t.

(c) If the computer costs £1000 initially, what is the value to the nearest £1 at 1 year, 2 years and 3 years old respectively?

(d) After what time does it drop to half its cost value?

14 Sketch the graphs of the following functions:

(a) $xe^{-x}, x > 0$

(b) $\dfrac{e^x}{e^x + 1}$ (What are the asymptotes?)

(c) $e^{-x} \sin \pi x, -1 \le x \le 2$

(d) $\ln(e^x + 1) - x, x > -3$.

 15 Determine x from the following equations:

(a) $2^x - 2^{-x} = 6$

(b) $9^x - 7 \times 3^x - 8 = 0$.

And determine x and y from the following simultaneous equations:

(c) $2^{x+y} = 32, 3^x + 3^y = 36$.

16 Simplify

(a) $\ln(2e^x)$

(b) $e^{5 \ln x}$

(c) $e^{(\ln x - z \ln y)}$.

17 Express $\dfrac{\log_{10}(e^x + 1)}{\log_{10} e}$ as a single logarithm.

18 The **hyperbolic sine** of x is defined to be

$$\sinh x = \frac{1}{2}(e^x - e^{-x})$$

likewise the **hyperbolic cosine** of x is defined to be

$$\cosh x = \frac{1}{2}(e^x + e^{-x}).$$

From the definitions show that sinh is an odd function, i.e.

$$\sinh(-x) = -\sinh x$$

and that cosh is an even function, i.e.

$$\cosh(-x) = \cosh x.$$

Use these properties to show that

(a) $\sinh x = 0$ when $x = 0$ only

(b) $\cosh x \geq 1$, all x

(c) $\cosh^2 x - \sinh^2 x \equiv 1$

(d) $\sinh 2x \equiv 2 \sinh x \cosh x$

(e) $\cosh 2x \equiv 2 \cosh^2 x - 1$.

Sketch the graphs of $\sinh x$ and $\cosh x$.

19   The **hyperbolic tangent** is defined by

$$\tanh x = \frac{\sinh x}{\cosh x}$$

(a)   Prove that $\tanh x = \dfrac{e^x - e^{-x}}{e^x + e^{-x}}$

(b)   When is $\tanh x = 0$?

(c)   Describe how $\tanh x$ behaves for
(i) large positive x (ii) large negative x.

 20   Use HYP and INV HYP on your calculator to find

(a) $\sinh (1)$

(b) $\cosh (2)$

(c) $\cosh (-2)$

(d) $\tanh (-5)$

(e) $\sinh^{-1} (-1)$

(f) $\cosh^{-1} (2)$

(g) $\tanh^{-1} \left(\dfrac{1}{2}\right)$.

21*   The hyperbolic functions are all exponential in type, which suggests their inverses are logarithmic. We can prove this for one example by letting

$$y = \sinh^{-1} x \quad \text{or} \quad x = \sinh y = (e^y - e^{-y})$$

Now let $z = e^y$, noting that $z > 0$, so that $x = \dfrac{1}{2}\left(z - \dfrac{1}{z}\right)$. Solving for z in terms of x gives $z = x \pm \sqrt{x^2 + 1}$.

Now $\sqrt{x^2 + 1} > x$ for any real x, and because $z > 0$ only the positive root is acceptable, so $y = \ln z = \ln(x + \sqrt{x^2 + 1})$, i.e. $\sinh^{-1} x \equiv \ln(x + \sqrt{x^2 + 1})$ for any real x.

(a)   Prove that $\tanh^{-1} x \equiv \dfrac{1}{2}\ln\left(\dfrac{1+x}{1-x}\right)$.

(b)   For any $x > 1$ prove that $x - \sqrt{x^2 - 1} \equiv \dfrac{1}{x + \sqrt{x^2 - 1}}$. Deduce that

$$\ln(x \pm \sqrt{x^2 - 1}) \equiv \pm \ln(x + \sqrt{x^2 - 1}).$$

(c)   It is known that $y = \cosh^{-1} x$ is double-valued in that $x = \cosh y = \cosh(-y)$. Prove using (b) that two values of $\cosh^{-1} x$ are

$$\cosh^{-1} x \equiv \pm \ln(x + \sqrt{x^2 - 1})$$

22* Use logarithms to verify parts (f) and (g) of Question 20.

23* Use the principle of reflection in the line $y = x$ to sketch curves on the same axes.

(a) $\sinh x$ and $\sinh^{-1} x$ (b) $\cosh x$ and $\cosh^{-1} x$ (c) $\tanh x$ and $\tanh^{-1} x$

Exercise 10.5: further applications and models

1   Draw the graph of $y = e^x$ and draw the tangents at the points $(0, 1)$ and $(1, e)$. Determine the intercepts of the tangents on the x-axis graphically and verify that the gradients are 1 and e respectively. What are the equations of the tangents?

2   Repeat the procedure of Question 1 for $y = 2^x$ at the points $(0, 1)$ and $(1, 2)$. For the tangents at these points verify graphically that

$$\frac{\text{gradient}}{\text{value of } y} = \ln 2.$$

3   For a radioactive isotope whose half-life is 3×10^{-4} s what is the law of exponential decay?

4   The build-up of voltage in a d.c. circuit with capacitance is modelled by $v = V(1 - e^{-t/CR})$ where v is the voltage measured in volts at any time, V is the final voltage, R is the resistance of the circuit measured in ohms (Ω) and C is its capacitance measured in microfarads ($1\ \mu\text{F} = 10^{-6}\ \text{F}$). The time constant of the system is the value of t, in seconds, such that $t = CR$.

(a)   If $R = 5 \times 10^6\ \Omega$ and $C = 10\ \mu\text{F}$ what the time constant (t_c)?

(b)   How long does the circuit voltage take to build up to 99% of its final value?

(c)   Show that no matter what values R and C happen to possess, $v \simeq 0.63V$ when $t = t_c$, and $v = 0.99\ V$ after $4.605 t_c$.

5 Beer's law of diminishing light intensity is based upon the idea that the light of intensity I_0 is reduced to an intensity I_1 where $I_1 = I_0 q$, $0 < q < 1$ on passing through a specified thickness of a translucent medium.

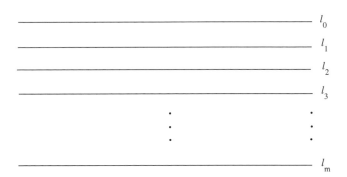

(a) If the medium is m units thick, m is a positive integer, given that

$$I_1 = I_0 q, \ I_2 = I_1 q, \ I_3 = I_2 q, \text{etc, establish that } I_m = I_0 q^m$$

(b) If $a = -\ln q$ prove that $I_m = I_0 e^{-am}$.

(c) The result in (b) holds for any $m > 0$, not just a positive integer, and for clear, still, fresh water $I = I_0 e^{-1.95x}$ where x is the depth in metres. Using this result determine
(i) the reduced light intensity at a depth of 1 m
(ii) the depth at which light intensity is reduced by a factor of $\left(\dfrac{1}{2}\right)$.

6 A cup of hot water is observed initially to be at a temperature of $60\,^{\circ}\text{C}$. Fifteen minutes later its temperature has dropped to $40\,^{\circ}\text{C}$ and after a further fifteen minutes this is $30\,^{\circ}\text{C}$.

(a) Write down the two equations which relate the observed temperatures according to Newton's law of cooling and show that

$$\frac{40 - \theta_s}{60 - \theta_s} = \frac{30 - \theta_s}{40 - \theta_s}.$$

Hence find θ_s.

(b) Show that the thermal constant is given by $k = \dfrac{\ln 2}{15}$.

(c) Assuming that the water had cooled down from boiling point at the time of the first observation, how long before the initial observation was it boiling ($100\,^{\circ}\text{C}$)?

7 In an electronic diode with two parallel plates the electric current (i) flowing at one plate varies with the voltage difference (V) between the two plates. The following observations are made.

V	1	4	8	12	17	20
i	0.030	0.240	0.679	1.247	2.103	2.683

Take logarithms and replot on a straight line graph. Determine the power law that operates between V and i. Alternatively use log-log graph paper and plot to obtain a straight line.

 8 When a gas expands or contracts without any supply or removal of heat, the changes that take place are adiabatic and governed by the law $PV^\gamma = \text{constant}$, where P is the pressure and V is the volume. For air $\gamma = 1.4$, approximately; use this value to plot P against V for $P = 1$ (1) 10 given that $V = 1$ when $P = 1$.

9* A weight suspended on a light helical spring vibrates up and down when impulsively jolted downwards.
The downward displacement x is a function of time t,

$$x: t \mapsto 10 \cos 2\pi ft$$

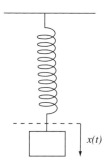

x is measured in centimetres and t in seconds.

(a) Sketch the graph of $x(t)$ when $f = 2$ and calculate the time when

(i) $x = 0$ for the first time

(ii) $x = 0$ for the third time

(iii) $x = -10$ for the second time

(b) What is the period of a complete oscillation?

10* An alternating voltage is a function of time, following a sinusoidal pattern of the form

$$v: t \mapsto v_0 \sin 100\pi t.$$

Find the width of the time interval when $v > 80v_0$ for the first time after $t = 0$.

11* In information theory the degree of disorder in a system is measured by its entropy. The entropy E of a system S with probability distribution $(p_1, p_2, \ldots, p_n)$ is given by

$$E(S) = -\sum_{i=1}^{n} p_i \log_2 p_i$$

where the sum of the probabilities $\sum_{i=1}^{n} p_i = 1$. Find $E(S)$ for a source with four elements having probability distribution $(0.02, 0.18, 0.35, 0.45)$.

12* In error-correcting code theory the use of n check digits produces a redundancy function $R(n) = \dfrac{n}{2^n - 1}, n \in \mathbb{N}$.

(a)   How many check digits are needed to ensure that the redundancy function is less than 0.01?

(b)   Plot the graph of $R: x \mapsto \dfrac{x}{2^x - 1}$ over $[1, \infty]$ and verify that it is decreasing for $x > 1$.

SUMMARY

- **Function**: a rule which associates each member of one set, the **domain**, with a unique member of a second set, the **co-domain**. The set of values taken by the function is its **range**. We write $y = f(x)$ where x is the **independent variable** and y is the **dependent variable**.

- **Transformation**: The graph of $y = f(x)$ may be translated, reflected or scaled.

- **Translation**: $f(x - a)$ represents a move to the right by a, but $f(x) + a$ represents a move upwards by a.

- **Reflection**: $-f(x)$ represents a reflection about the x-axis, but $f(-x)$ represents a reflection about the y-axis.

- **Scaling**: $af(x)$ represents a scaling in the y-direction, an expansion if $a > 1$, a compression if $a < 1$; but $f(ax)$ represents a scaling in the x-direction, an expansion if $a < 1$, a compression if $a > 1$.

- **Function of a function**: the output of one function becomes the input of a second function. The notation $f(g(x))$ implies that the function f is applied to the output of the function g.

- **The inverse of a function**: $f^{-1}(x)$ reverses the action of the function $f(x)$; its graph is obtained by reflecting the graph of $f(x)$ in the line $y = x$. It is a function if the original function is such that no two inputs are mapped to the same output. The domain of the inverse function is the range of the original function and vice-versa.

- **The exponential function**: is written e^x, where $e \approx 2.718$; its inverse is $\ln x$.

- **Inverse trigonometric functions**: the domains of the trigonometric functions are restricted to an interval of length π; the function $\sin^{-1} x$ has domain $-1 \leq x \leq 1$.

Answers

1 (a) $-\infty < x < 0$ and $0 < x < \infty$

(b) $-1 \leq x < 0, 0 < x < \sqrt{2}$ and $x > \sqrt{2}$

(c) $x \leq -4$ and $x \geq 4$

(d) $-5 < x < -4.8$

2 (a) $f(x) = \dfrac{x}{2} + 4, \ f : x \mapsto \dfrac{x}{2} + 4$ (b) $f(x) = \sqrt{x-5}, f : x \mapsto \sqrt{x-5}$

(c) $f(x) = x^2 + 1, \ f : x \mapsto x^2 + 1$ (d) $f(x) = (x+1)^2, f : x \mapsto (x+1)^2$

(e) $f(x) = x^2 - 1, \ f : x \mapsto x^2 - 1$

3

	Domain	Range
(a)	All real numbers	All real numbers
(b)	Real numbers ≥ 5	Non-negative real numbers
(c)	All real numbers	Real numbers ≥ 1
(d)	All real numbers	Non-negative real numbers
(e)	All real numbers	Real numbers ≥ -1

4

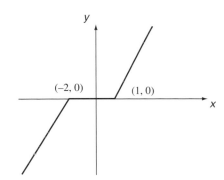

The domain and range are each the set of all real numbers

5

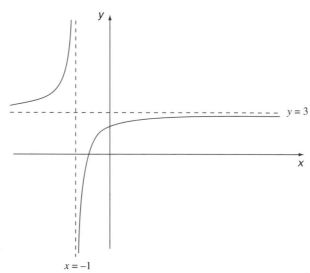

$x = -1$

Domain is the set of real numbers except -1
Range is the set of real numbers except 3

6 (a,b)

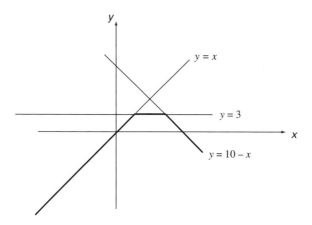

(c)   Range is the set of real numbers ≤ 3

(d) $a = 7$

7 (a)

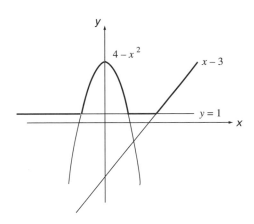

(b) $f(x) = \begin{cases} 1 & x < 1 \\ 4 - x^2 & -1 \leq x < 1 \\ 1 & 1 \leq x \leq 4 \\ x - 3 & x \geq 4 \end{cases}$

Equality has been taken at the changeover points from left to right; this was our choice.

8

	Domain	Range
(a)	All real numbers	All positive real numbers
(b)	All real numbers except -2	All real numbers ≤ 2
(c)	Not a function: for values of $x > 0$ there are *two* values of y	
(d)	Not a function: for values of x such that $-a < x < 0$ or $0 < x < a$ there are *two* values of y	

9

	Function	Domain	Range
(a)	$f(x) = 2x^2$	All real numbers	Non-negative real numbers
(b)	$f(x) = x^3$	All real numbers	All real numbers
(c)	$f(x) = \dfrac{1}{x^3}$	Real numbers $\neq 0$	Real numbers $\neq 0$
(d)	$f(x = \dfrac{1}{(x+2)}$	Real numbers $\neq -2$	Real numbers $\neq 0$
(e)	$f(x) = \cos x$	$-\dfrac{\pi}{2} \leq x \leq \dfrac{\pi}{2}$	Real numbers $0 \leq y \leq 1$

	Function	Domain	Range
(f)	$f(x) = \tan x$	$-\dfrac{\pi}{2} < x < \dfrac{\pi}{2}$	All real numbers
(g)	$f(x) = |2x + 3|$	All real numbers	Real numbers ≥ 0
(h)	$f(x) = \sin\left(\dfrac{1}{x}\right)$	Real numbers $\neq 0$	Real numbers $-1 \leq y \leq 1$

10 (a) $f(x) \geq 5$ (b) $\dfrac{1}{20} \leq x \leq 5$

     (c) $1 \leq f(x) < 2$ (d) $-7 < x < -5$ and $2 < x < 4$

     (e) $\dfrac{1}{\sqrt{2}} \leq f(x) \leq 1$ (f) $\dfrac{\pi}{6} < x < \dfrac{\pi}{3}$

11 (a)

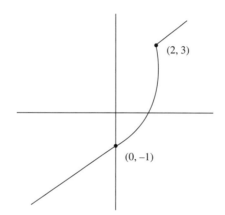

(b) $x = 3$

(c)   Yes, the graph has a U shape at $x = 0$ but it does not break since $|0| = 0$.

(d)

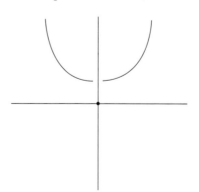

The graph breaks at $x = 0$; it is not continuous

12 (a,b)

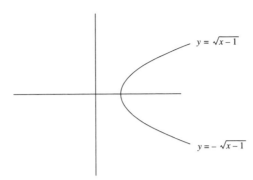

$y^2 = x - 1$ does *not* define a function because a simple input x can lead to two outputs y

(c) $x = g(y) = y^2 + 1$; domain all real numbers, range all real numbers ≥ 1

(d) $x \geq 1 \quad f(x) \geq 0$

13 (a)

n	-4	-3	-2	-1	0	1	2	3	4
$f(n)$	$-\dfrac{3}{5}$	-1	-1	0	$\dfrac{1}{5}$	$\dfrac{1}{5}$	$\dfrac{3}{17}$	$\dfrac{2}{13}$	$\dfrac{5}{37}$

(b) $f(n) \to 0$ (i) through positive values and (ii) through negative values

14 (b)

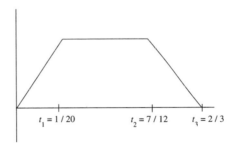

(i) Let t be measured in hours and

$$
s(t) = \begin{cases}
3600t & 0 \leq t < \dfrac{1}{20} \\[2mm]
180 & \dfrac{1}{20} \leq t < \dfrac{7}{12} \\[2mm]
180 - 2160\left(t - \dfrac{7}{12}\right) & \dfrac{7}{12} \leq t \leq \dfrac{2}{3}
\end{cases}
$$

(ii) A and B are $108\,$km apart

Exercise 10.2

1

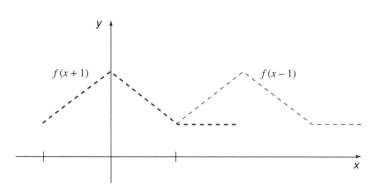

Domains change to $x \geq 1$ and $x \geq -1$ but the ranges remains the same

2 (a) (i) odd (ii) even (iii) odd (iv) even (v) odd
  (b) odd
  (c) (i) even (ii) even (iii) neither (iv) even
     (v) neither (vi) even

3 Domains of g and h are both $[-L, L]$

4

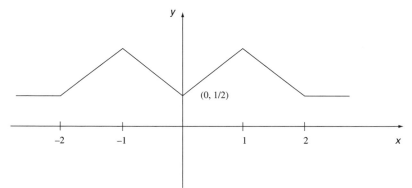

The domain is $\mathbb{R}$

5 (a) $f(2x)$ (b) $f(x/2)$

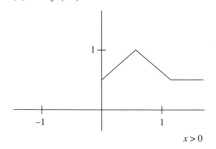

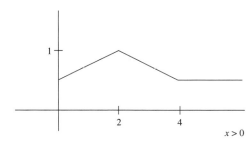

(c) $f(-2x)$

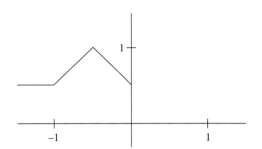

Range unchanged

6 (a)
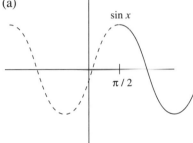

(b)
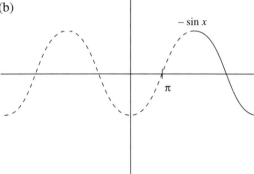

(c)
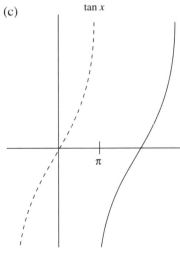

(d)
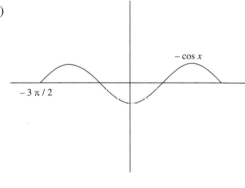

(e)
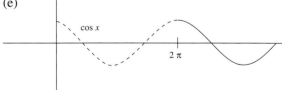

7

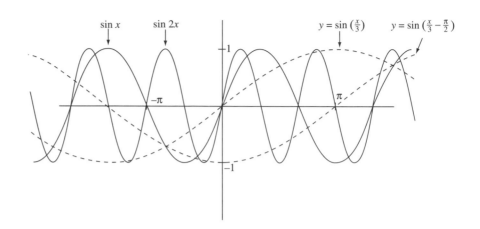

8 (a) $f(x) - \dfrac{1}{2}$ (b) $2f(x+1)$ (c) $\dfrac{1}{2} - f(-x)$

(d) $2f(4-x) - 3$ (e) $2(1 - f(2x+1))$

9 (a)

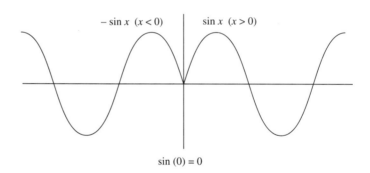

(b) Yes

10

	Domain	Range
(a) (b) (d)	$\mathbb{R}$	$0 < f(x) \le M$
(c)	$\mathbb{R}$	$4 < f(x) \le 3M + 4$

11 [domain], [range]

(a) $[3, 5]$, $[8, 14]$ (b) $[1, 2]$, $[5, 11]$

(c) $[-4, -3]$, $[-37, -28]$ (d) $[1, 4]$, $[2, 11]$

(e)   [13, 25], [38, 74]   (f)   [−2, −1], [−7, −4]

(g)   [a + b, 2a + b], [3(a + b) − 1, 3(2a + b) − 1]

(h)   ℝ, ℝ

Each of the input 'functions', $2x + 1, x − 5, 2x^2 + 6x + 5$ are defined over ℝ if x is defined over ℝ. f is a purely linear function and maps ℝ to itself. Note that [3, 5] means $3 \leq x \leq 5$, etc.

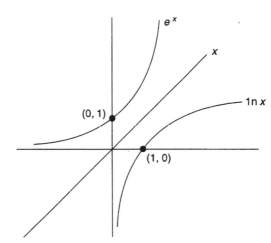

12   (a) $\dfrac{2}{x^2 + 3}$ (b) $\dfrac{2}{5x − 3}$ (c) $\dfrac{2}{3}$, constant

     (d) $\dfrac{2(6 + \sin x + \cos x)}{(\sin x + 3)(\cos x + 3)}$

13   (c)   Yes

14   $p = 1, \qquad q = \dfrac{b}{2a}, \qquad r = a, \qquad s = \dfrac{4ac − b^2}{4a}$

15   (a) $\dfrac{\pi}{3} + n\pi \qquad \dfrac{2\pi}{3} + n\pi$ (b) $\dfrac{n\pi}{3}$

Exercise 10.3

1   The order is inverse; domain, range. ℝ/{1} means real numbers except 1.

     (a) $f^{-1} : y \mapsto y − 1$; ℝ, ℝ   (b) $f^{-1} : y \mapsto y^{1/5}$; ℝ, ℝ

     (c) $f^{-1} : y \mapsto 1 − \dfrac{1}{y}$; ℝ/{1}, ℝ/{0}   (d) $f^{-1} : y \mapsto (y + 1)^{1/3}$; ℝ, ℝ

     (e) $f^{-1} : y \mapsto 1 + y^2$; (1, ∞), ℝ   (f) $f^{-1} : y \mapsto (y − b)/a$; ℝ, ℝ

     (g) $f^{-1} : y \mapsto y^{-1/3}$, ℝ/{0}, ℝ/{0}

2   (a) $f : x \mapsto$ constant, all x, not invertible

(b) $f^{-1} : y \mapsto \dfrac{b - dy}{cy - a}$ ($\mathbb{R}/\{-d/c\}$, $\mathbb{R}/\{a/c\}$)

If $ad = bc$, $\dfrac{ax + b}{cx + d} = \dfrac{a}{c}$, constant, not invertible as in (a)

3   (a) $f(x) > 0$ (b) $f^{-1} : x \mapsto \sqrt{x}$ (d) $x \geq 1$ both cases

4   (a) $f_1^{-1} : y \to -1 + \sqrt{y - 4}$, $f : (x \geq -1) \mapsto (y \geq 4)$
$f_2^{-1} : y \to -1 - \sqrt{y - 4}$, $f : (x < -1) \mapsto (y > 4)$

(b) $f_1^{-1} : y \to 6 + \sqrt{y + 86}$, $f : (x \geq 6) \mapsto (y \geq -86)$

$f_2^{-1} : y \to 6 - \sqrt{y + 86}$, $f : (x < 6) \mapsto (y > -86)$

(c) $f_1^{-1} : y \to -\dfrac{1}{3} + \sqrt{\dfrac{y}{3} + \dfrac{34}{9}}$, $f : \left(x \geq -\dfrac{1}{3}\right) \mapsto \left(y \geq -\dfrac{34}{3}\right)$

$f_2^{-1} : y \to -\dfrac{1}{3} - \sqrt{\dfrac{y}{3} + \dfrac{34}{9}}$, $f : \left(x < -\dfrac{1}{3}\right) \mapsto \left(y > -\dfrac{34}{3}\right)$

(d) $f_1^{-1} : y \to -\dfrac{b}{2a} + \sqrt{\dfrac{y - c + \dfrac{b^2}{4a}}{a}}$, $f : \left(x \geq -\dfrac{b}{2a}\right) \mapsto \left(y \geq c - \dfrac{b^2}{4a}\right)$

$f_2^{-1} : y \to -\dfrac{b}{2a} - \sqrt{\dfrac{y - c + \dfrac{b^2}{4a}}{a}}$, $f : \left(x < -\dfrac{b}{2a}\right) \mapsto \left(y > c - \dfrac{b^2}{4a}\right)$

(e)   (i) $f^{-1} : y \mapsto \dfrac{y - c}{b}$, $f : \mathbb{R} \to \mathbb{R}$ (ii) f^{-1} does not exist

5   (a)

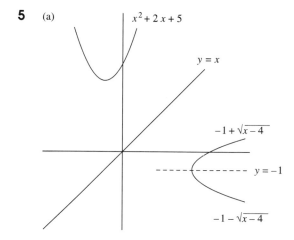

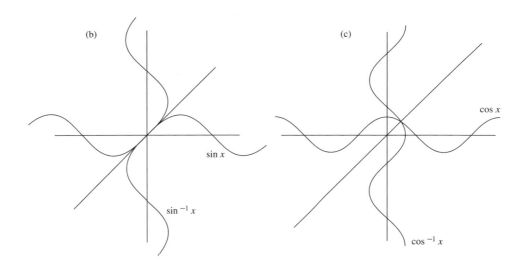

6

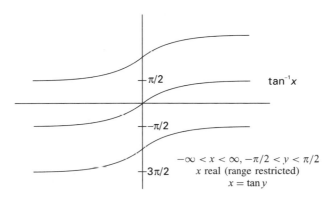

$-\infty < x < \infty, -\pi/2 < y < \pi/2$
x real (range restricted)
$x = \tan y$

7 (a)   30°   (b)   120°   (c)   89.43°

    (d)   9.879°   (e)   75°

8 (a)

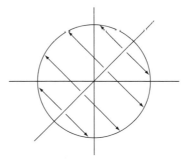

    (b) $y = \sqrt{1 - x^2}$, i.e. $f : x \mapsto (1 - x^2)^{1/2}$ upper half $[-1, 1] \mapsto [0, 1]$

           $y = -\sqrt{1 - x^2}$, i.e. $f : x \mapsto -(1 - x^2)^{1/2}$ lower half $[-1, 1] \mapsto [0, 1]$

9 (In each quadrant, f, f^{-1} are given.)

I $\quad x \mapsto b\sqrt{1 - x^2/a^2},\, [0, a] \to [0, b];$ $\qquad x \mapsto a\sqrt{1 - x^2/b^2},\, [0, b] \to [0, a]$

II $\quad x \mapsto b\sqrt{1 - x^2/a^2},\, [-a, 0] \to [0, b];$ $\qquad x \mapsto -a\sqrt{1 - x^2/b^2},\, [0, b] \to [-a, 0]$

III $x \mapsto -b\sqrt{1 - x^2/a^2},\, [-a, 0] \to [-b, 0];$ $\quad x \mapsto -a\sqrt{1 - x^2/b^2},\, [-b, 0] \to [-a, 0]$

IV $x \mapsto -b\sqrt{1 - x^2/a^2},\, [0, a] \to [-b, 0];$ $\qquad x \mapsto a\sqrt{1 - x^2/b^2},\, [-b, 0] \to [0, a]$

10 **I** $x \mapsto b\sqrt{x^2/a^2 - 1},\, (x \geq a),\, (y \geq 0)$ $\qquad x \mapsto a\sqrt{1 + x^2/b^2},\, (x \geq 0),\, (y \geq a)$

II $x \mapsto b\sqrt{x^2/a^2 - 1},\, (x \leq -a),\, (y \geq 0)$ $\qquad x \mapsto -a\sqrt{1 + x^2/b^2},\, (x \geq 0),\, (y \leq -a)$

III $x \mapsto -b\sqrt{x^2/a^2 - 1},\, (x \leq -a),\, (y \leq 0)$ $\quad x \mapsto -a\sqrt{1 + x^2/b^2},\, (x \leq 0),\, (y \leq -a)$

IV $x \mapsto -b\sqrt{x^2/a^2 - 1},\, (x \geq a),\, (y \leq 0)$ $\qquad x \mapsto a\sqrt{1 + x^2/b^2},\, (x \leq 0),\, (y \geq a)$

11 (a) (i) $fg : x \mapsto \dfrac{1}{1 - x}, \quad gh : x \mapsto 1 - x^3, \quad hf : x \mapsto 1/x^3$

(ii) $f^{-1} : x \mapsto 1/x, \quad g^{-1} : x \mapsto 1 - x, \quad h^{-1} : x \mapsto x^{1/3}$

(b) $gf : x \mapsto 1 - 1/x\, (\neq fg), \quad fh : x \mapsto 1/x^3\, (= hf,$ but this is an exception) $hg : x \mapsto (1 - x)^3$

(c) $g^{-1}f^{-1} : x \mapsto 1 - \dfrac{1}{x}(= fg)^{-1}, \quad h^{-1}g^{-1} : x \mapsto (1 - x)^{1/3}, \quad f^{-1}h^{-1} : x \mapsto x^{-1/3}$

(d) $ff = I,\, gg = I,\, hh : x \mapsto (x^3)^3 = x^9,\, fff \ldots = f,$ odd, n times and $= I,$ even $g,$ is as $f.\ h \ldots h$ (n times), $x \to x^{3n}$

(e) $f(x) + g(x) = \dfrac{1}{x} + 1 - x = \dfrac{1 + x - x^2}{x}$

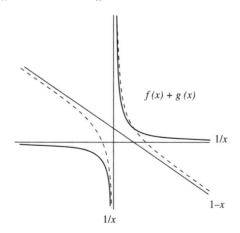

$f(x) + g(x)$

$1/x$

$1-x$

$1/x$

A vertical ruler shows that $f + g$ is a function defined for $x \in \mathbb{R}/\{0\}$ but $(f + g)^{-1}$ must be defined for $x > 0$ and $x < 0$.

$$(f + g)^{-1} : y \mapsto \frac{1 - y \pm \{(y - 1)^2 + 4\}^{1/2}}{2}$$

(f) $\dfrac{1-x}{x}$, $x^3(1-x)$, x^2

12   (b) $f(n) = \{f(1)\}^n = k^n$ if $f(1) = k$ (e)   Any integer, $f(1) = a$

    (f) a^x exists if $a > 0$ for a general $x \in \mathbb{R}$.

13   (e)

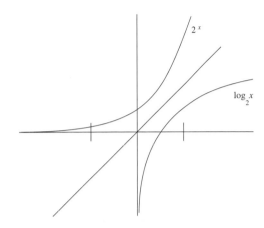

14   (a)   No inverse over $\mathbb{N}$

    (b)   No inverse over $\mathbb{N}$

    (c)   No inverse; $2 \to 2, 4 \to 2$, etc.   (d) $f^{-1} = f$

    (e)   No inverse   (f)   No inverse

    (g)   No inverse; $1 \to 1, 3 \to 1$, the only exception to a 1–1 mapping

15

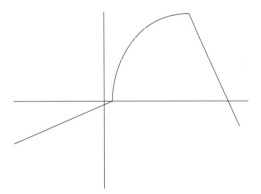

$$f^{-1}: x \mapsto \dfrac{x-1}{2}, (-\infty, 1); \qquad x \mapsto \sqrt{x-1}, [1, 10]; \qquad x \mapsto 13 - x, (10, \infty)$$

Exercise 10.4

1 $2^{10} = 1024 \sim 10^3$ so that $(2^{10})^n \sim (10^3)^n$

2   (a) $7 = 2^2 + 2 + 1 = 111,$ $19 = 2^4 + 2 + 1 = 10011$
$41 = 2^5 + 2^3 + 1 = 101001,$ $59 = 2^5 + 2^4 + 2^3 + 2 + 1 = 111011$
$113 = 2^6 + 2^5 + 2^4 + 1 = 1110001,$ $520 = 2^9 + 2^3 = 1000001000$

   (b)   Since $2^{10} \sim 10^3$, ten bits correspond roughly to three decimal digits.

3   (a)   (i) $S\left(1 + \dfrac{p}{100}\right)$ (ii) $\left(1 + \dfrac{p}{100}\right)^2$ (iii) $S\left(1 + \dfrac{p}{100}\right)^n$

   (b)   (i) $£100 \left(1 + \dfrac{5}{100}\right)^5 = 127.63$

      (ii)   Just over 14 years; with $n = 14$ the figure is £197.99.

4   (a)   4   (b) -2 (c)   6

   (d) $-\dfrac{4}{3}$ (e) $\dfrac{7}{2}$ (f) $-2x - 3y$

   (g) $-\dfrac{3}{2}$ (h) $\dfrac{1}{4}$

5   (a) $xyz = (xyz)^{xyz}$, so $xyz = 1$

7 $n > \dfrac{\ln 0.05}{\ln 0.9}$, i.e. $n \geq 29$

8   (a) $x = 0.695$ (b) $x = 2$ (c) $x = 1.664, \quad y = 0.171$

9

Each function on e^{-x} and e^x, is a reflection of the other in the vertical axis

10 $x = p/\ln 3$

11

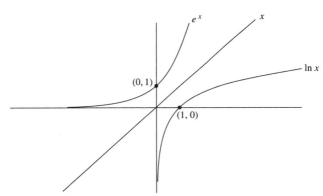

Each function, e^x and $\ln x$, is a reflection of the other in the line $y = x$

12 (a)   (i) e^4 (ii) $2\ln 3$ (iii) $\sqrt{2}$ (iv) $2\ln 2$

(b)   (i) true   (ii) false   (iii) true

(c) $\ln \ln x$ has domain $x > 1$. $\ln \ln \sin x$ cannot be formed because $\sin x \leq 1$; the same is true for $\cos x$.

13 (a) $C = x(0)$, i.e. the value of x at $t = 0$ (b) $x = Ce^{-0.336t}$ $(0.336 = \ln 1.4)$

(c)

Year	Value (£)
1	714
2	510
3	364

(d)   2.06 years

14

(a)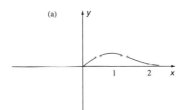

(b)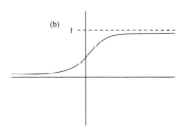

(c)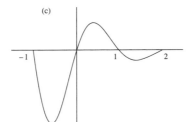

(d)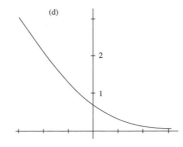

15 (a) $\quad x = \dfrac{\ln(3\sqrt{10})}{\ln 2} = 2.623$ (b) $\quad 3^x = 8, x = 1.893$

    (c) $\quad x = 2, y = 3,$ or $x = 3, y = 2$

16 (a) $\quad \ln 2 + x$ (b) $\quad x^5$ (c) $\quad x/y^z$

17 $\ln(e^x + 1)$, which cannot be reduced further

18

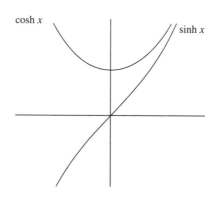

19 (b) $\quad \tanh x = 0$ if $x = 0$, i.e. only when $\sinh x = 0$

    (c) $\quad \tanh x = \dfrac{e^{2x} - 1}{e^{2x} + 1} \sim 1, x \gg 0; = \dfrac{1 - e^{-2x}}{1 + e^{-2x}} \sim -1, x \ll 0.$

20 (a) 1.175   (b) 3.762   (c) 3.762   (d) -1.000

    (d) -1.000 (e) -0.881 (f) 1.317   (g) 0.549

23

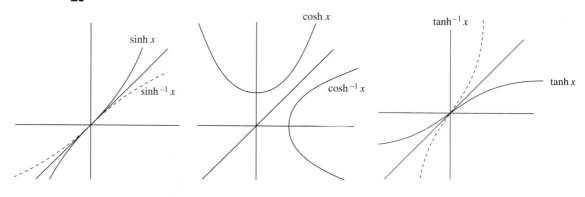

Exercise 10.5

1   Tangents $y = x + 1, y = ex$

3 $m = m_0 e^{-kt}, k = \dfrac{\ln 2}{3 \times 10^{-4}} = 2.310 \times 10^3$

4   (a) $t_c = 50\,\text{s}$ (b)   230.3 s

5   (c)   (i) 14.2% of surface value   (ii) 0.355 m

6   (c)   15 min

7 $i = 3 \times 10^{-2} V^{3/2}$

8

P	1	2	3	4	5	6	7	8	9	10
V	1	0.6095	0.4562	0.3715	0.3168	0.2781	0.2491	0.2264	0.2081	0.1931

9   (a) (i) 0.125 s   (b) (ii) 0.625 s   (c) (b) 0.75 s

10   0.0041 s

11   1.607

12   (a) $n = 10$
       (b)

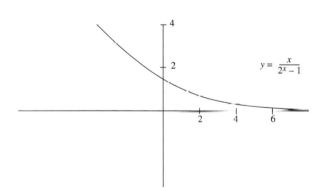

$y = \dfrac{x}{2^x - 1}$

11 DIFFERENTIATION

INTRODUCTION

The need to calculate accurately the rate of change of a variable quantity is crucial in many areas of activity. Often we have a formula which describes the behaviour of the quantity, and differential calculus provides us with the means of obtaining the rate of change as accurately as the formula will allow. We can calculate the acceleration of an object if we know its speed as a function of time. The process of differentiation can be used to calculate maximum and minimum values; for example, we can find the dimensions of the box of maximum volume which can be cut from a given area of sheet metal.

OBJECTIVES

After working through this chapter you should be able to

- understand the terms *derived function* and *derivative*
- relate the derivative of a function to the gradient at a point on its graph
- differentiate standard functions
- differentiate a linear combination of standard functions
- identify from its derived function where a given function is increasing or decreasing
- identify the stationary points of a function
- use the first derivative test to determine the nature of a stationary point
- obtain the second derivative of a function by repeated differentiation
- use the second derivative test to classify the stationary points of a function
- locate the points of inflection of a function
- know the derived function of the standard functions

11.1 RATES OF CHANGE

A function expresses the dependence of one variable quantity on another. Often we are interested in the *rate of change* of one quantity with the other. In many cases we are concerned with the rate of change of a quantity with time; for example, the temperature of a cooling liquid, the current in an electrical circuit, the distance travelled by a vehicle. Examples which involve other rates of change are sales revenue with number of units produced and the strength of a signal with distance from the source. In order to keep the discussion general we shall mostly use x and y as the variables.

Sometimes it is clear from the function what the rate of change is. If $y = 2x$ then y increases at twice the rate of x. More generally, if $y = mx + c$ then y increases at m times the rate of x. This constant rate of change is the *gradient* of the straight line $y = mx + c$. If m is negative then y *decreases* as x increases and we speak of a negative rate of change. Refer to Figure 11.1.

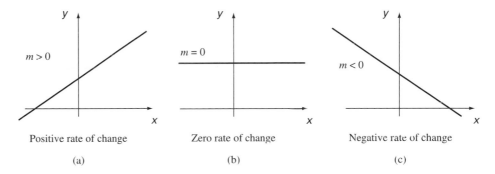

Positive rate of change   Zero rate of change   Negative rate of change

(a)   (b)   (c)

Figure 11.1   Rate of change for a linear relationship

Average rate of change

If we consider the curve $y = x^2$, shown in Figure 11.2(a), then we see that the rate of change of y with x is not constant. For $x < 0$ the rate of change is negative and for $x > 0$ it is positive. As x increases from zero the rate of change becomes larger. For example, between $x = 0$ and $x = 1$ the value of y increases by 1 whereas between $x = 1$ and $x = 2$ the value of y increases by 3 (from 1 to 4). As x decreases from zero the rate of change also becomes larger but is always a decrease.

We can define the **average rate of change** of y over an interval of x by reference to Figure 11.2(b). As x increases from x_1 to x_2, y changes from y_1 to y_2. The average rate of change of y in this interval is $\dfrac{y_2 - y_1}{x_2 - x_1}$.

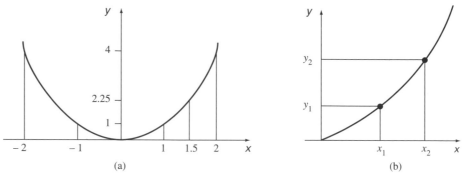

Figure 11.2 Average rate of change of a function

If $y_2 > y_1$ this average rate of change is positive, indicating an increase in y; if $y_2 < y_1$ then y has decreased.

Example

What are the average rates of change of $y = x^2$ between the following values: (a) $x_1 = 1$ and $x_2 = 1.5$, (b) $x_1 = 1.5$ and $x_2 = 2$, (c) $x_1 = -2$ and $x_2 = -1$, (d) $x_1 = -2$ and $x_2 = 2$?

Notice that in all cases we have chosen $x_2 > x_1$.

(a) At $x_1 = 1, y_1 = 1$ and at $x_2 = 1.5, y_2 = 2.25$; therefore

$$\text{Average rate of change } = \frac{2.25 - 1}{1.5 - 1} = \frac{1.25}{0.5} = 2.5$$

(b) At $x_1 = 1.5, y_1 = 2.25$ and at $x_2 = 2, y_2 = 4$; therefore

$$\text{Average rate of change} = \frac{4 - 2.25}{2 - 1.5} = \frac{1.75}{0.5} = 3.5$$

(c) $$\text{Average rate of change} = \frac{4 - 1}{-2 - (-1)} = \frac{3}{-1} = -3$$

(d) $$\text{Average rate of change} = \frac{4 - 4}{2 - (-2)} = \frac{0}{4} = 0$$ ∎

Notice that in (d), although the values of y change as x goes from -2 to 2, the *average* rate of change is zero.

Instantaneous rate of change

Suppose we wish to find the rate of change of $y = x^2$ at the *instant* when $x = 1$. Referring to Figure 11.3, we can calculate the average rate of change of y in the intervals $x_1 = 1$ to $x_2 = 2, x_1 = 1$ to $x_2 = 1.5$ and $x_1 = 1$ to $x_2 = 1.1$

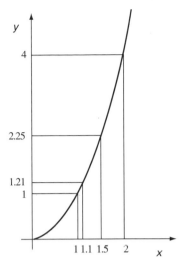

Figure 11.3 Instantaneous rate of change

They are, respectively, $\dfrac{4-1}{2-1} = 3, \dfrac{2.25-1}{1.5-1} = 2.5, \dfrac{1.21-1}{1.1-1} = 2.1.$

The values of the average rate of change are decreasing but what happens as x_2 moves closer to $x_1 = 1$? In the tables below, the first table shows the results of some calculations while the second table shows similar calculations based on $x_1 = 2$.

Table 11.1

x_2	2	1.1	1.01	1.001	1.0001
Average rate of change	3	2.1	2.01	2.001	2.0001

Table 11.2

x_2	3	2.1	2.01	2.001	2.0001
Average rate of change	5	4.1	4.01	4.001	4.0001

It *looks* as though the average rate of change in the first case is edging towards the value 2 as x_2 closes to x_1, whereas the second case seems to approach the value 4. Rather than carry out a similar set of calculations for other values of x_1, we choose a general value $x_1 = a$ and consider a nearby value $x = a + h$, as in Figure 11.4(a).

The average rate of change of y over the interval $a \le x \le a + h$ is

$$\frac{(a+h)^2 - a^2}{(a+h) - a} = \frac{a^2 + 2ah + h^2 - a^2}{a + h - a} = \frac{2ah + h^2}{h} = 2a + h$$

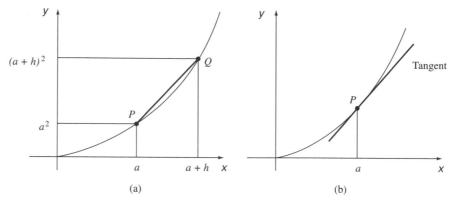

Figure 11.4 Instantaneous rate of change and slope of a tangent

This gives the gradient of the **chord** PQ shown in the diagram. As Q moves down the curve towards P, the chord is getting shorter in length; when Q coincides with P this chord is effectively replaced by the **tangent** to the curve at P, as in Figure 11.4(b).

As Q moves towards P the value of h gets progressively closer to zero and the expression $(2a + h)$ gets closer to the value $2a$. This suggests that the instantaneous rate of change of $y = x^2$ at $x = 1$ is 2 and at $x = 2$ is 4.

You *could* try to draw the tangents to the curve at $x = 1$ and at $x = 2$ and estimate the gradient of each but it would be a somewhat inaccurate procedure.

We now put these ideas on a formal footing. The **derivative** of the function $f(x) = x^2$ at $x = a$ is $2a$. Note that the derivative is a number. Since the result is true for *any* value a we can define the **derived function** of $f(x) = x^2$ to be the function $g(x) = 2x$.

The derived function of $f(x)$ is written $f'(x)$ which we read as 'f dashed x'. The value of the derived function at $x = a$ is the derivative of $f(x)$ at $x = a$, written $f'(a)$. Hence, in this example $f'(x) = 2x$. As special cases, $f'(1) = 2$ and $f'(2) = 4$.

Example

Show that the derived function of $f(x) = mx + c$ is the constant function $f'(x) = m$, hence find the derivative of $f(x)$ at $x = 2$ and $x = -3$.

Let $x_1 = a$ and $x_2 = a + h$. Now $y = f(a) = ma + c$ and $y_2 = f(a + h) = m(a + h) + c$, so the average rate of change of y over the interval $a \leq x \leq a + h$ is

$$\frac{y_2 - y_1}{x_2 - x_1} = \frac{m(a + h) + c - ma - c}{(a + h) - a} = \frac{mh}{h} = m$$

This result is independent of h, so the derived function is given by $f'(x) = m$. Hence the derivatives $f'(2)$ and $f'(-3)$ both have value m. ∎

Note the special case that if $f(x) = x$ then $f'(x) = 1$.

In general, the **derivative** of the function $f(x)$ at $x = a$ is found by considering the fraction $\dfrac{f(a+h) - f(a)}{h}$. If this fraction tends to a limit as h approaches zero, we say that $f(x)$ is **differentiable** at $x = a$ and the limiting value of the fraction is the derivative $f'(a)$. We write

$$f'(a) = \lim_{h \to 0} \left\{ \frac{f(a+h) - f(a)}{h} \right\} \tag{11.1}$$

where $\lim\limits_{h \to 0}$ means 'the limit as h approaches zero'. The word 'differentiable' simply means 'can be differentiated', i.e. the derivative exists at that point. The process of obtaining the derived function is called **differentiation**.

An alternative notation which will be more useful in certain contexts is illustrated in Figure 11.5. P is the point (x, y) and Q is a point *close by* with coordinates $(x + \delta x, y + \delta y)$. We take the symbol δx to mean 'a small increase in x' and the symbol δy to mean 'a small increase in y'. Note that the symbol δ by itself has no meaning.

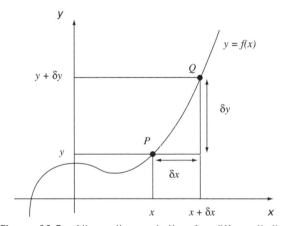

Figure 11.5 Alternative notation for differentiation

The gradient of the chord PQ is therefore $\dfrac{\delta y}{\delta x}$. For the curve $y = x^2$ this is

$$\frac{(y + \delta y) - y}{(x + \delta x) - x} = \frac{(x + \delta x)^2 - x^2}{\delta x} = \frac{2x\delta x + (\delta x)^2}{\delta x} = 2x + \delta x$$

As Q approaches P, $\delta x \to 0$ and the expression $2x + \delta x \to 2x$. Hence the derived function is $2x$.

When we take the limit of $\dfrac{\delta y}{\delta x}$ as x approaches zero, we write the result as $\dfrac{dy}{dx}$. Hence for $y = x^2$, $\dfrac{dy}{dx} = 2x$.

Note that $\dfrac{dy}{dx}$ is a *single entity* not a fraction; for example, the d's do *not* cancel.

Example

Find $\dfrac{dy}{dx}$ for the curve $y = cx^2$ where c is a constant.

The coordinates of P are (x, cx^2) and of Q are $(x + \delta x, c(x + \delta x)^2)$. The gradient of the chord PQ is

$$\frac{\delta y}{\delta x} = \frac{c(x + \delta x)^2 - cx^2}{(x + \delta x) - x} = \frac{2cx(\delta x) + c(\delta x)^2}{\delta x} - = 2cx + c(\delta x)$$

The limit of this expression as $\delta x \to 0$ is given by $\dfrac{dy}{dx} = 2cx$. ∎

Hence for the curve $y = -x^2$, $\dfrac{dy}{dx} = -2x$. Sketch the curve and check this result is reasonable. The table below summarises the main ideas in the two notations with an example.

Table 11.3

Function	$f(x)$	y	x^2
Derived function	$f'(x)$	$\dfrac{dy}{dx}$	$2x$
Derivative	$f'(a)$	$\dfrac{dy}{dx}$ at $x = a$	$2a$

Note that where $\dfrac{dy}{dx} = f'(x)$ is positive the values of $y = f(x)$ *increase* as x increases; where $\dfrac{dy}{dx} = f'(x)$ is negative the values of $y = f(x)$ *decrease* as x increases.

Exercise 11.1

 1 The data below are taken from the graph of $y = x^3$.

x	0	1	2	3	4	5
y	0	1	8	27	64	125

(a) Determine the average rate of change (a.r.o.c.) in each unit interval, e.g. from $x = 2$ to $x = 3$, written $[2, 3]$.

(b) Take $x = 2$, 2.1 and 2.2. Determine $y(2.1)$ and $y(2.2)$ and the average rate of change in the intervals [2.0, 2.1] and [2.1, 2.2].

(c) Repeat (b) but with $x = 2.01$ and 2.02 and the intervals [2.0, 2.01] and [2.01, 2.02].

(d) Set $x = 2 + h$ and determine $y(2 + h)$. If $h \to 0$ what is the limiting value of the average rate of change (a.r.o.c) at $x = 2$?

(e) Repeat (d) but for a general value of x, i.e. x_0. What is the rate of change at $x = x_0$?

2 (a) Expand

(i) $(x_0 + h)^2$ (ii) $(x_0 + h)^3$

(iii) $(x_0 + h)^4$ (iv) $(x_0 + h)^5$

(b) If h is small enough to neglect terms of order h^2 and higher, written $O(h^2)$, write the appropriate forms for the expansions in (a).

(c) Guess at an approximate form for $(x_0 + h)^n$, where n is a positive integer greater than 4. Multiply this form by $x_0 + h$ and ignore terms of $O(h^2)$ to verify that $(x_0 + h)^{n+1} \approx x_0^n(x_0 + (n + 1)h)$.

3 (a) With $x_0 = 1$ and $h = 0.01$ determine the approximate rate of change (a.r.o.c.) of $y = x^6$ in the interval [1.00, 1.01]. Repeat with $h = 0.001$.

(b) What is the exact value of the rate of change at $x_0 = 1$?

4 For $x_0 = 1, 2, 3$ and $h = 0.001$ estimate the approximate rate of change (a.r.o.c.) for $y = 1/x$ at those points.

5 (a) If terms of $O(h^2)$ can be ignored, verify that $(x_0 + h)^{-1} \approx \dfrac{1}{x_0}\left(1 - \dfrac{h}{x_0}\right)$.

(b) The average rate of change (a.r.o.c.) of $y = 1/x$ in $[x_0, x_0 + h]$ is given by

$$\frac{\dfrac{1}{x_0 + h} - \dfrac{1}{x_0}}{h}$$

Simplify this expression. What is the limiting value, i.e. the derivative at $x = x_0$, as $h \to 0$?

6 It is known that the actual rate of change of the function $f(x) = \sin x$ is the derived function $f'(x) = \cos x$. Tabulate $\sin x$ for $x = 0$ (0.1) 0.5 and determine the average rate of change (a.r.o.c.) to 3 s.f. in each interval. Compare it to the cosine value at the midpoint of the interval. Remember x is in radians.

7 Estimate the average rate of change (a.r.o.c.) for $f(x) = e^x$ for $x = 0$ (1) 3. Divide this estimate by e^x itself. Take $\dfrac{f(x+h) - f(x)}{h}$ with $h = 0.01$, then $h = 0.001$. Retain 4 s.f. What conclusion do you reach?

8 Repeat Question 7 for $f(x) = 2^x$, dividing the estimates by 2^x. Use the x^y button on your calculator with $h = 0.01$. Retain 4 s.f.

9 (a) Draw the graph of $f(x) = \ln x$ and the tangents at the points $x = \dfrac{1}{3}, \dfrac{1}{2}, 1, 2, 3$.

(b) Estimate the rate of change (3 s.f.) at these points, taking $h = 0.01$. What conclusion do you reach about $f'(x)$? *Hint*: $\ln 2 = 0.693$ to 3 d.p.

10 The average rate of change can be readily estimate from data tabulated at equally spaced intervals even if the actual function is not known. Given the tabulated data, plot $f(x)$ and $f'(x)$ using the average rate of change estimates.

x	0.0	0.1	0.2	0.3	0.4	0.5
$f(x)$	1.000	0.794	0.725	0.697	0.693	0.707

In this case you should assume that the average rate of change estimates for $f'(x)$ are evaluated at the midpoints, e.g. $f'(0.05)$ is estimated in the interval [0.0, 0.1].

11 The distance D covered by a train departing from a station platform is measured accurately over the first five minutes of its journey.

t (min)	1	2	3	4	5
D (km)	0.7	2.1	4.0	6.5	9.2

(a)   Estimate the mean speed over each minute of the journey.

(b)   What is the mean of all the mean speeds?

(c)   Sketch a graph showing speed and distance.

11.2 SIMPLE DIFFERENTIATION

In this section we extend the idea of differentiation to some of the functions we have already met and to sums, differences and multiples of them.

From the previous section we know the derivative of the functions x, x^2, x^3, x^4 and x^5. The results are summarised in the table below.

Table 11.4

$f(x)$, or y	x	x^2	x^3	x^4	x^5
$f'(x)$, or $\dfrac{dy}{dx}$	1	$2x$	$3x^2$	$4x^3$	$5x^4$

It seems reasonable to suppose that if n is a positive integer then the derived function of $f(x) = x^n$ is $f'(x) = nx^{n-1}$.

What happens if we allow the power of x to be a negative integer? Consider the following example.

Example

Find the derived function of $f(x) = x^{-1}$.

If $f(x) = x^{-1} = \dfrac{1}{x}$ then $f(a) = \dfrac{1}{a}$ and $f(a+h) = \dfrac{1}{(a+h)}$. Therefore

$$\frac{f(a+h)-f(a)}{h} = \frac{1}{h}\left[\frac{1}{a+h} - \frac{1}{a}\right] = \frac{1}{h}\left[\frac{a-(a+h)}{(a+h)a}\right] = \frac{-h}{h(a+h)a} = \frac{-1}{(a+h)a}$$

As $h \to 0$, $(a+h) \to a$ and the whole expression has a limiting value of $-\dfrac{1}{a^2}$.

Since this is true for any value of a (except $a = 0$ for which $f(a)$ is not defined) then the derived function of $x^{-1} = \dfrac{1}{x}$ is $-\dfrac{1}{x^2} = -x^{-2}$. ∎

The result is consistent with the formula above for differentiating x^n.

We show in Chapter 5 of *Mathematics in Engineering and Science* that the formula nx^{n-1} holds true when n is a fraction. In fact we can quote the following *for any real number n*.

$$\text{If } y = x^n \quad \text{then} \quad \frac{dy}{dx} = nx^{n-1} \tag{11.2}$$

We quote three rules without proof that can be used generally, *provided* the functions concerned can be differentiated. Here $f(x)$ and $g(x)$ are two functions and α and β are two real numbers.

1. **Scalar multiple**: the derivative of a scalar multiple of a function is that scalar multiplied by the derivative of the function, i.e. the derivative of $\alpha f(x)$ is $\alpha f'(x)$.

2. **Sum**: the derivative of the sum of two functions is the sum of their derivatives, i.e. the derivative of $f(x) + g(x)$ is $f'(x) + g'(x)$.

3. **Difference**: the derivative of the difference of two functions is the difference of their derivatives, i.e. the derivative of $f(x) - g(x)$ is $f'(x) - g'(x)$.

Note that these three rules can be covered by a single general rule:

4. **Linear combination**: the derivative of a linear combination of two functions is the same linear combination of their derivatives, i.e. the derivative of $\alpha f(x) + \beta g(x)$ is $\alpha f'(x) + \beta g'(x)$.

Putting $\beta = 0$ gives Rule 1, putting $\alpha = \beta = 1$ gives Rule 2 and putting $\alpha = 1, \beta = -1$ gives Rule 3.

Note that when we use the term 'derivative' this means the results are true for a particular value, $x = a$ say. If there are no restrictions on a then we should strictly replace 'derivative' by 'derived function'.

In the alternative notation we let $u = f(x)$ and $v = g(x)$. Then the rules become

$$
\begin{aligned}
&1. \quad \frac{d}{dx}(\alpha u) = \alpha\frac{du}{dx} \\[2mm]
&2. \quad \frac{d}{dx}(u + v) = \frac{du}{dx} + \frac{dv}{dx} \\[2mm]
&3. \quad \frac{d}{dx}(u - v) = \frac{du}{dx} - \frac{dv}{dx} \\[2mm]
&4. \quad \frac{d}{dx}(\alpha u + \beta v) = \alpha\frac{du}{dx} + \beta\frac{dv}{dx}
\end{aligned}
\tag{11.3}
$$

Examples   ✓ 1. Differentiate the following using (11.2) and Rule 1.

   (a) x^9 (b) x^{-6} (c) $x^{1/2}$ (d) $x^{-2/3}$

   (e) $5x^7$ (f) $-2x^3$ (g) $4x^{1/5}$ (h) $-3x^{-3/4}$

Solution

(a) Here $n = 9$ and the answer is $9x^8$.

(b) With $n = -6$ the answer becomes $-6x^{-7}$.

(c) Since $n = \dfrac{1}{2}$ the answer is $\dfrac{1}{2}x^{-1/2}$

(d) With $n = -\dfrac{2}{3}$ the answer becomes $-\dfrac{2}{3}x^{-5/3}$.

(e) $5 \times 7x^6 = 35x^6$ (f) $(-2) \times 3x^2$.

(g) $4 \times \dfrac{1}{5}x^{-4/5} = \dfrac{4}{5}x^{-4/5}$ (h) $(-3) \times \left(-\dfrac{3}{4}x^{-7/4}\right) = \dfrac{9}{4}x^{-7/4}$

 2. Differentiate the following using Rules 2, 3 and 4

   (a) $x^2 + 1$ (b) $x^2 - x$ (c) $4x^2 - 3x$

   (d) $x^2 + 4x + 3$ (e) $-2x^3 + 4x^2 + 3x - 5$

Solution

(a) Via Rule 2 we obtain $2x + 0 = 2x$.

(b) Via Rule 3 we obtain $2x - 1$.

(c) The result is $4(2x) - 3(1) = 8x - 3$.

(d) The result is $1(2x) + 4(1) + 3(0) = 2x + 4$

(e) The result is $-2(3x^2) + 4(2x) + 3(1) - 5(0) = -6x^2 + 8x + 3$.   ■

After practice you will be able to write down the answers to such problems by carrying out the multiplications mentally.

Note that when we differentiate a polynomial of degree n,
i.e. $ax^n + a_{n-1}x^{n-1} + \cdots + a_0$, where $a_n, a_{n-1}, \ldots, a_0$ are real numbers and a_n is not zero, we obtain a polynomial of degree $n - 1$.

Trigonometric functions

Figure 11.6 shows the graph of $y = \sin x$ marked with nine points, A to I. On the lower graph we have marked corresponding points A' to I'. The lower graph has values at each x which are the gradients at the corresponding points on the first graph.

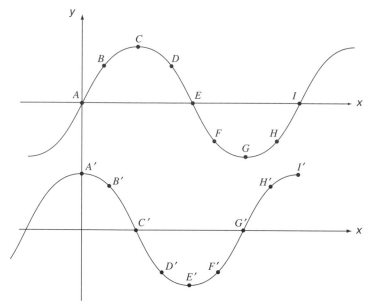

Figure 11.6   Gradients of the sine function

The value of the second function at A' is 1, which is the value of the *gradient* on the first graph at A; the gradient on the first graph at B is $\frac{1}{2}$, which is the value of the second function at B'. It is clear that the gradient on the first graph at C is zero and this is the value on the second graph at C', and so on. From the periodic nature of the function $f(x) = \sin x$ we may also expect the gradients to be periodic, e.g. the situation at I should be identical to the situation at A.

The lower graph looks very similar to $y = \cos x$; in fact this is exactly what the curve is. Without proof we state that

$$\text{If } y = \sin x \quad \text{then} \quad \frac{dy}{dx} = \cos x. \tag{11.4}$$

Similar consideration of the graph of $y = \cos x$ leads to the following result

$$\text{If } y = \cos x \quad \text{then} \quad \frac{dy}{dx} = -\sin x. \tag{11.5}$$

We state without proof two further results.

$$\text{If } y = e^x \quad \text{then} \quad \frac{dy}{dx} = e^x. \qquad (11.6)$$

$$\text{If } y = \ln x \quad \text{then} \quad \frac{dy}{dx} = \frac{1}{x}. \qquad (11.7)$$

It would be instructive for you to verify the plausibility of results (11.5), (11.6) and (11.7) by consideration of the graphs concerned.

Example   A stone is thrown vertically upwards with initial speed u. Its height above the point of projection at subsequent times t is given by

$$y = ut - \frac{1}{2}gt^2$$

where g is the acceleration due to gravity. Find its velocity at any time t and determine where the speed is zero. What do you conclude about the motion of the stone at this time?

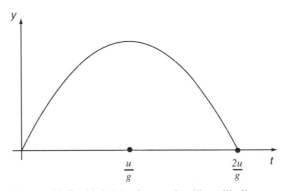

Figure 11.7 Height of a projectile with time

The velocity is given by

$$v = \frac{dy}{dt} = u - \frac{1}{2}g \times 2t = u - gt \quad \text{(positive upwards)}$$

When $t = \dfrac{u}{g}$, the speed is zero. The stone is then at its highest point. Subsequently $\dfrac{dy}{dt} = v < 0$, indicating that the stone is descending again.

Figure 11.7 shows the graph of y against t. Note that $y = 0$ when $ut - \dfrac{1}{2}gt^2 = 0$, i.e.

$2ut - gt^2 = 0$ or $t(2u - gt) = 0$. Therefore, $y = 0$ at $t = 0$ or at $t = \dfrac{2u}{g}$. ∎

In general, if a parabola crosses the horizontal axis at two distinct points then its highest (or lowest) value is taken midway between the two crossing values.

Scaled variables

Consider the graphs of $y = \sin x$ and $y = \sin 2x$ in Figure 11.8(a) and (b) respectively. Because the graph of $y = \sin 2x$ can be obtained from the graph of $y = \sin x$ by compressing it horizontally by a factor of 2, the gradients on the graph of $y = \sin 2x$ are steeper.

These ideas make *plausible* the following result

$$\text{If } y = \sin kx \quad \text{then} \quad \frac{dy}{dx} = k \cos kx. \tag{11.8}$$

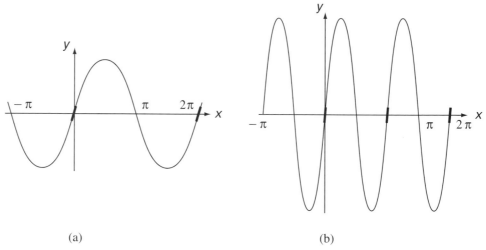

(a)   (b)

Figure 11.8 Scaling the sine function: (a) $y = \sin x$ and (b) $y = \sin 2x$

The factor $\cos kx$ implies that the gradients repeat with the same periodicity as the function and the factor k emphasises the relative change in steepness of the gradients. A formal justification will be given in Chapter 5 of *Mathematics in Engineering and Science*.

Another useful result which we state without proof is as follows.

$$\text{If } y = e^{kx} \quad \text{then} \quad \frac{dy}{dx} = ke^{kx}. \tag{11.9}$$

Examples

1. Newton's law of cooling states that the rate at which a 'hot' liquid cools is directly proportional to the temperature difference between the liquid and the surrounding air. Write the mathematical equivalent of this law using θ for the temperature of the liquid at time t, θ_s for the temperature of the air (assumed constant) and k as a constant of proportionality (assumed positive).

    Show that $\theta = \theta_s + (\theta_0 - \theta_s)e^{-kt}$ satisfies the equation you produce where θ_0 is the initial temperature of the liquid (i.e. the temperature at $t = 0$). (Note that the derived function of e^{-kt} is $(-k)e^{-kt}$).

    Solution

    The rate of change of temperature with time is $\dfrac{d\theta}{dt}$. The statement 'is proportional to' means 'is equal to a constant times' and the temperature difference is $(\theta - \theta_s)$. We cannot equate $\dfrac{d\theta}{dt}$ and $k(\theta - \theta_s)$ because $k(\theta - \theta_s)$ is positive since $k > 0$ and $\theta > \theta_s$ for the liquid to cool. If $\dfrac{d\theta}{dt} > 0$ then θ *increases* with time, which is clearly wrong. We require $\dfrac{d\theta}{dt} < 0$, so we write $\dfrac{d\theta}{dt} = -k(\theta - \theta_s)$, which is the required equation. (In Chapter 11 of *Mathematics in Engineering and Science* we see how to solve this equation. For the moment, we take the given 'solution' and verify that it does satisfy the equation.)

    Using the given derived function of e^{-kt} we apply Rule 1 so that

    $$\frac{d}{dt}\{(\theta_0 - \theta_s)e^{-kt}\} = -k(\theta_0 - \theta_s)e^{-kt}$$

The derivative of the constant θ_s is zero and by Rule 2 we have

$$\frac{d\theta}{dt} = 0 - k(\theta_0 - \theta_s)e^{-kt}$$

But from the given 'solution'

$$\theta - \theta_s = (\theta_0 - \theta_s)e^{-kt}$$

therefore $\dfrac{d\theta}{dt} = -k(\theta - \theta_s)$, as required. (Note that when $t = 0$,

$\theta = \theta_s + (\theta_0 - \theta_s) \times 1 = \theta_0$ and as $t \to \infty$, $e^{-kt} \to 0$ and $\theta \to \theta_s$ from above.)

2. In the previous example at what time is the rate of cooling *half* the initial rate?

Solution

The initial rate of cooling is $-k(\theta_0 - \theta_s)$, by putting $\theta = \theta_0$ in the formula for $\dfrac{d\theta}{dt}$. And when

$$\frac{d\theta}{dt} = -\frac{1}{2}k(\theta_0 - \theta_s)$$

we have

$$-k(\theta - \theta_s) = \frac{1}{2}k(\theta_0 - \theta_s)$$

or $\qquad\qquad \theta - \theta_s = \frac{1}{2}(\theta_0 - \theta_s)$

i.e. $\qquad (\theta_0 - \theta_s)e^{-kt} = \frac{1}{2}(\theta_0 - \theta_s)$

hence $\qquad\qquad e^{-kt} = \frac{1}{2}$

and then $\qquad\qquad e^{kt} = 2$

and $\qquad\qquad kt = \ln 2$

so that $\qquad\qquad t = \frac{1}{k}\ln 2$ ■

In general, we can say the derived function of $f(kx)$ is $kf'(kx)$. For example, the derived function of $\cos 4x$ is $4(-\sin 4x) = -4\sin 4x$. The table below lists some important functions and their derivatives. Together with the rules for linear combination, this allows you to differentiate a wide range of functions.

Table 11.5

Function $f(x)$	Derived function $f'(x)$
C, constant	0
x^n	nx^{n-1}
$\sin x$	$\cos x$
$\cos x$	$-\sin x$
$\tan x$	$1/(\cos^2 x)$
e^x	e^x
$\ln x, x > 0$	$1/x$
$\sin kx$	$k \cos kx$
$\cos kx$	$-k \sin kx$
$\tan kx$	$k/(\cos^2 kx)$
e^{kx}	ke^{kx}

Examples $\longrightarrow$ 1. If $y = \ln(kx), k > 0, x > 0$ find $\dfrac{dy}{dx}$ by writing y as the sum of two logarithms. $\ln kx = \ln k + \ln x$ and $\ln k$ is a constant with zero derivative. Hence $\dfrac{dy}{dx} = 0 + \dfrac{1}{x} = \dfrac{1}{x}$.

2. If $y = (x + b)^2$ find $\dfrac{dy}{dx}$ by expanding the power on the right-hand side. Repeat the process for $y = (x + b)^3$.

First $y = (x + b)^2 = x^2 + 2bx + b^2$.

Hence

$$\frac{dy}{dx} = 2x + 2b = 2(x + b).$$

Next

$$y = (x + b)^3 = x^3 + 3bx^2 + 3b^2x + b^3.$$

Hence

$$\frac{dy}{dx} = 3x^2 + 6bx + 3b^2 = 3(x^2 + 2bx + b^2) = 3(x + b)^2.$$

We show in Chapter 5 of *Mathematics in Engineering and Science* that if $y = (x + b)^n$ then $\dfrac{dy}{dx} = n(x + b)^{n-1}$.

3. Repeat the ideas of the previous example for $y = (ax + b)^2$ and $y = (ax + b)^3$.

Solution

First

$$y = (ax + b)^2 = a^2x^2 + 2abx + b^2$$

so that

$$\frac{dy}{dx} = 2a^2x + 2ab = 2a(ax + b)$$

Second

$$y = (ax + b)^3 = a^3x^3 + 3a^2bx^2 + 3ab^2 + b^3$$

so that

$$\frac{dy}{dx} = 3a^3x^2 + 6a^2bx + 3ab^2$$

$$= 3a(a^2x^2 + 2abx + b^2x^2)$$

$$= 3a(ax + b)^2$$

In general, if $y = (ax + b)^n$ then $\dfrac{dy}{dx} = na(ax + b)^{n-1}$.

Exercise 11.2

1   From first principles differentiate

    (a) x^3 (b) x^5 (c) $\dfrac{1}{x^2}$

2   Find the derivatives with respect to x of the following functions of x:

    (a) $3x^2$ (b) $-4x^2$ (c) $8x^3$

    (d) $\dfrac{x^3}{2}$ (e) $x^2 - 7x$ (f) $5x^2 + 6$

    (g) $2x^3 + 5x$ (h) $x^3 - 3x^2$ (i) $3x - 5$

    (j) $7 - 4x$ (k) $5(x + x^2)$ (l) $3(2x - 3x^2)$

    (m) $3x^2 + 5x + 6$ (n) $4x^2 - 7x + 2$ (o) $3 + 6x - 8x^2$

    (p) $2x^3 - 7x^2 + 5$ (q) $5x^3 + 8x - 9$ (r) $\dfrac{x^3}{2} - 4x^2 + 3x$

    (s) $x + \dfrac{1}{x}$ (t) $\dfrac{3}{x} - 4x$ (u) $4x^3 - \dfrac{2}{x}$

    (v) $\dfrac{5}{x} - 8 - 9x^2$

3 Sketch the curve $y = x^2 - 5x + 4$. Find $\dfrac{dy}{dx}$ and evaluate it at the following points:

(a)   the two places where the x-axis is crossed (i.e. the roots of $x^2 - 5x + 4 = 0$)

(b)   the single place where the y-axis is crossed

Where is $\dfrac{dy}{dx} = 0$? What does this mean?

4 Sketch the cubic $y = (x + 3)(x - 2)(x - 5)$.

(a)   Multiply out the right-hand side, determine $\dfrac{dy}{dx}$ and evaluate it at each of the three roots of the equation $y = 0$.

(b)   Solve the quadratic equation $\dfrac{dy}{dx} = 0$ for x. Determine y for these values of x and mark the points on the curve.

5 The velocity v (m s^{-1}) of a particle constrained to move in a straight line is defined to be $v = \dfrac{ds}{dt}$, where s is the displacement (m) and t is the time (s). If $s = 3t - t^3 (t > 0)$ determine the value of v when $t = 0.5$. When is the particle instantaneously at rest?

6 A weight falling freely under gravity obeys the law $s = \dfrac{1}{2}gt^2$, where s is the downward vertical displacement (m), t the time (s) and g is the acceleration due to gravity (9.81 m s^{-2}); air resistance ignored.

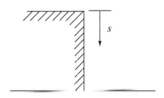

(a) (i) A small object is dislodged from the roof of a building 100 m tall. How long will it take to reach the ground below?

(ii) At what speed will it be travelling then?

(iii) How far has the object fallen at half the time it takes to reach the bottom?

(b) In a similar problem, if the object reached the ground after 6 s, how tall was the building and what would the velocity be on impact?

7 Differentiate the following with respect to the variable concerned:

(a) $\dfrac{(x-3)(x+4)}{x}$

(b) $\dfrac{(x+1)(x-1)(x-2)}{x^2}$

(c) $x^{1/2} + x^{-1/2}$

(d) $\dfrac{(6x-5)(2x+1)}{\sqrt{x}}$

(e) $9 - z^{5/3}$

(f) $\dfrac{p-1}{\sqrt{p}}$

(g) $ax^3 + bx^2 + cx + d \ (x)$

(h) $y^{1/4} - y^{1/5}$

(i) $(2x)^{1/3}$

(j) $\sin 10x$

(k) e^{-11x}

(l) $(e^x + e^{-x})(e^x - 3e^{-x})$

(m) $\ln(-x), x < 0$

(n) $\tan x(\sin^2 x + \cos^2 x)$

8 Differentiate the following with respect to x:

(a) $\ln x^2 \ (= 2\ln x)$ (b) $\ln x^k$ (c) 10^x

In case (c) express 10^x as a power of e.

9 Differentiate $y = ax^2 + bx + c$ to determine the coordinates of the vertex, i.e. where $\dfrac{dy}{dx} = 0$.

10 A cubic function of the form $f(x) = ax^3 + bx^2 + cx + d$ may possess a graph with a variety of shapes, as illustrated. In the second case the slope of $y = f(x)$ is positive for all x. By finding $\dfrac{dy}{dx}$ determine the conditions satisfied by a, b and c for the slope to change sign. In this event what are the implications for the cubic?

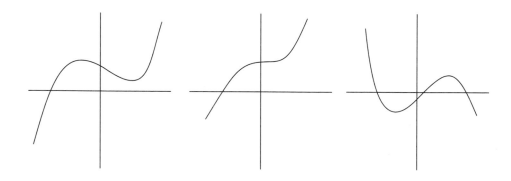

11* (a) If $y = |x|$ determine $\dfrac{dy}{dx}$ for both positive and negative x and draw its graph. Is $\dfrac{dy}{dx}$ defined at $x = 0$?

(b) Given that $y = x|x|$, prove that $\dfrac{dy}{dx} = 2|x|$ $(x \neq 0)$. By considering $\dfrac{\delta y}{\delta x}$ as $\delta x \to 0$ through positive or negative values, prove there exists a common limit equal to zero. Draw the graphs of y and $\dfrac{dy}{dx}$.

12* A road bridge is constructed so the approach roads up to points A and B are of constant gradient and the joining span between A and B is a parabolic arc.

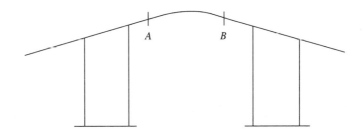

The construction model takes the form illustrated. The slope of each approach road is $1/50$. Determine the equation of each approach road and the parabolic span, noting that its slope is equal to the slope of the approach roads where they join.

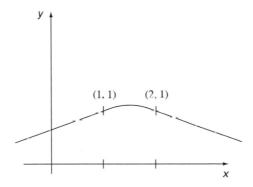

11.3 TANGENTS AND STATIONARY POINTS

We know that $\dfrac{dy}{dx}$ measures the gradient of the tangent to a curve $y = f(x)$ at a given point.

Knowing the coordinates (x_1, y_1) of a point on the curve allows us to determine the equation of the tangent to the curve at that point using the equation $y - y_1 = m(x - x_1)$ where m is the gradient.

The **normal** to the curve at a point is a straight line that is perpendicular to the tangent there, as shown in Figure 11.9.

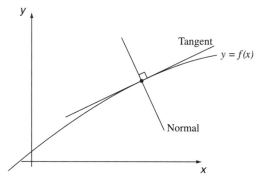

Figure 11.9   Tangent and normal to a curve

Examples

1. Find the equations of the tangent and normal to the curve $y = x^2 - 4x + 3$ at $x = 0$ and at $x = 4$. Find where these tangents meet and where these normals meet.

Solution

For the given curve $\dfrac{dy}{dx} = 2x - 4$. At $x = 0$, $y = 3$ and $\dfrac{dy}{dx} = -4$.

The equation of the tangent is $y - 3 = (-4)(x - 0)$, i.e.

$$y - 3 = -4x \quad \text{or} \quad y = -4x + 3$$

The normal has a gradient of $\dfrac{-1}{-4} = \dfrac{1}{4}$; its equation is

$$y - 3 = \frac{1}{4}(x - 0) \quad \text{or} \quad y = \frac{1}{4}x + 3$$

At $x = 4$, $y = 3$ and $\dfrac{dy}{dx} = 4$. The equation of the tangent is

$$y - 3 = 4(x - 4) \quad \text{i.e. } y = 4x - 13$$

The gradient of the normal is $-\dfrac{1}{4}$ and its equation is

$$y - 3 = -\frac{1}{4}(x - 4) \quad \text{i.e. } y = -\frac{1}{4}x + 4$$

The tangents meet where

$$-4x + 3 = 4x - 13 \quad \text{i.e. } x = 2$$

The equation of either tangent shows that when $x = 2, y = -5$, so the point of intersection is $(2, -5)$. The normals meet where

$$\frac{1}{4}x + 3 = -\frac{1}{4}x + 4 \quad \text{i.e. } x = 2$$

When $x = 2, y = -\dfrac{1}{4} \times 2 + 4 = 3\dfrac{1}{2}$. The point of intersection is $\left(2, 3\dfrac{1}{2}\right)$. Refer to Figure 11.10(a).

2. Find $\dfrac{dy}{dx}$ for the curve $y = \dfrac{1}{x}$. Find the two points on the curve from which normals can be drawn to pass through the origin. Is the gradient on the curve ever zero?

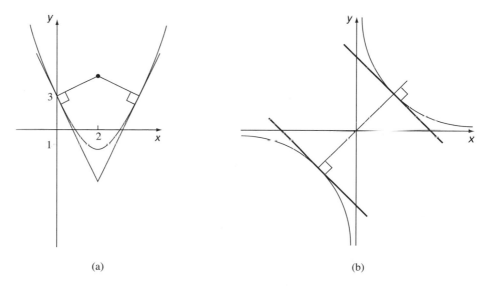

(a)   (b)

Figure 11.10 Tangents and normals for (a) $x^2 - 4x + 3$ and (b) $y = \dfrac{1}{x}$

Solution

On the curve, $\dfrac{dy}{dx} = -\dfrac{1}{x^2}$. The gradient of the normal at (x_1, y_1) is therefore $-1 \Big/ \left(\dfrac{1}{x_1^2}\right) = x_1^2$. The equation of the normal is $y - y_1 = x_1^2(x - x_1)$.

But $y_1 = \dfrac{1}{x_1}$ so that

$$y - \frac{1}{x_1} = x_1^2(x - x_1).$$

If this line passes through the origin then

$$0 - \frac{1}{x_1} = x_1^2(0 - x_1)$$

i.e. $\qquad -\dfrac{1}{x_1} = -x_1^3$

or $\qquad x_1^4 = 1.$

Hence $x_1 = 1$ or -1 and the two points are $(1, 1)$ and $(-1, -1)$.

The gradient is *never* zero but approaches zero as x takes increasingly large values, positive or negative; see Figure 11.10(b). ∎

Horizontal tangents

Consider the function $y = x^2 - 4x + 3$ shown in Figure 11.10(a). We saw that $\dfrac{dy}{dx} = 2x - 4$. When $x < 2, \dfrac{dy}{dx} < 0$ and the function is **decreasing**; when $x > 2, \dfrac{dy}{dx} > 0$ and the function is **increasing**. At $x = 2, \dfrac{dy}{dx} = 0$ and the function is **stationary**.

We can generalise these results to the following statements using the other notation.

> If $f'(x) > 0$ in an interval of x then $f(x)$ is **increasing** in the interval.
> If $f'(x) < 0$ in an interval of x then $f(x)$ is **decreasing** in the interval. $\qquad$ (11.9)
> If $f'(x) = 0$ at a point then $f(x)$ is **stationary** there.

In this example the curve is stationary at the point $(2, -1)$, which is called a **stationary point**. The tangent is horizontal at a stationary point.

Examples

1. $f(x) = \tan x$ is an increasing function everywhere since $f'(x) = \dfrac{1}{\cos^2 x} > 0$, for any x. Refer to Figure 7.26 to remind yourself of the graph of $y = \tan x$.

2. $f(x) = \dfrac{1}{x}$ has no stationary points. $f'(x) = \dfrac{-1}{x^2}$ and is always negative so that on each branch $f(x)$ is decreasing; see Figure 11.10(b).

3. The function $f(x) = x^3 - 3x$ has two stationary points. Since $f'(x) = 3x^2 - 3$ the stationary points occur where

$$3x^2 - 3 = 0 \quad \text{or} \quad x = \pm 1$$

When $x = 1, y = 1^3 - 3 = -2$ and when $x = -1, y = (-1)^3 + 3 = 2$. Hence $A = (-1, 2)$ and $B = (1, -2)$ are stationary points.

Figure 11.11 shows the graph of the function; notice the regions where the function is increasing and the region where it is decreasing. ∎

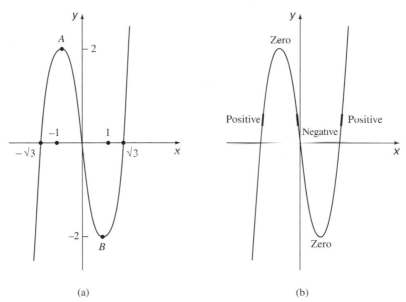

(a)   (b)

Figure 11.11   Gradients on the curve $f(x) = x^3 - 3x$

Types of stationary point

The graph in Figure 11.11 is typical of a cubic function with a positive coefficient for x^3. The feature at A is a *local maximum*; the function is at its highest *in the neighbourhood*; it does take higher values for large positive x, in fact for $x > 2$. The feature at B is a *local minimum*; the function is at its lowest *in the neighbourhood*; it does take lower values for large negative x, in fact for $x < -2$.

We make a formal definition as follows.

> The function $f(x)$ has a **local minimum** at $x = a$ if $f(x) \geq f(a)$ for all x near a.
> The function $f(x)$ has a **local maximum** at $x = a$ if $f(x) \leq f(a)$ for all x near a.
>
> (11.10)

Local minima and local maxima are both examples of stationary points, but there is a third kind. Consider the function $f(x) = x^3$, whose graph is depicted in Figure 11.12.

Now $f'(x) = 3x^2$, and for almost every value of $x, f'(x) > 0$, showing that $f(x)$ is increasing. At $x = 0$, however, $f'(x) = 0$ and the function is stationary there. Yet this point is clearly neither a local minimum nor a local maximum.

If you imagine driving along the curve from negative values of x to positive values, i.e. in a general left-to-right direction, then at $x = 0$ the sense of bending would change from clockwise to anticlockwise.

The point of change in the sense of bending is called a **horizontal point of inflection**. (Some older books call it a *point of contraflexure* to emphasise the change.)

At a local maximum the sense of bending is clockwise, at a local minimum it is anticlockwise. Local maxima and minima are points at which the function changes from decreasing to increasing or vice versa; in other words, the derivative changes sign. They

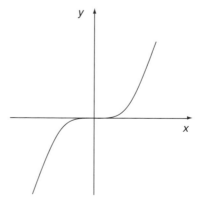

Figure 11.12 Graph of the function $f(x) = x^3$

are also known as **turning points** since the curve effectively changes direction from upwards to downwards or vice versa.

Testing for stationary points

A test to determine the nature of the stationary point employs the first derivative of a function. Table 11.6 shows a pictorial representation of the test where the stationary point is at $x = a$.

Examples

1. Determine the location and nature of the stationary points of $y = x^4 - 6x^2 + 8x + 9$.

Solution

Now $\dfrac{dy}{dx} = 4x^3 - 12x + 8 = 4(x^3 - 3x + 2) = 4(x - 1)^2(x + 2)$. Stationary points occur

when $x = 1$ and $x = -2$.

The only places where $\dfrac{dy}{dx}$ *might* change sign are at $x = -2$ and $x = 1$. Hence we choose as typical values some simple values of x to the left of -2, between -2 and 1, and to the right of 1.

Table 11.6

Sign of $f'(x)$ to the left of $x = a$	Sign of $f'(x)$ to the right of $x = a$	Picture	Decision or definition
Negative	Positive		Local minimum
Positive	Negative		Local maximum
Positive	Positive		Horizontal point of inflection
Negative	Negative		

Consider the sign of $\dfrac{dy}{dx}$ at $x = -3, 0, 2$.

$$\text{At} \quad x = -3, \frac{dy}{dx} = 4(-4)^2(-1) < 0.$$

$$\text{At} \quad x = 0, \quad \frac{dy}{dx} = 4(-1)^2(2) > 0.$$

$$\text{At} \quad x = 2, \quad \frac{dy}{dx} = 4(1)^2(4) > 0.$$

Using the table above we see there is a local minimum at $x = -2$ and a horizontal point of inflection at $x = 1$. The graph of the function is sketched in Figure 11.13.

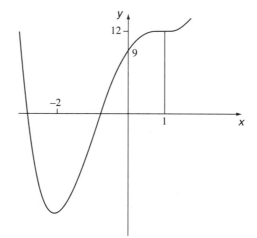

Figure 11.13 Graph of $y = x^4 - 6x^2 + 8x + 9$

2. Locate the stationary points of the function $f(x) = \sin x$ and determine their nature.

Solution

Now $f'(x) = \cos x$, therefore $f'(x) = 0$ where $\cos x = 0$, i.e. at

$$x = \pm\frac{\pi}{2}, \qquad \pm\frac{3\pi}{2}, \qquad \pm\frac{5\pi}{2}, \text{ etc.}$$

There are infinitely many stationary points.

Between $x = -\dfrac{\pi}{2}$ and $x = \dfrac{\pi}{2}$ the derivative does not change sign; if we take $x = 0$ for simplicity we see that $f'(0) = 1 > 0$. Hence the function increases between $x = \dfrac{\pi}{2}$ and $x = \dfrac{\pi}{2}$.

Between $x = \dfrac{\pi}{2}$ and $x = \dfrac{3\pi}{2}$ the derivative does not change sign; if we take $x = \pi$ for simplicity we see that $f'(\pi) = -1 < 0$. Hence the function decreases from $x = \dfrac{\pi}{2}$ to $x = \dfrac{3\pi}{2}$.

Since the derivative changes from positive to negative at $x = \dfrac{\pi}{2}$, this point must be a *local maximum*. At $x = 2\pi, f'(x) = 1 > 0$, so the derivative changes from negative to positive at $x = \dfrac{3\pi}{2}$, indicating a *local minimum*. The periodic nature of $\sin x$ suggests an alternative of the local minima, of value -1, and local maxima, of value 1. Figure 11.14 repeats the graph of $y = \sin x$ to illustrate these ideas.

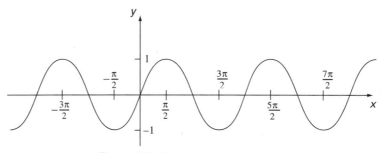

Figure 11.14 Graph of $y = \sin x$

3. Find the stationary points of the function $f(x) = x + \dfrac{4}{x}$.
Here

$$f'(x) = 1 - \frac{4}{x^2} = \frac{x^2 - 4}{x^2}.$$

Hence $f'(x) = 0$ when $x^2 - 4 = 0$, i.e. $x = \pm 2$.

These are the only places where the derivative *might* change sign. For convenience we take $x = -4, 1$ and 4 in turn:

$$f'(-4) = 1 - \frac{4}{16} > 0, \qquad f'(1) = 1 - 4 < 0, \qquad f'(4) = 1 - \frac{4}{16} > 0.$$

Furthermore, when $x = -2, y = -2 - 2 = -4$, and at $x = 2, y = 4$. Hence there is a local maximum at $(-2, -4)$ and a local minimum at $(2, 4)$. The graph of $f(x)$ is shown in Figure 11.15.

The dotted lines $y = x$ and $y = -x$ are not part of the graph; they are **asymptotes** of the curve, as are the x and y axes. Note that the function is not defined at $x = 0$, as demonstrated by the break in its graph. Also, $f'(0)$ does not exist. ■

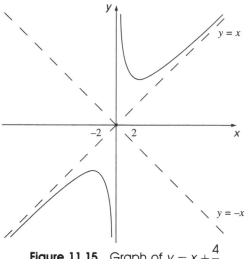

Figure 11.15 Graph of $y = x + \dfrac{4}{x}$

Special cases

The function $f(x) = |x|$ has no derivative at $x = 0$ yet it clearly has a local minimum there.

It is important to distinguish between *local* maximum and minimum values and *global* maximum and minimum values. First note that maxima and minima are known collectively as **extrema**, or extreme points, the singular is **extremum**. The function depicted in Figure 11.16 has a local maximum at P and a local minimum at Q, but in the interval $a \le x \le b$ it has a global minimum at $x = a$ and a global maximum at b.

Sometimes a local extreme value coincides with its global counterpart. For example, the function $f(x) = x^2$ has a local and global minimum at $x = 0$, but if the interval of interest is $1 \le x \le 3$, say, the global minimum is at $x = 1$; see Figure 11.17(a).

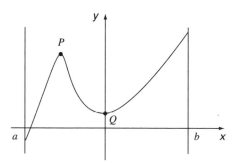

Figure 11.16 Local and global maxima and minima

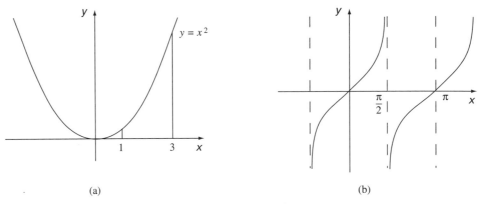

Figure 11.17 Graphs of (a) $y = x^2$ and (b) $y = \tan x$

Figure 11.17(b) depicts part of the graph of $y = \tan x$. In the interval $0 \le x \le \dfrac{\pi}{4}$ the function has a minimum at $x = 0$ and a maximum at $x = \dfrac{\pi}{4}$. Over the interval $0 \le x \le \pi$ it has neither a maximum nor a minimum.

In some situations we do not need calculus at all.

Example

Suppose that we wish to find the global maximum and minimum values of the function

$$f(x) = \frac{1}{[(x-4)^2 + 10]^2}.$$

This looks formidable, but the function clearly takes its greatest value when the expression $(x-4)^2 + 10$ is least, i.e. when $x = 4$; there the expression has a value of 10 and $f(x)$ is $\dfrac{1}{100}$.

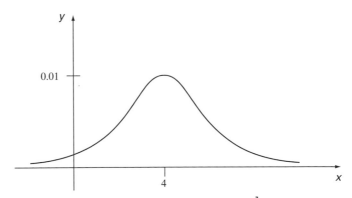

Figure 11.18 Graph of $y = \dfrac{1}{[(x-4)^2 + 10]^2}$

However, the expression $(x-4)^2 + 10$ has no greatest value since its graph is a U-shaped parabola. Since $f(x)$ is the square of some quantity, it can never be negative and for very large (positive and negative) values of x it is approximately given by $\dfrac{1}{x^2}$. The larger the value of x^2, the smaller the value of $f(x)$. Therefore the values of $f(x)$ approach zero but never achieve it. The graph of the function is shown in Figure 11.18. ■

Exercise 11.3

1 Find the tangent and normal to the curves indicated at the given point.

(a) $y = 3 + 5x - x^2$, $x = 2$ (b) $y = 2 - x^2, x = 0$

(c) $y = x^3$, $x = -1$ (d) $y = \dfrac{2}{x}$, $x = \sqrt{2}$

(e) $y = \dfrac{3x+1}{x^2}$, $x = -1$ (f) $y = \sin x$, $x = \dfrac{\pi}{3}$.

Draw a sketch for case (b).

2 On the following curves, determine the coordinates of the points where the gradient is 1.

(a) $y = e^x$ (b) $y = \ln x$.

Show that the curves share a common normal at these points. Draw a sketch.

3 At what points on the curve $y = \tan x$ does the normal have gradient $-\dfrac{1}{2}$?

4 The parabola $y = x^2 - 8x + 12$ has two real zeros.

(a) Determine the equation of the tangent to the parabola at each of the zeros.

(b) Likewise determine the equations of the normals at the zeros.

(c) Deduce that the point where the tangents intersect, the vertex and the point where the normals intersect, all three lie on the axis of symmetry of the parabola.

5 Draw a parabola of your choice which possesses real zeros. At the zeros draw in the tangents and normals. Mark the respective meeting points, as in part (c) of Question 4 and observe that the result holds in general.

6   Draw the parabola $y = x^2 + 3x + 1$.

    (a)   Determine the gradient of the tangent and normal at the point $P_1(1, 5)$.

    (b)   Determine the coordinates of point P_2 on the parabola where the normal and tangent respectively have the same slope as the tangent and normal at $(1, 5)$.

    (c)   Determine the coordinates of the points when the tangent at P_1 meets the normal at P_2 and vice versa.

    (d)   Determine the area of the rectangle comprised of the two points on the parabola and the two intersection points. Draw a sketch.

7   Show that the tangent to the parabola $y = x^2 + x + k$ at the point $x = 1$ always has gradient 3. If it passes through the origin, determine k.

8*   The straight line $y + 4x = k$ is a tangent to the rectangular hyperbola $y = 1/x$ at a point P on the first quadrant, as illustrated. Determine the coordinates of P and the value of k. Verify that $y + 4x = -k$ is a tangent at the point P', vertically opposite P in the third quadrant.

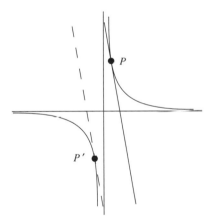

9*   A straight line $y = mx$ is drawn tangentially to the curve $y = \sin x$, meeting it at $x = \alpha$ just to the left of the second positive maximum at $x = \dfrac{5\pi}{2}$.

    (a)   Determine m.

    (b)   Prove that α satisfies the equation $\alpha = \tan \alpha$. Do not attempt to solve the equation.

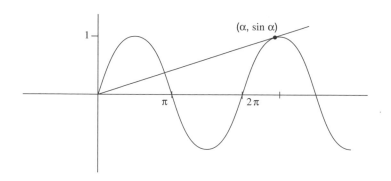

10. If $f(x) = \sqrt{x} - \dfrac{1}{\sqrt{x}}$, $x > 0$ prove that $f'(x) > 0$ for all $x > 0$ and determine the equation of the normal to $f(x)$ at the point $(1, 0)$.

11.4 SECOND DERIVATIVES AND STATIONARY POINTS

The **second derivative** of a function is obtained by applying the process of differentiation to the derived function. We use the notation $f''(x)$ and $\dfrac{d^2y}{dx^2}$.

Example   Find the second derivative of (a) $f(x) = x^6$, (b) $y = \sin x$ and (c) $y = 5\sin x - 3\cos x$.

(a) $f'(x) = 6x^5$ and $f''(x) = 6(5x^4) = 30x^4$

(b) $\dfrac{dy}{dx} = \cos x$ and $\dfrac{d^2y}{dx^2} = \dfrac{d}{dx}(\cos x) = -\sin x$

(c) $\dfrac{dy}{dx} = 5\cos x - 3(-\sin x) = 5\cos x + 3\sin x$

$\dfrac{d^2y}{dx^2} = -5\sin x + 3\cos x$ ∎

Note that in parts (b) and (c) $\dfrac{d^2y}{dx^2} = -y$.

Geometrical interpretation of the second derivative

The quantity $f''(x) = \dfrac{d^2 y}{dx^2}$ measures the rate of change of the gradient $\dfrac{dy}{dx}$.

If $f(x) = x^2$ then $f'(x) = 2x$ and $f''(x) = 2$

If $f(x) = -x^2$ then $f'(x) = -2x$ and $f''(x) = -2$

Figure 11.19(a) shows that the gradient of $y = x^2$ is *increasing* from left to right. From large negative values it goes through small negative values to zero to small positive values and to large positive values. For this curve $\dfrac{d^2 y}{dx^2} > 0$.

Conversely, in Figure 11.19(b) the gradient of $y = -x^2$ is *decreasing* from left to right; for this curve $\dfrac{d^2 y}{dx^2} < 0$.

Curves like $y = x^2$ are said to be **concave upwards** and those like $y = -x^2$ are **concave downwards**. The graphs of many functions are concave upwards in some parts and concave downwards in others. Look at a graph of $y = \sin x$ for an example.

In the region of a local minimum the graph is concave upwards and $\dfrac{d^2 y}{dx^2} > 0$. In the region of a local maximum the graph is concave downwards and $\dfrac{d^2 y}{dx^2} < 0$. What happens at points where $\dfrac{d^2 y}{dx^2} = 0$?

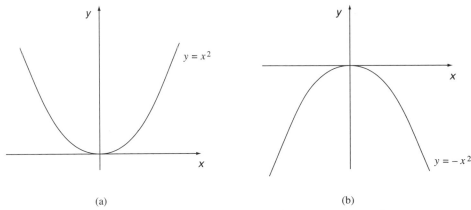

(a)   (b)

Figure 11.19 Graphs of (a) $y = x^2$ and (b) $y = -x^2$

Points of inflection

Consider the graphs of $y = x^3$ and $y = x^3 - 3x$ in Figure 11.20. If $y = x^3$ then $\dfrac{dy}{dx} = 3x^2$ and $\dfrac{d^2y}{dx^2} = 6x$.

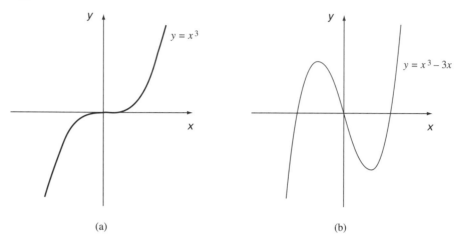

Figure 11.20 Graphs of (a) $y = x^3$ and (b) $y = x^3 - 3x$

When $x < 0, \dfrac{d^2y}{dx^2} < 0$ and the curve is concave downwards. When $x > 0, \dfrac{d^2y}{dx^2} > 0$ and the curve is concave upwards. At $x = 0, \dfrac{d^2y}{dx^2} = 0$ and the curve has a horizontal point of inflection. The sense of bending has changed at $x = 0$ and this is reflected by the change in sign of $\dfrac{d^2y}{dx^2}$ as we pass through the origin.

The graph of $y = x^3 - 3x$ indicates that the curve is concave downwards for $x < 0$ and concave upwards for $x > 0$. At the origin there is again a change in the sense of bending. For this function

$$\frac{dy}{dx} = 3x^2 - 3 \quad \text{and} \quad \frac{d^2y}{dx^2} = 6x$$

Clearly $\dfrac{d^2y}{dx^2} = 0$ when $x = 0$ but the gradient of the function is not horizontal. The feature at $x = 0$ is known simply as a **point of inflection**.

One word of warning. Simply because $\dfrac{d^2y}{dx^2} = 0$ at a point, it does not immediately follow there is a point of inflection. Consider the graph of $y = x^4$ shown in Figure 11.21. It is obvious that the curve has a local (and global) minimum at the origin, but $\dfrac{d^2y}{dx^2} = \dfrac{d}{dx}(4x^3) = 12x^2$, which is zero at $x = 0$.

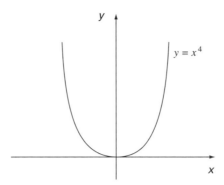

Figure 11.21   Graph of $y = x^4$

We can base a test for stationary points and points of inflection on the second derivative as follows.

If $f'(a) = 0$ and $f''(a) > 0$ then $f(x)$ has a local minimum at $x = a$.

If $f'(a) = 0$ and $f''(a) < 0$ then $f(x)$ has a local maximum at $x = a$.

If $f'(a) \neq 0$ and $f''(a) = 0$ then $f(x)$ has a point of inflection at $x = a$.

If $f'(a) = 0$ and $f''(a) = 0$ then further investigation is needed.

Example

Locate the points of inflection on the following curves:

(a) $y = x^3 - x^2$ (b) $y = \sin x$

(c) $y = 3x^4 - 8x^3 + 6x^2 + 24x$ (d) $y = e^{-x}$

(a) $\dfrac{dy}{dx} = 3x^2 - 2x, \dfrac{d^2y}{dx^2} = 6x - 2$; when $x = \dfrac{1}{3}, \dfrac{d^2y}{dy^2} = 0$ but $\dfrac{dy}{dx} = 3 \times \dfrac{1}{9} - \dfrac{2}{3} \neq 0$. Hence *the* point of inflection occurs when $x = \dfrac{1}{3}$.

(b) $\dfrac{dy}{dx} = \cos x, \dfrac{d^2y}{dx^2} = -\sin x$. When x is a multiple of π then $\dfrac{d^2y}{dx^2} = 0$ but $\cos x = 1$ or -1 at such points, hence the points of inflection occur infinitely often, sandwiched between the alternate local maxima and minima. Refer to a graph of $y = \sin x$.

(c) $\dfrac{dy}{dx} = 12x^3 - 24x^2 + 12x + 24,$

$\dfrac{d^2y}{dx^2} = 36x^2 - 48x + 12 = 12(3x^2 - 4x + 1) = 12(3x - 1)(x - 1).$

When $x = 1$, $\dfrac{d^2y}{dx^2} = 0$ but $\dfrac{dy}{dx} = 24 \neq 0$. When $x = \dfrac{1}{3}$, $\dfrac{d^2y}{dx^2} = 0$ but $\dfrac{dy}{dx} = 25\dfrac{7}{9} \neq 0$.

Hence there *are* points of inflection, at $x = 1$ and $x = \dfrac{1}{3}$.

(d) $\dfrac{dy}{dx} = -e^{-x}$, $\dfrac{d^2y}{dx^2} = +e^{-x}$ and $\dfrac{d^2y}{dx^2}$ is never zero, so that there are no points of inflection on the curve. ∎

For locating and identifying stationary points, we now have two tests: one is based on the first derivative at the stationary point and on either side of it, the other is based on the first and second derivatives at the stationary point.

Other than the case where both $f'(x)$ and $f''(x)$ are zero for the same value of x, the second derivative test is probably easier to apply, provided the differentiations do not become too complicated.

It might be argued that the first derivative test is foolproof so it should always be used. Remember, though, it will only pick out points of inflection which have a horizonal gradient. The choice depends partly on what is sought and partly on your preference.

Examples

1. A rectangular plot of land is to be enclosed by brick walls. If the area enclosed is $400\,\text{m}^2$, what dimensions of the plot will require the least perimeter and what is this least value? Refer to Figure 11.22.

Solution

Let the dimensions of the plot be x m by y m. Then the area is $xy\,\text{m}^2$ so that

$$xy = 400 \quad \text{or} \quad y = 400/x$$

The perimeter is given by $p = 2x + 2y$. Substituting $y = 400/x$ we obtain $p = 2x + \dfrac{800}{x}$. Differentiating twice in succession produces

$$\frac{dp}{dx} = 2 - \frac{800}{x^2}, \qquad \frac{d^2p}{dx^2} = \frac{1600}{x^3}.$$

Now $\dfrac{dp}{dx} = 0$ when $2 - \dfrac{800}{x^2} = 0$, i.e. $x^2 = 400$; the only relevant solution is $x = 20$. Then $y = \dfrac{400}{x} = 20$, too, so the rectangle is a square and $p = 80$ m. Since $\dfrac{d^2p}{dx^2} > 0$ this solution represents a minimum value of p.

2. An object moves so its displacement from a fixed point is given by $s = \sin 3t + \cos 3t$. When does s take its maximum and minimum values? Interpret your results.

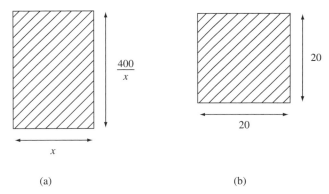

Figure 11.22 Diagrams to show the rectangular plot of land

Solution

Differentiating twice we obtain

$$\frac{ds}{dt} = 3\cos 3t - 3\sin 3t, \quad \frac{d^2s}{dt^2} = 3(-3\sin t) - 3(3\cos 3t) = -9\sin 3t - 9\cos 3t$$

$$\left(\text{Note that } \frac{d^2s}{dt^2} = -(3)^2 \times s \right).$$

Now $\frac{ds}{dt} = 0$ when $3\cos 3t - 3\sin 3t = 0$ i.e.

$$\sin 3t = \cos 3t \quad \text{or} \quad \tan 3t = 1.$$

This equation is satisfied when $3t = \dfrac{\pi}{4}, \dfrac{5\pi}{4}, \dfrac{9\pi}{4}$, etc., so $t = \dfrac{\pi}{12}, \dfrac{5\pi}{12}, \dfrac{9\pi}{12}$, etc. When

$$3t = \frac{\pi}{4}, \quad \frac{d^2s}{dt^2} = -9 \times \frac{1}{\sqrt{2}} - 9 \times \frac{1}{\sqrt{2}} < 0.$$

When

$$3t = \frac{5\pi}{4}, \quad \frac{d^2s}{dt^2} = -9\left(-\frac{1}{\sqrt{2}} \right) - 9\left(-\frac{1}{\sqrt{2}} \right) > 0.$$

When

$$3t = \frac{9\pi}{4}, \quad \frac{d^2s}{dt^2} = -9 \times \frac{1}{\sqrt{2}} - 9 \times \frac{1}{\sqrt{2}} < 0, \text{ etc.}$$

Hence at $t = \dfrac{\pi}{12}, \dfrac{9\pi}{12}, \dfrac{17\pi}{12}$, etc., s is at its maximum of $\dfrac{1}{\sqrt{2}} + \dfrac{1}{\sqrt{2}} = \dfrac{2}{\sqrt{2}} = \sqrt{2}$. At $t = \dfrac{5\pi}{12}$, $\dfrac{13\pi}{12}, \dfrac{21\pi}{12}$, etc., s is at its minimum of $-\dfrac{1}{\sqrt{2}} - \dfrac{1}{\sqrt{2}} = -\sqrt{2}$.

This displacement describes a (simple harmonic) oscillation about the fixed point. The maximum represents the greatest distance travelled in the direction of positive s and the minimum represents the greatest distance travelled in the opposite direction; together they define the points between which the object oscillates.

3. A firm mass-produces a computer component. The unit price p (in thousands of £) is related to the demand x (in hundreds of units) by the equation

$$x = \left(\frac{90}{p}\right)^2.$$

The cost C (in £) producing the components is given by

$$C = 5(x + 24)$$

(a) Find the level of demand x for which the profit is maximum.

(b) Find the unit price p at which this occurs.

(c) Find the maximum profit.

Solution

The revenue R gained from selling x units is $R = xp$. Now $x = \dfrac{8100}{p^2}$ so that $p = \left(\dfrac{8100}{x}\right)^{1/2} = \dfrac{90}{x^{1/2}}$. Hence the revenue R is $x\left(\dfrac{90}{x^{1/2}}\right) = 90x^{1/2}$. The profit

$$P = \text{revenue} - \text{costs} = R - C$$
$$= 90x^{1/2} - 5(x + 24) = 90x^{1/2} - 5x - 120.$$

Now

$$\frac{dP}{dx} = \frac{1}{2}\frac{90}{x^{1/2}} - 5 = \frac{45}{x^{1/2}} - 5 \quad \text{and} \quad \frac{d^2P}{dx^2} = -\frac{1}{2}\frac{45}{x^{3/2}}.$$

Hence $\dfrac{dP}{dx} = 0$ when $\dfrac{45}{x^{1/2}} = 5$ or $x^{1/2} = 9$, i.e. $x = 81$. At this value of x, $\dfrac{d^2P}{dx^2} < 0$, so this represents a maximum profit. When $x = 81, p = \dfrac{90}{(81)^{1/2}} = \dfrac{90}{9} = 10$ and $P = 90(9) - 5(81) - 120 = 285$.

Hence the maximum profit of £285 000 is achieved when 8100 units are produced at a selling price of £10 000 per 100 items. ■

Exercise 11.4

1 Which of the following functions have maximum or minimum values? State the values of x, if any, which gives those values of the functions.

(a) $x^2 - 8x$ (b) $7x - x^2$ (c) $3x^2 + 4x$

(d) $1 - x^3$ (e) $(x - 1)(x - 4)$ (f) $(x + 2)(x - 5)$

(g) $7 + 5x - 2x^2$.

2 For the following functions identify any maxima, minima or points of inflection, stating the values of x for which they occur. Sketch a graph for each function.

(a) $x^3 - 3x^2 + 3x$ (b) $x^2(x - 6)$ (c) $x(x^2 - 12)$

(d) $x(x - 5)(x - 8)$ (e) $(x + 1)(x - 1)^2$ (f) $(1 + x)^2(2 - x)$

(g) $x + \dfrac{9}{x}$ (h) $x^2 + 2x$.

3 Find the maximum and minimum values of $f(x) = (x - 1)(x - 2)/x$ and illustrate your result by sketching the graph of the function between $x = \pm 3$.

4 Find the area of the largest rectangular piece of ground that can be enclosed by 100 m of fencing if part of an existing wall can be used.

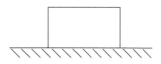

5 A rectangular sheet of cardboard is 8 m long and 5 m wide. Equal shares are cut away at each corner and the remainder is folded to form an open box. Find the maximum volume.

6 A particle which moves in a straight line relative to a fixed origin obeys the law

$$s = \frac{2t^3}{3} - \frac{9t^2}{2} + 10t$$

where s is the distance (m) and t the time (s).

(a) Find an expression for the *velocity*. When is the particle instantaneously at rest?

(b) What is the acceleration?

(c) What is the maximum displacement?

(d) What ultimately happens to the particle?

7   A stone is thrown vertically upwards from the ground. Neglecting air resistance, it obeys the law $s = ut - \dfrac{1}{2}gt^2$ where s is the upward displacement (m), u the initial upward projection velocity ($\mathrm{m\,s^{-1}}$) and g the acceleration due to gravity ($\mathrm{m\,s^{-2}}$) measured downwards, hence the negative sign.

(a)   Differentiate s and verify that $u = \dfrac{ds}{dt}$ when $t = 0$.

(b)   When does the stone reach its maximum height?

(c)   What is the maximum height?

8   (a)   A person drops a stone into the sea from a vertical cliff 100 m high. How long does it take for the stone to reach the sea?

(b)   Suppose, the person threw the stone upwards at $10\,\mathrm{m\,s^{-1}}$.
(i)   To what height would it rise?
(ii)   How much longer would it take the stone to reach the sea?
(Take g to be $10\,\mathrm{m\,s^{-2}}$)

9   Prove that $f(x) = \dfrac{x^2 - 4x - 1}{x}$ can have no maxima or minima when $x > 0$.

10   The function f is defined as follows:

$$f(x) = \begin{cases} 2x(x-1) & x < 1 \\ (x-1)(x-2)(x-3) & x \geq 1 \end{cases}$$

Find the maximum and minimum values of $f(x)$ and identify its least value. Sketch $f(x)$ and $f'(x)$ together on the same axes.

11   Given that $f(x) = x^3 + ax^2 + bx + c$ has a maximum at $x = -2$ and a minimum when $x = \dfrac{2}{3}$, determine a, b, c.

12   The volume of a cylinder is $16\pi\,\mathrm{cm}^3$. If its total surface area is the least possible, find its radius.

13   The amount, W litres, of fuel used by an aircraft flying a certain distance at a speed $v\,\mathrm{km\,min^{-1}}$ is given by $W = 25v^2 + \dfrac{400}{v^2}$.

(a)   Find the rate of change of W with respect to v when $v = 4$.

(b)   Find the smallest number of litres required for the distance and the economical cruising speed.

14* A ship is $10\,\mathrm{km}$ north of A and another is $9\,\mathrm{km}$ east of A. Starting at the same time, the two ships move south and west respectively at $6\,\mathrm{kph}$. If the distance between the two ships after t hours is $x\,\mathrm{km}$, show that $x^2 = 72t^2 - 228t + 181$. Find the time when the ships are nearest to one another by finding the value for t where x^2 is least. What is the shortest distance apart?

15* Which of the following functions are concave downwards, upwards or neither?

(a) x^2 (b) x^3 (c) $-x^4$ (d) e^x

(e) $\ln x$ (f) $x - 1/x^2$ (g) $x^3 + 3x^2 + 6x + 11$

16* The positive function f is defined for $x > 0$. It is both decreasing and concave upwards.

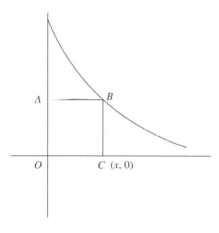

(a)   What conditions must be fulfilled for the rectangle $ABCO$ to have maximum area?

(b)   For $f(x) = e^{-x}$, defined for $x > 0$, find the rectangle of maximum area. You may assume that $\dfrac{d}{dx}(xe^{-x}) = (x - 1)e^{-x}$.

17 Determine the derivative of the following functions to the order stated.

(a) $\sin x$ (3)   (b) x^7 (4)   (c) $\ln x$ (5)

(d) $x^6 - 19x^5 + 26x^4 - 5x^3 + 3x^2 - 11x + 1$ (6)

(e)   As (d) (7)   (f) $x^{1/3}$ (5)

18   Keep differentiating e^{-x}, recording the result each time. What do you observe?

19   Repeat Question 18 for $\sin x$.

20   Differentiate x^k a number of times. Consider the cases

(a) k a positive integer   (b) k not a positive integer

What do you observe?

21*   Sketch the graph of $f(x) = 1 + (x - 1)^4$. Show that $f^{(k)}(1) = 0, k \leq 3$. Set $x = 1 + \varepsilon$, where ε is small, positive or negative, and by finding $f(x)$ near $x = 1$ prove that f has a minimum at $x = 1$.

SUMMARY

- **Differentiation of a function**: a function $f(x)$ is differentiated to give the derived function $f'(x)$. The value $f'(a)$ is the derivative of $f(x)$ at $x = a$. The derivative measures the instantaneous rate of change of the function.

$$f'(a) = \lim_{h \to 0} \frac{f(a+h) - f(a)}{h}$$

- **Linear combination**:

$$\frac{d}{dx}\{\alpha f(x) + \beta g(x)\} = \alpha f'(x) + \beta g'(x)$$

- **Gradient of the curve**: $y = f(x)$ has gradient $\dfrac{dy}{dx} = f'(x)$.

- **Increasing and decreasing**: are properties of a function $f(x)$ revealed by its derivative.

$f'(a) > 0$	$f(x)$ is increasing at $x = a$
$f'(a) < 0$	$f(x)$ is decreasing at $x = a$
$f'(a) = 0$	$f(x)$ has a stationary point at $x = a$

- **There is a local minimum** at $x = a$ if $f'(x) < 0$ for $x < a$ and $f'(x) > 0$ for $x > a$.

- **There is a local maximum** at $x = a$ if $f'(x) > 0$ for $x < a$ and $f'(x) < 0$ for $x > a$.

- **There is a point of inflection** at $x - a$ if $f'(x)$ has the same sign on both sides of a stationary point.

- **The second derivative** of a function is the derivative of the first derivative, written $f''(x)$ or $\dfrac{d^2y}{dx^2}$.

- **Second derivative test** for a stationary point at $x = a$

  If $f''(a) > 0$ the point is a local minimum
  If $f''(a) < 0$ the point is a local maximum
  If $f''(a) = 0$ we need further investigation

- **Point of inflection**: if $f''(a) = 0$ then when $f'(a) \neq 0$ the function has a point of inflection at $x = a$; if $f'(a) = 0$ we need to investigate further.
- **Some derived functions**

$f(x)$	x^n	e^{kx}	$\sin kx$	$\cos kx$	$\ln kx$
$f'(x)$	nx^{n-1}	ke^{kx}	$k \cos kx$	$-k \sin kx$	$\dfrac{1}{x}$

Answers

Exercise 11.1

1 (a)

x	0	1	2	3	4	5
y	0	1	8	27	64	125
a.r.o.c.		1	7	19	37	61

(b)

x	2		2.1		2.2
y	8		9.261		10.648
a.r.o.c.		12.61		13.87	

(c)

x	2.00		2.01		2.02
y	8		8.120 601		8.242 408
a.r.o.c.		12.0601		12.1807	

(d)   a.r.o.c. $= 12 + 6h + h^2 \to 12$ as $h \to 0$

(e)   At $x = x_0$, a.r.o.c. $= 3x_0^2 + 3x_0 h + h^2 \to 3x_0^2$ as $h \to 0$

2 (a,b) $x_0^2 + 2x_0 h + h^2 \approx x_0^2 + 2x_0 h$

$x_0^3 + 3x_0^2 h + 3x_0 h^2 + h^3 \approx x_0^3 + 3x_0^2 h$

$x_0^4 + 4x_0^3 h + 6x_0^2 h^2 + 4x_0 h^3 + h^4 \approx x_0^4 + 4x_0^3 h$

(c) $(x_0 + h)^n \approx x_0^n + nx_0^{n-1}h = x_0^{n-1}(x_0 + nh)$

3 (a)   a.r.o.c. $(h = 0.01) = 6.152(3$ d.p.$)$
a.r.o.c. $(h = 0.001) = 6.015$ $(3$ d.p.$)$

(b)   6

4

x	1	2	3
$1/x$	1.000	0.500	0.333
a.r.o.c. $(4$ d.p.$)$	-0.9901	-0.2488	-0.1107

5 (b)   a.r.o.c. $= \dfrac{-1}{x_0(x_0 + h)}$; limiting value $= \dfrac{-1}{x_0^2}$

6

x	0.0		0.1		0.2		0.3		0.4		0.5
$\sin x$	0.000		0.100		0.199		0.296		0.389		0.479
a.r.o.c.		1.00		0.99		0.97		0.93		0.90	
$\cos x$		1.00		0.99		0.97		0.94		0.90	

7

x	0	1	2	3	h
e^x	1.000	2.718	7.389	20.09	
a.r.o.c.	1.005	2.732	7.426	20.19	0.01
a.r.o.c.$/e^x$	1.005	1.005	1.005	1.005	0.01
a.r.o.c.	1.001	2.719	7.393	20.10	0.001
a.r.o.c.$/e^x$	1.001	1.001	1.001	1.001	0.001

Note that for $h = 0.001$ the ratio a.r.o.c.$/e^x$ is 1.0005, rounded 1.001. You can see that (a.r.o.c.$/e^x$) $\to 1$ as $h \to 0$ and that the relative error for $h = 10^{-n}$ is approximately $\frac{1}{2} \times 10^{-n}$.

8

x	0	1	2	3
2^x	1	2	4	8
a.r.o.c.	0.696	1.391	2.782	5.564
a.r.o.c.$/2^x$	0.696	0.696	0.696	0.696

In fact a.r.o.c.$/2^x \simeq 0.695\,555$ in each case. Notice that if $f(x) = 2^x$ then $f'(x) = \ln 2.2^x$, where $\ln 2 \simeq 0.693$.

9 (b)

x	0.333	0.500	1.000	2.000	3.000
$\ln x$	-1.098	-0.693	0	0.693	1.098
a.r.o.c.	2.96	1.98	1.00	0.50	0.33

$$\left(f'(x) = \frac{1}{x} \right)$$

9 (a)

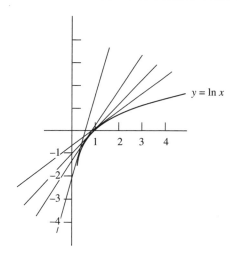

10

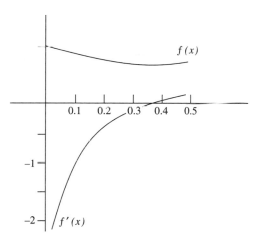

11 (a)

t (min)	1	2	3	4	5
S (kph)	42	84	114	150	162

(b) Mean of five speeds = 110.4 kph.

(c)

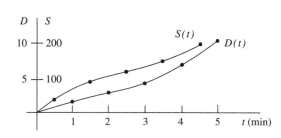

Exercise 11.2

1 (a) $3x^2$ (b) $5x^4$ (c) $-2/x^3$

2 (a) $6x$ (b) $-8x$ (c) $24x^2$ (d) $\dfrac{3x^2}{2}$

(e) $2x-7$ (f) $10x$ (g) $6x^2+5$ (h) $3x^2-6x$

(i) 3 (j) -4 (k) $5+10x$ (l) $6-18x$

(m) $6x+5$ (n) $8x-7$ (o) $6-16x$ (p) $6x^2-14x$

(q) $15x^2+8$ (r) $\dfrac{3x^2}{2}-8x+3$ (s) $1-\dfrac{1}{x^2}$

(t) $-\dfrac{3}{x^2}-4$ (u) $12x^2+\dfrac{2}{x^2}$ (v) $-\dfrac{5}{x^2}-18x$

3

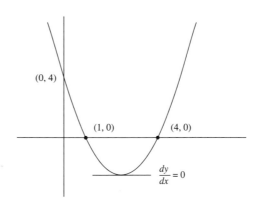

(a) $\dfrac{dy}{dx}=2x-5$; -3 at (1, 0), 3 at (4, 0)

(b) $\dfrac{dy}{dx}=-5$ at (0, 4)

$\dfrac{dy}{dx}=0$ at the vertex (2.5, -2.25) where the slope is zero.

4

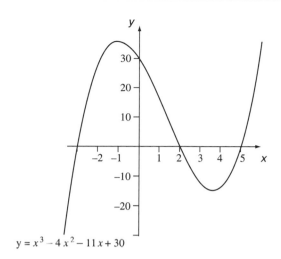

$$y = x^3 - 4x^2 - 11x + 30$$

(a) $\dfrac{dy}{dx} = 3x^2 - 8x - 11 = (3x - 11)(x + 1)$

$\dfrac{dy}{dx} = 40$ at $(-3, 0)$, $= 15$ at $(2, 0)$, 24 at $(5, 0)$

(b) $\dfrac{dy}{dx} = 0$ at $(-1, 36)$ and $\left(\dfrac{11}{3}, -\dfrac{400}{27}\right)$

5 $v - 2.25, t = 1$

6   (a)   (i) 4.52 s   (ii) 44.3 m s^{-1} (iii) 25 m

(b)   (i) 176.6 m   (ii) 58.9 m s^{-1}

7   (a) $1 + 12/x^2$ (b) $1 + 1/x^2 - 4/x^3$ (c) $\dfrac{1}{2\sqrt{x}}\left(1 - \dfrac{1}{x}\right)$

(d) $\dfrac{36x^2 - 4x + 5}{2x\sqrt{x}}$ (e) $-\left(\dfrac{5}{3}\right)z^{2/3}$ (f) $\dfrac{p + 1}{2p\sqrt{p}}$

(g) $3ax^2 + 2bx + c$ (h) $\dfrac{1}{4}y^{-3/4} - \dfrac{1}{5}y^{-4/5}$ (i) $\dfrac{1}{3}\left(\dfrac{2}{x^2}\right)^{1/3}$

(j) $10 \cos 10x$ (k) $-11e^{-11x}$ (l) $2e^{2x} + 6e^{-2x}$

(m) $1/x$ (n) $\sec^2 x$

8   (a) $2/x$ (b) k/x (c) $\ln 10 \times 10^x$

9 $\left(-\dfrac{b}{2a}, c - \dfrac{b^2}{4a}\right)$

10 $b^2 > 3ac$; cubic has a local maximum and a local minimum

11   (a) $\dfrac{d}{dx}(|x|)$

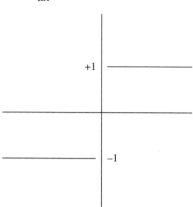

(a)

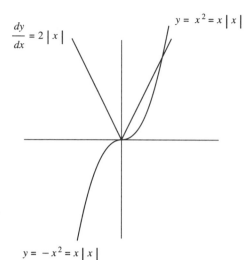

(b)

12   Left span: $y - 1 = \dfrac{1}{50}(x - 1)$

Right span: $y - 1 = -\dfrac{1}{50}(x - 2)$

Join: $y = \dfrac{1}{50}(-x^2 + 3x + 48)$

Exercise 11.3

1   (a) $y = x + 7, x + y = 11$

(b) $y = 2, x = 0$

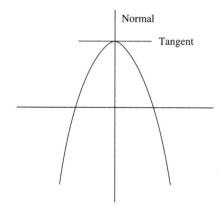

(c) $y = 3x + 2, 3y + x + 4 = 0$

(d) $x + y = 2\sqrt{2}, y = x$

(e) $x + y + 3 = 0, y = x - 1$

(f) $y = \dfrac{x}{2} + \dfrac{\sqrt{3}}{2} - \dfrac{\pi}{6}, \dfrac{y}{2} + x = \dfrac{\sqrt{3}}{4} + \dfrac{\pi}{3}$

2

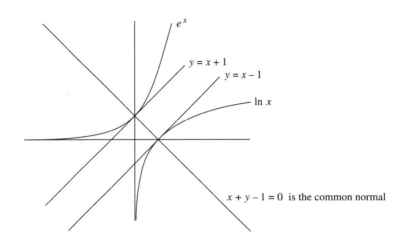

$x + y - 1 = 0$ is the common normal

3 $x = \pm\pi/4$

4   Tangents $y + 4x = 8$ at $(2, 0)$

                $y = 4x - 24$ at $(6, 0)$ meet at $(4, -8)$

     Normals $4y = x - 2$ at $(2, 0)$

                $4y + x = 6$ at $(6, 0)$ meet at $\left(4, \dfrac{1}{2}\right)$

5

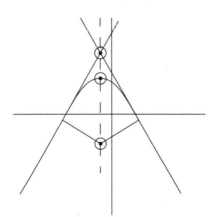

6 (a)   Tangent gradient 5, normal gradient $-\dfrac{1}{5}$ at (1, 5)

    (b) $P_2\left(-\dfrac{8}{5}, -\dfrac{31}{25}\right)$ i.e. $(-1.6, -1.24)$

    (c)   Tangents/normals meet at $(-3, -1.5)$ and $(-0.3, 5.26)$

    (d)   8.788

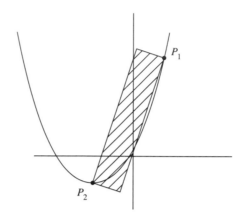

7   Tangent at $(1, 2+k)$, $k=1$

8 $P\left(\dfrac{1}{2}, 2\right)$, $k = 4$

9 (a) $\dfrac{\sin \alpha}{\alpha}$

10 $f'(x) = \dfrac{1}{2\sqrt{x}}\left(1 + \dfrac{1}{x}\right) > 0$ if $x > 0$; normal is $x + y = 1$.

Exercise 11.4

1 (a)   min 4   (b)   max $\dfrac{7}{2}$ (c)   min $-\dfrac{2}{3}$

    (d)   infl 0   (e)   min $\dfrac{5}{2}$ (f)   min $\dfrac{3}{2}$

    (g)   max $\dfrac{5}{4}$.

2 (a)   infl 1   (b)   max 0, min 4, infl 2

    (c)   max -2, min 2, infl 0   (d)   max 2, min $\dfrac{20}{3}$, infl $\dfrac{13}{3}$

    (e)   max $-\dfrac{1}{3}$, min 1, infl $\dfrac{1}{3}$ (f)   min -1, max 1, infl 0

    (g)   max -3, min 3   (h)   min -1

3 max $-(3 + 2\sqrt{2})$, min$(2\sqrt{2} - 3)$

4 $1250\,\text{m}^2$

5 $18\,\text{m}^3$

6 (a) $v = 2t^2 - 9t + 10$; $t = 2.5$ and 2 (b) $a = 4t - 9$ (c) $7\dfrac{1}{3}$

(d)

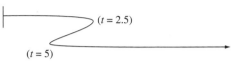

ultimately accelerates away

7 (b) $t = u/g$ (c) $u^2/2g$

8 (a) $4.47\,\text{s}$ (b) (i) $5\,\text{m}$ (ii) $1.11\,\text{s}$

9 $f'(x) = x + \dfrac{1}{x^2} > 0$ when $x > 0$. Must have $f'(x) = 0$ for a local maximum or minimum.

10 Least and minimum $\left(\dfrac{1}{2}, -\dfrac{1}{2}\right)$, max $\left(2 - \dfrac{\sqrt{3}}{3}, \dfrac{2}{9}\sqrt{3}\right)$, min $\left(2 + \dfrac{\sqrt{3}}{3}, \dfrac{2}{9}\sqrt{3}\right)$.

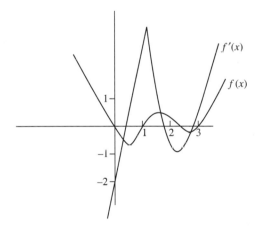

11 $a = 2, b = -4, c = -8$

12 $2\,\text{cm}$

13 (a) 187.5 litres per kilometre (b) 200 litres, $2\,\text{km}$ per minute

14 1 h 35 min, $\sqrt{0.5}$ km ($=0.707$ km)

15 (a) concave up (b) neither (c) concave down

(d) concave up (e,f) concave down (g) neither

16 (a) (i) $\dfrac{d}{dx}(xf(x)) = 0, \dfrac{d^2}{dx^2}(xf(x)) < 0$

(b) $x = 1, \dfrac{d^2}{dx^2}(xe^{-x}) = (x-2)e^{-x} = -e^{-1} < 0$ at $x = 1$

Area $= e^{-1}$, maximum

17 (a) $-\cos x$ (b) $840x^3$ (c) $\dfrac{24}{x^5}$

(d) 720 (e) 0 (f) $\dfrac{880}{243}x^{-14/3}$

18 $\dfrac{d^2}{dx^2}(e^{-x}) = e^{-x}$; alternates $\pm e^{-x}$, even/odd

19 Successive differentiations give $\cos x, -\sin x, -\cos x, \sin x, \ldots$; the pattern repeats every fourth time.

20 $k(k-1)\ldots(k-n+1)x^{k-n}$ n times

(a) derived function $= 0$ after $k+1$ differentiations and subsequently

(b) derived function never zero

21

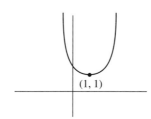

12 INTEGRATION

INTRODUCTION

Integration is a versatile tool in applied mathematics. It can be regarded as the reverse of differentiation; for example, we can find the distance travelled by an object knowing its speed at every instant during the journey. To find the position of the object, we would need to know where it started from. On the other hand, integration can be applied to finding plane areas, volumes of non-regular solids, the location of centres of gravity and the mean value of a function over a given interval. The link between the two types of application is provided by a powerful theorem: the fundamental theorem of calculus.

OBJECTIVES

After working through this chapter you should be able to

- understand the distinction between indefinite and definite integration
- obtain the indefinite integrals of simple functions by reversing differentiation
- use definite integration to find the area under a given curve
- deal with cases where the area crosses the horizontal axis
- appreciate the importance of the fundamental theorem of calculus
- calculate the mean value of a function over a specified interval
- calculate the root-mean-square value of a function over a specified interval
- find the volume of a solid of revolution by integration
- calculate the moment of an area about either axis
- find the position of the centre of gravity of a plane area
- use the trapezium rule to find the approximate value of a definite integral

12.1 REVERSING DIFFERENTIATION

If we know (at any instant) the rate at which a liquid is cooling, can we determine its temperature at any subsequent time? If we know the instantaneous speed of an object moving in a straight line path, can we determine its position? The answer to both questions is a partial yes. The process requires the reverse of differentiation, but this leads to some uncertainty, as the following example shows.

Example

If $f'(x) = 2x$ then what is $f(x)$?

From our work on differentiation we know that *one* answer is $f(x) = x^2$. However, other possible answers include $f(x) = x^2 + \frac{1}{2}$, $f(x) = x^2 - 3$ and $f(x) = x^2 - 4.7$. In each of these three cases $f'(x) = 2x$ because the constant differentiates to 0.

To cover all possibilities we write $f(x) = x^2 + C$ where C is an **arbitrary constant**, known as the **constant of integration**.

In the alternative notation, if $\frac{dy}{dx} = 2x$ then $y = x^2 + C$. ■

In solving the equation $\frac{dy}{dx} = 2x$ we have produced a **family** of **solution curves** with the general formulation $y = x^2 + C$. Each value of C gives a different member of the family and no two members of the family intersect; see Figure 12.1(a). In this example the family cover the entire x–y plane in that every point in the plane lies on one (and therefore

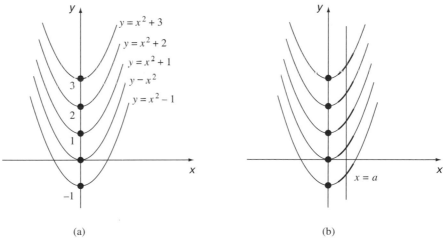

(a)   (b)

Figure 12.1 The family of curves $y = x^2 + C$

only one) curve in the family. Specifying a point through which the solution curves passes selects one of the curves uniquely.

In Figure 12.1(b) we see that for any value of x, say $x = a$, the gradients on the member curves are all equal, in this case $2a$.

In our example, if we specify that when $x = 0, y = 3$, i.e. the solution curve passes through the point (0, 3), then $y = x^2 + 3$ is *the* solution.

This process of reversing differentiation is known as **indefinite integration**. The resulting function is a **primitive**, or more popularly, an **indefinite integral**, sometimes denoted $F(x)$.

For $f(x) = 2x$, examples of primitives are $F(x) = x^2, F(x) = x^2 - \dfrac{1}{2}, F(x) = x^2 + 3$.

Example

A stone is thrown vertically upwards with initial speed u. Its speed at subsequent times t is given by

$$v = \frac{dy}{dt} = u - gt$$

where y is its height above the point of projection at time t and g is the acceleration due to gravity. Find the height as a function of time.

Solution

A function y for which $\dfrac{dy}{dt} = u - gt$ is $y = ut - \dfrac{1}{2}gt^2$, so the general solution is $y = ut - \dfrac{1}{2}gt^2 + C$ where C is an arbitrary constant.

We know that when $t = 0, y = 0$ so that

$$0 = 0 - 0 + C.$$

Hence $C = 0$ and *the* solution is

$$y = ut - \frac{1}{2}gt^2. \qquad \blacksquare$$

Notation

To help us later on we introduce some notation for integrals. The process of integrating the function $2x$ to get $x^2 + C$ is written

$$\int 2x \, dx = x^2 + C.$$

The left-hand-side is read as 'the integral of $2x$ (with respect to x)'.

The symbol $\int$ is a corruption of S (for sum, as we shall see in the next section). You should regard it and dx as forming part of a package into which a function is placed, i.e. $\int(\ )dx$, where the formula for the function replaces the brackets. The function being integrated is known as the **integrand**. Note that x is the variable used to convey information about the process of integration. Other letters could be used. Hence

$$\int 2t\,dt = t^2 + C, \qquad \int 2v\,dv = v^2 + C, \text{ etc.}$$

In general, if $F'(x) = f(x)$ then $F(x) = \int f(x)dx$.

To allow us to integrate a range of functions we make use of rules similar to those in Chapter 11 and apply parts of Table 11.5 in reverse.

In the following, $f(x)$ and $g(x)$ are two functions and α and β are two real numbers.

1. **Scalar multiple**: the integral of a scalar multiple of a function is that scalar multiplied by the integral of the function.

2. **Sum**: the integral of the sum of two functions is the sum of their integrals.

3. **Difference**: the integral of the difference of two functions is the difference of their integrals.

4. **Linear combination**: the integral of a linear combination of two functions is the same linear combination of their integrals.

We have used the term 'integral' to mean 'indefinite integral'.

These rules may be summarised as follows.

$$
\begin{aligned}
&1. \quad \int \alpha f(x)dx = \alpha \int f(x)dx \\[1em]
&2. \quad \int \{f(x) + g(x)\}dx = \int f(x)dx + \int g(x)dx \\[1em]
&3. \quad \int \{f(x) - g(x)\}dx = \int f(x) - \int g(x)dx \\[1em]
&4. \quad \int \{\alpha f(x) + \beta g(x)\}dx = \alpha \int f(x)dx + \beta \int g(x)dx
\end{aligned}
\tag{12.1}
$$

In the following table we have denoted the constant of integration by C. Note the use of $|x|$ in the integral of $\dfrac{1}{x}$. This is because the ln function requires a positive number as input.

Table 12.1

Function $f(x)$	Indefinite integral $\int f(x)dx$
0	C
1	$x + C$
$2x$	$x^2 + C$
$x^n, \quad n \neq -1$	$\dfrac{x^{n+1}}{n+1} + C$
$\cos x$	$\sin x + C$
$\sin x$	$-\cos x + C$
e^x	$e^x + C$
$\dfrac{1}{x}, \quad x \neq 0$	$\ln|x| + C$
$\cos kx$	$\dfrac{1}{k}\sin kx + C$
$\sin kx$	$-\dfrac{1}{k}\cos kx + C$
e^{kx}	$\dfrac{1}{k}e^{kx} + C$

Examples

1. Given the following derivatives, find the function y in each case.

(a) $\dfrac{dy}{dx} = 5x^4$

(b) $\dfrac{dy}{dx} = 2x^3 - 3x^2 + 2$

(c) $\dfrac{dy}{dx} = \dfrac{2}{x^3}$

(d) $\dfrac{dy}{dx} = 3x^2 + 2 + \dfrac{3}{x^4}$.

Solution

(a) $y = 5\left(\dfrac{x^5}{5}\right) + C = x^5 + C$

(b) $y = 2\left(\dfrac{x^4}{4}\right) - 3\left(\dfrac{x^3}{3}\right) + 2(x) + C = \dfrac{x^4}{2} - x^3 + 2x + C$

(c) $\dfrac{dy}{dx} = 2x^{-3} \quad y = 2\left(\dfrac{x^{-2}}{-2}\right) + C = -x^{-2} + C = -\dfrac{1}{x^2} + C$

(d) $\dfrac{dy}{dx} = 3x^2 + 2 + 3x^{-4} \quad y = \dfrac{3x^3}{3} + 2x + \dfrac{3x^{-3}}{(-3)} + C = x^3 + 2x - \dfrac{1}{x^3} + C$.

2. Integrate the following with respect to x:

(a) x^3

(b) $x^2 + 2x^3 + 3 + \dfrac{5}{x^2}$

(c) $2\cos 3x + 5\sin 2x$

(d) $\dfrac{1}{2}(e^{2x} + e^{-2x})$.

Solution

(a) $\int x^3 dx = \dfrac{x^4}{4} + C$

(b) $\int \left(x^2 + 2x^3 + 3\dfrac{5}{x^2} \right) dx = \int x^2 dx + 2\int x^3 dx + 3\int 1 \, dx + 5\int x^{-2} dx$

$$= \dfrac{x^3}{3} + \dfrac{2x^4}{4} + 3x + \dfrac{5x^{-1}}{(-1)} + C$$

$$= \dfrac{x^3}{3} + \dfrac{1}{2}x^4 + 3x - \dfrac{5}{x} + C$$

(c) $\int (2\cos 3x + 5\sin 2x) dx = 2\int \cos 3x \, dx + 5\int \sin 2x \, dx$

$$= 2\left(\dfrac{\sin 3x}{3} \right) + 5\left(-\dfrac{\cos 2x}{2} \right) + C = \dfrac{2}{3}\sin 3x - \dfrac{5}{2}\cos 2x + C$$

(d) $\int \left(\dfrac{1}{2}e^{2x} + \dfrac{1}{2}e^{-2x} \right) dx = \dfrac{1}{2}\int e^{2x} dx + \dfrac{1}{2}\int e^{-2x} dx$

$$= \dfrac{1}{2}\left(\dfrac{e^{2x}}{2} \right) + \dfrac{1}{2}\left(\dfrac{e^{-2x}}{-2} \right) + C = \dfrac{1}{4}e^{2x} - \dfrac{1}{4}e^{-2x} + C \qquad \blacksquare$$

Note the following points:

1. Although we carried out several separate integrations in parts (b), (c) and (d) of the second example, we need only add *one* arbitrary constant at the end. The ' $+ C$' is not to be treated as a part of the arithmetic, rather as a statement of uncertainty about the answer. Once all integrations have been completed it is sufficient to put the single '$+ C$' after the expression.
2. It is always possible, if time allows, to check the accuracy of your integration by differentiating your answer and checking that the result matches the original integrand.

Exercise 12.1

1   By reversing the process of differentiation, write down integrals of the following functions. Include the arbitrary constant.

(a) x^6

(b) $5x^2$

(c) $-\dfrac{19}{x}$

(d) $\dfrac{\pi}{x^4}$

(e) $\sqrt{x}$

(f) $(x^2 - 1)^2$

(g) $(x+1)(x+2)(x+3)$

(h) $\left(x - \dfrac{1}{x}\right)^2$

(i) $\dfrac{9}{x^2} - \dfrac{5}{x^5} - \dfrac{7}{x^7}$

(j) $\dfrac{(x-1)^2}{x^3}$.

2 Repeat Question 1 for the following, by determining the indefinite integrals of the functions. Ignore the arbitrary constant.

(a) $x^{3/2}$

(b) $\sqrt{x}(x-1)(x-2)$

(c) $(6x^{2/3} - 5x^{-1/3})(1 - 2/x)$

(d) $(x-1)^2(x-2) + (x-3)^2(x-4)$

(e) $\dfrac{\{(x-5)(x-6) + (x-2)^2(x-1)\sqrt{x}\}}{x^{5/4}}$

(f) $(x^{7/2} - x^{2/11})(x^{19/6} - x^{3/7})$.

3 By reversing the process of differentiation, write down integrals of the following; this time include the arbitrary constant.

(a) $6e^x$

(b) $11e^{-x} + 15e^{2x}$

(c) $\sin 5x$

(d) $3\sin 5x - 4\cos 4x - 11/x$

(e) $a\sin ax + b\cos bx$ with a, b constant.

4 $\ln \sqrt{x} + C$ is the indefinite integral of one of the functions given below. Which one?

(a) $\dfrac{2}{x}$

(b) $\dfrac{1}{2x}$

(c) $\dfrac{1}{\sqrt{x}}$

(d) $\dfrac{1}{2\sqrt{x}}$.

5 By writing $3^x = e^{x\ln 3}$, determine $\displaystyle\int 3^x\,dx$.

6* Determine $\int(x^a + x^b)(x^c + x^d)dx$ where a, b, c, d are constants. Consider different cases, depending on the values of the constants.

7* Determine $\int((x-a)(x-b) + (x-c)^2(x+d)\sqrt{x})dx$ where a, b, c, d are constants.

8 Given that $\int \sqrt{x}(x+a)dx = \dfrac{2}{5}x^{3/2}(x+1) + C$ determine a.

12.2 DEFINITE INTEGRATION AND AREA

Suppose that we wish to find the area between the line $y = x$, the x-axis and the ordinates $x = a$ and $x = b$, as in Figure 12.2(a). We can use elementary geometry. The required area is

$$ABDC = OBD - OAC$$

$$= \frac{1}{2}OB \times BD - \frac{1}{2}OA \times AC$$

$$= \frac{1}{2}b \times b - \frac{1}{2}a \times a$$

$$= \frac{1}{2}b^2 - \frac{1}{2}a^2 = F(b) - F(a)$$

where $F(x) = \frac{1}{2}x^2$

Note that $F'(x) = x$. And in the special case when $a = 0$, the area is $\frac{1}{2}b^2$.

Figure 12.2(b) represents a velocity–time graph; the velocity of an object is given by the formula $v = u + at$, where u is the initial velocity (i.e. at $t = 0$) and a is the (constant) acceleration of the object.

The area under the graph of $v = u + at$, the t-axis and the ordinates $t = t_1$ and $t = t_2$ represents the distance travelled by the object in the interval between times t_1 and t_2. The ordinate $AC = u + at_1$ and the ordinate $BD = u + at_2$. The area $ABDC$ is given by

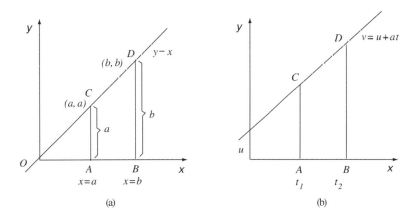

Figure 12.2   Area under a straight line graph

$$ABDC = \frac{1}{2}\{(u + at_2) + (u + at_2)\} \times (t_2 - t_1)$$

$$= \left\{u + \frac{a}{2}(t_1 + t_2)\right\}(t_2 - t_1)$$

$$= u(t_2 - t_1) + \frac{a}{2}(t_2^2 - t_1^2)$$

$$= \left(ut_2 + \frac{a}{2}t_2^2\right) - \left(ut_1 + \frac{a}{2}t_1^2\right)$$

$$= F(t_2) - F(t_1)$$

where $\quad F(t) = ut + \frac{1}{2}at^2$

Note that $F'(t) = u + at = v$. And in the special case where $t_1 = 0$, the distance travelled is $ut_2 + \frac{1}{2}at_2^2$. Therefore the distance travelled at any time $t \geq 0$ is $ut + \frac{1}{2}at^2 = F(t)$.

Suppose now that we find the area between the graph of $y = x^2$, the x-axis and the ordinates $x = a$ and $x = b$; this is the shaded area in Figure 12.3.

We *could* try to draw the graph accurately on squared paper and *estimate* the area by counting the number of squares in the region concerned, *estimating* the fraction of each square to be included in the count when the graph cuts through it. This would be tedious and not very accurate.

In Figure 12.3(b) we have taken the simpler case where $a = 0$. The area under the curve between $x = 0$ and $x = b$ is clearly less than the area of the rectangle $OPQR$; this rectangle has area $OR \times QR = b \times b^2 = b^3$. We could say that $0 < \text{Area} < b^3$. Now we introduce the ordinate $x = \frac{b}{2}$.

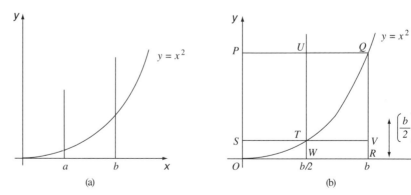

Figure 12.3 Area under the curve $y = x^2$

The area we want has a value greater than the area of the rectangle *WTVR* and less than the sum of the areas of *OSTW* and *WUQR*, i.e.

$$\frac{b}{2} \times \left(\frac{b}{2}\right)^2 < \text{Area} < \frac{b}{2} \times \left(\frac{b}{2}\right)^2 = \frac{b}{2} \times b^2$$

$$\frac{b^3}{8} < \text{Area} < \frac{5b^3}{8}.$$

Notice that the **overestimate** of the area has *decreased* (because we have omitted the rectangle *SPUT*) and the **underestimate** of the area has *increased* (because we have included the rectangle *WTVR*).

Figure 12.4 shows the process continuing with the interval $0 \le x \le b$ divided into 4 **strips** of width $b/4$ and 8 strips of width $b/8$. The sum of the areas of the shaded rectangles represents the difference between the values of the overestimate and the underestimate in each case.

The table on the next page shows the results as the number of strips increases over the interval $0 \le x \le b$. Results are quoted to 5 d.p. As the number of strips increases and therefore their width decreases, the underestimate and the overestimate get closer together. To what value are they converging?

Question 3 in Exercise 12.2 shows that for n strips, of width $\frac{b}{n}$, the underestimate and the overestimate are given respectively by

$$\frac{(n-1)(2n-1)b^3}{6n^2} \quad \text{and} \quad \frac{(n+1)(2n+1)}{6n^2}b^3$$

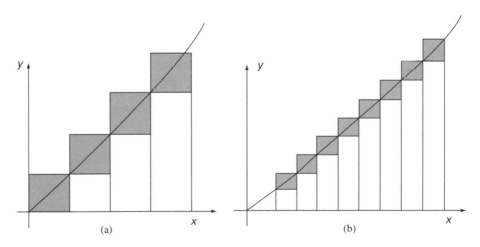

Figure 12.4 Improving the accuracy of area estimates

i.e. $\left(1 - \dfrac{1}{n}\right)\left(2 - \dfrac{1}{n}\right)\dfrac{b^3}{6}$ and $\left(1 + \dfrac{1}{n}\right)\left(2 + \dfrac{1}{n}\right)\dfrac{b^3}{6}$

As n increases, both $\left(1 + \dfrac{1}{n}\right)$ and $\left(1 + \dfrac{1}{n}\right)$ both converge to the value 1 and both $\left(2 - \dfrac{1}{n}\right)$ and $\left(2 + \dfrac{1}{n}\right)$ converge to the value 2.

Table 12.2

	Area/b^3	
Number of strips, n	Underestimate	Overestimate
1	0	1
2	0.125	0.625
4	0.218 75	0.468 75
8	0.273 44	0.398 44
16	0.283 20	0.365 23
32	0.317 87	0.349 12
64	0.325 56	0.341 19
128	0.329 44	0.337 25
256	0.331 39	0.335 29

Each estimate therefore converges to the value

$$1 \times 2 \times \frac{b^3}{6} \quad \text{i.e.} \quad \frac{b^3}{3}.$$

If we look back to Figure 12.3(a) it is therefore reasonable to argue that the shaded area has the value $\dfrac{b^3}{3} - \dfrac{a^3}{3}$.

Even for this simple function it is a tedious process to obtain the result. It is much more tedious to find that the area between the curve $y = x^3$, the x-axis and the ordinates $x = a$ and $x = b$ is $\dfrac{b^4}{4} - \dfrac{a^4}{4}$. Fortunately the next section provides a simpler means of finding these areas.

Fundamental theorem of calculus

The clue is provided by the results obtained so far; see the following table. It would be tempting to suggest that the result for the function $f(x) = x^4$ is $\dfrac{b^5}{5} - \dfrac{a^5}{5}$, since $\dfrac{x^5}{5}$ differentiates to x^4. In fact this is the correct result.

Table 12.3

Function $f(x)$	Area between $x = a$ and $x = b$, $g(b) - g(a)$	$g(x)$	$g'(x)$
x	$\dfrac{b^2}{2} - \dfrac{a^2}{2}$	$\dfrac{x^2}{2}$	x
x^2	$\dfrac{b^3}{3} - \dfrac{a^3}{3}$	$\dfrac{x^3}{3}$	x^2
x^3	$\dfrac{b^4}{4} - \dfrac{a^4}{4}$	$\dfrac{x^4}{4}$	x^3

In general, for a function $f(x)$, when the underestimates and the overestimates of area converge to the same value as the number of strips increases, the value to which they converge is called the **definite integral** of $f(x)$ from $x = a$ to $x = b$. We often call it 'the integral of $f(x)$ from a to b'.

In each case in the above table the definite integral could be expressed in the form $F(b) - F(a)$ where $F'(x) = f(x)$, i.e. $F(x)$ is an indefinite integral of $f(x)$. Which indefinite integral is chosen does not matter.

For example, with the function $f(x) = x^2$ we have chosen $F(x) = \dfrac{x^3}{3}$, then $F(b) - F(a) = \dfrac{b^3}{3} - \dfrac{a^3}{3}$; if we had chosen $F(x) = \dfrac{x^3}{3} + 2$ then

$$F(b) - F(a) = \left(\frac{b^3}{3} + 2\right) - \left(\frac{a^3}{3} + 2\right) = \frac{b^3}{3} - \frac{a^3}{3}$$

Whatever constant we add to $\dfrac{x^3}{3}$ will cancel out in this way; therefore we choose $F(x)$ without a constant of integration.

Because of the link between the area and the primitive we write the definite integral of $f(x)$ from $x = a$ to $x = b$ as

$$\int_a^b f(x)dx.$$

Notice the position of the **lower limit of integration** $x = a$ and the **upper limit** $x = b$.

To evaluate the definite integral we find an indefinite integral of $f(x)$, evaluate it at $x = a$ and $x = b$ and subtract the first value from the second. This is expressed in the **fundamental theoreom of calculus**.

If $F'(x) = f(x)$ then $\displaystyle\int_a^b f(x)dx = F(b) - F(a)$. $\qquad$ (12.2)

Evaluating definite integrals

Examples

1. Find the values of

(a) $\displaystyle\int_2^4 3x^2\,dx$ (b) $\displaystyle\int_0^{\pi/2} \cos x\,dx$ (c) $\displaystyle\int_0^{\pi/2} \cos\theta\,d\theta$

(a) The first step is to find an indefinite integral of $3x^2$; x^3 will do. The notation
for   evaluating x^3 at $x = 2$, at $x = 4$ and subtracting is

$$[x^3]_2^4$$

Notice again the position of the limits of integration. We therefore write

$$\int_2^4 3x^2\,dx = [x^3]_2^4 = 4^3 - 2^3 = 64 - 8 = 56$$

(b) A suitable indefinite integral of $\cos x$ is $\sin x$. Hence

$$\int_0^{\pi/2} \cos x\,dx = [\sin x]_0^{\pi/2} = \sin\frac{\pi}{2} - \sin 0 = 1$$

(c) Here the problem is described in terms of the variable θ. This does not matter;
it is the sine function and the limits of 0 and $\dfrac{\pi}{2}$ that determine the answer, which is
therefore 1.

2. Find the area under the curve $y = \sin x$ between $x = 0$ and (a) $\dfrac{\pi}{2}$, (b) $x = \pi$.

(a) The area is given by

$$\int_0^{\pi/2} \sin x\,dx = [-\cos x]_0^{\pi/2} = \left(-\cos\frac{\pi}{2}\right) - (-\cos 0) = (-0) - (-1) = 1$$

This is the same as for the cosine function between 0 and $\dfrac{\pi}{2}$. Why is this?

(b) $\displaystyle\int_0^{\pi} \sin x\,dx = [-\cos x]_0^{\pi} = (-\cos\pi) - (-\cos 0) = (+1) - (-1) = 2$

Since the cosine curve is symmetrical about $x = \dfrac{\pi}{2}$ we would expect the second
answer to be twice the first. ■

When a curve crosses the x-axis between the limits of integration we can obtain
misleading results, as the following example shows.

Example

Find

(a) $\displaystyle\int_1^3 4x^3\,dx$ (b) $\displaystyle\int_{-1}^3 4x^3\,dx$

(c) $\displaystyle\int_{-2}^3 4x^3\,dx$ (d) $\displaystyle\int_{-3}^3 4x^3\,dx$

and hence find the area between the curve $y = 4x^3$, the x-axis and the values of x shown in each case.

Solution

An indefinite integral of $4x^3$ is x^4. Then

(a) $\displaystyle\int_1^3 4x^3\,dx = [x^4]_1^3 = 81 - 1 = 80$

(b) $\displaystyle\int_{-1}^3 4x^3\,dx = [x^4]_1^3 = 81 - (-1)^4 = 80$

(c) $\displaystyle\int_{-2}^3 4x^3\,dx = [x^4]_{-2}^3 = 81 - (-2)^4 = 65$

(d) $\displaystyle\int_{-3}^3 4x^3\,dx = [x^4]_{-3}^3 = 81 - 81 = 0.$

We clearly cannot interpret all these integrals as areas between the curves and the x-axis. As we extend the range of x further to the left, we expect the area to increase. This does not appear to have happen in cases (b) and (c). The result in (d) is alarming; what does it mean?

The problems are resolved if we accept the convention that areas below the x-axis are given a negative sign. This is plausible since in going from the axis to the curve we move in the *negative* y-direction. We then make use of the following result.

If $a \le c \le b$ then

$$\int_a^b f(x)dx = \int_a^c f(x)dx + \int_c^b f(x)dx \qquad (12.3)$$

Hence $\displaystyle\int_{-1}^3 4x^3\,dx = \int_{-1}^0 4x^3\,dx + \int_0^3 4x^3\,dx$

$$= [x^4]_{-1}^0 + [x^4]_0^3$$
$$= 0 - (-1)^4 + 81 - 0 = -1 + 81.$$

Recognising that the first integral represents an area *below* the x-axis then the *area* we require is $+1+81=82$. Similarly,

$$\int_{-2}^{3} 4x^3 \, dx = \int_{-2}^{0} 4x^3 \, dx + \int_{0}^{3} 4x^3 \, dx$$
$$= [x^4]_{-2}^{0} + [x^4]_{0}^{3}$$
$$= 0 - (-2)^4 + 81$$
$$= -16 + 81.$$

The area is $+16+81=97$. Finally,

$$\int_{-3}^{3} 4x^3 \, dx = [x^4]_{-3}^{0} + [x^4]_{0}^{3} = -81 + 81.$$

The area is $+81+81=162$. ■

The lesson to learn is that we must be careful to understand exactly what we are looking for in this problem. If it is a physical area then we should split the integral into the sum of two or more integrals, each of which has one or both limits where the function is zero. The negative values which result are multiplied by -1 before adding to find the true total area.

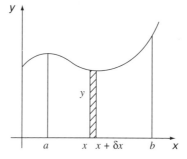

Figure 12.5 Finding the area under the curve $y = f(x)$ from $x = a$ to $x = b$

More formally, consider the problem of finding the area under the curve $y = f(x)$ from $x = a$ to $x = b$. We divide the area into strips of thickness $h = \delta x$, as shown in Figure 12.5. The strip shown shaded is approximately a rectangle if δx is small and its area is approximately $y \, \delta x$ or $f(x)\delta x$. The approximation improves as δx approaches zero. The total area required is approximately the sum of the areas of such strips and may be written

$$A \simeq \sum f(x)\delta x.$$

If A converges to a fixed value as $\delta x \to 0$ then $f(x)$ is **integrable** over $a \leq x \leq b$ and the limiting value is the definite integral. In our notation

$$\int_a^b f(x)dx = \lim_{\delta x \to 0} \{\sum f(x) \times \delta x\}.$$

Note that, although we have used area as a means of illustrating the process of definite integration, we shall see that definite integrals can represent volumes, mean values of functions, etc. These will have appropriate units attached. With this proviso, a definite integral is a *value*, not a *function* like an indefinite integral.

Exercise 12.2

1 Obtain overestimates and underestimates (as illustrated) for the area under the curve $y = x^3$ over the interval [0, 1]. Use the following steps:

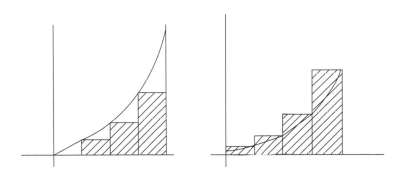

(a) $\dfrac{1}{2}$ (b) $\dfrac{1}{4}$ (c) $\dfrac{1}{8}$

How do these estimates compare to the true answers?

2 Repeat the process with $y = x^3$, but this time take the area to that formed by the trapeziums linking the values of y at $x = \dfrac{1}{4}, \dfrac{1}{2}$ as illustrated. How does the value compare to the true value of $\dfrac{1}{4}$.

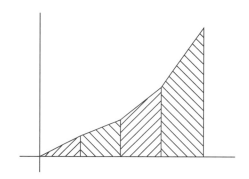

3* The interval $0 \leq x \leq 6$ is divided into n equal subintervals of width b/n. Noting that $1^2 + 2^2 + \cdots + n^2 = \frac{1}{6}n(n+1)(2n+1)$, show that the underestimate and overestimate of the area between the curve $y = x^2$ and the x-axis over this interval are respectively

$$\frac{(n-1)(2n-1)b^3}{6n^2} \quad \text{and} \quad \frac{(n+1)(2n+1)b^3}{6n^2}.$$

4 Evaluate the following definite integrals:

(a) $\displaystyle\int_0^1 x^2 \, dx$

(b) $\displaystyle\int_1^2 3x^2 \, dx$

(c) $\displaystyle\int_1^4 \sqrt{x} \, dx$

(d) $\displaystyle\int_{-2}^{-1} \frac{dx}{x^2}$

(e) $\displaystyle\int_{-1}^1 (2x^2 - 1) dx$

(f) $\displaystyle\int_0^2 (3x^3 - x) dx$

(g) $\displaystyle\int_2^3 \left(x^2 - \frac{1}{x^2} \right) dx$

(h) $\displaystyle\int_{-1}^2 (3x-1)(2x+1) dx$

(i) $\displaystyle\int_1^2 x(x-1)(x-2) dx$

(j) $\displaystyle\int_1^9 \frac{x+1}{\sqrt{x}} dx.$

5 Sketch the curve with equation $y = 4x^2$. Find the area included by the curve, the axis of x and the ordinates at $x = 0$ and $x = 3$.

6 The sketch represents the curve $y = 2x(2 - x)$. What are the coordinates of P? Evaluate the area above the x-axis.

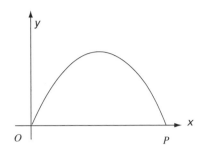

7   Sketch the curve $y = (1 - x)(2 + x)$. Find the area above the x-axis.

8   Find the individual areas of the segments cut off by the x-axis from the curves with equations (a) $y = x(x + 1)(x - 1)$ and (b) $y = x(x + 1)(x + 2)$.

9   Sketch the curve $y = 3x - x^2$ for values of x from 0 to 5. Evaluate $\int_0^5 (3x - x^2)dx$ and interpret the result.

10   Evaluate $\int_0^3 (2x - x^2)dx$ and explain the result.

11   Find the area of the segment cut off on the curve $y = 5x - x^2$ by the line $y = 6$.

12   If $\int_1^a y^3 dy = 20$, find the value of a.

13*   Sketch the curve $y = 1/x$ and mark off the points where $x = -b, -a, a, b$ where a and b are two arbitrary positive values satisfying $b > a > 0$.
   What is the value of

(a) $\int_a^b \dfrac{dx}{x}$

(b) $\int_{-b}^{-a} \dfrac{dx}{x}$ (shade the areas)

(c)   Put $a = 1$ and prove that $\ln|b| = \begin{cases} \displaystyle\int_1^b \dfrac{dx}{x}, & b > 0 \\ \displaystyle\int_{-1}^{-b} \dfrac{dx}{x}, & b < 0 \end{cases}$

(This is why the indefinite integral of $1/x$ is taken to be $\ln|x|$ or $\ln|Kx|$, incorporating the arbitrary constant into the logarithm.)

14   Determine the following definite integrals:

(a) $\int_0^{\pi/2} \sin x \, dx$

(b) $\int_0^1 e^x \, dx$

(c) $\displaystyle\int_0^\pi (\sin 2x + 3\cos 3x)dx$

(d) $\displaystyle\int_0^1 \sqrt{x}\{3x + 4) - (x - 2)^2(x + 3)\}dx$

(e) $\displaystyle\int_1^2 e^{-x}\,dx$ (f) $\displaystyle\int_e^{e^2} \frac{dx}{x}$ (g) $\displaystyle\int_1^X \frac{dx}{x^2}$

In part (g) consider what happens as $X \to \infty$.

15*   Consider $\displaystyle\int_a^b x^N dx$ where a, b are real constants, and N is an integer. Write down the value of the integral in the following cases:

(a) $N > 0$ (b) $N = 0$

(c) $N = -1, b > a > 0$ (d) $N < -1, b > a > 0$

Now answer the following questions:

(e)   Does a or b need to be restricted in cases (a) and (b)?

(f)   Can we take $a = 0, b > 0$ in case (c)?

(g)   Can we take $a = 0, b > 0$ in case (d)?

(h)   Accepting that $b > a$, will be any other values of a and b, apart from $b > a > 0$, suffice in cases (c) and (d)?

16   Determine the numerical values of the following definite integrals:

(a) $\displaystyle\int_e^\pi \frac{(x + 3)}{x^{5/2}}dx$ (b) $\displaystyle\int_{1.327}^{6.121} 4x^3(1 - 1/x^4)dx$

(c) $\displaystyle\int_{0.5}^{2.5} 2^x\,dx \quad (2^x = e^{x\ln 2})$

(d) $\displaystyle\int_0^1 (6.525\sin 3.16x - 1.327\cos 2.66x)dx.$

17   If $m \le f(x) \le M$ over $[a, b]$ then the area enclosed between $f(x)$ and the x-axis must lie between the areas of the rectangles formed by a, b and m, M respectively, i.e.

$$m(b - a) \le \int_a^b f(x)dx \le M(b - a)$$

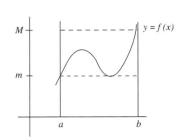

(a)   Observe that $f(x) = \dfrac{1}{1+x^2}$ is a decreasing function of x for $x \geq 0$. By taking its minimum and maximum values over the intervals in question, obtain lower and upper bounds for

(i) $\displaystyle\int_0^1 f(x)dx$ (ii) $\displaystyle\int_1^2 f(x)dx$

(b)   Using the same function obtain bounds for $\displaystyle\int_0^{1/2} f(x)dx$ and $\displaystyle\int_{1/2}^1 f(x)dx$. Use these results to give refined bounds for $\displaystyle\int_0^1 f(x)dx$.

18   The function $f(x) = (x+1)e^{-x^2}$ has a single stationary point in the interval $[0, 1]$.

(a)   Verify that $f(x)$ has both a local maximum and a global maximum at the stationary point.

(b)   Assuming that the global minimum of $f(x)$ is at the endpoint, obtain bounds for $\displaystyle\int_0^1 f(x)dx$.

19   The function $f(x)$ is to be integrated over the range $[a, b]$ and the area under the curve is approximated by four trapeziums. Prove that the combined area of the trapeziums is

$$\left(\frac{b-a}{8}\right)(f_0 + 2f_1 + 2f_2 + 2f_3 + f_4)$$

where $f_i, 0 \leq i \leq 4$ are the values of $f(x)$ at the node points. This is the trapezium rule.

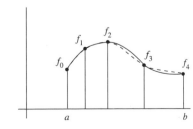

Determine the approximate value of $\int_1^2 \dfrac{dx}{x^{3/2}}$ using the above trapezium rule and compare it to the exact value of the integral (3 s.f.).

20 Use the method of Question 19 to determine an approximate value for

$$\int_1^2 \frac{dx}{\sqrt{x^3 + 0.2}} \, (3 \text{ s.f.})$$

12.3 APPLICATIONS TO AREA, VOLUME AND MEAN VALUES

Area between two curves

The area between two curves can be found by one of two methods, as the following example shows.

Example

Find the area between the curves $y = x^2$ and $y = x^3$.
The two curves intersect at $x = 0$ and $x = 1$ as shown in Figure 12.6.

(i) The required area can be found as the difference between the area under $y = x^2$ and that under $y = x^3$, i.e.

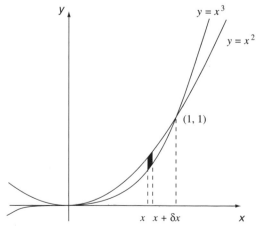

Figure 12.6 Area between the curves $y = x^2$ and $y = x^3$

$$A = \int_0^1 x^2 \, dx - \int_0^1 x^3 \, dx$$

$$= \left[\frac{x^3}{3}\right]_0^1 - \left[\frac{x^4}{4}\right]_0^1$$

$$= \frac{1}{3} - 0 - \left(\frac{1}{4} - 0\right) = \frac{1}{3} - \frac{1}{4} = \frac{1}{12}$$

(ii) The length of the strip highlighted in Figure 12.6 is $(x^2 - x^3)$ at the left-hand edge and if the strip is thin its area is approximately

$$(x^2 - x^3)\delta x.$$

The total area required, A, is approximately $\sum(x^2 - x^3)\delta x$. As $\delta x \to 0$ then

$$A = \int_0^1 (x^2 - x^3) dx$$

$$= \left[\frac{x^3}{3} - \frac{x^4}{4}\right]_0^1$$

$$= \left(\frac{1}{3} - \frac{1}{4}\right) - (0) = \frac{1}{12} \text{ (as before).} \qquad \blacksquare$$

Mean value of a function

In Figure 12.7 the rectangle $PQRS$ is constructed so its area is equal to the area under the curve $y = f(x)$ from $x - a$ to $x = b$.

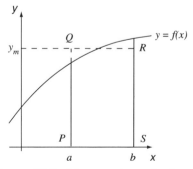

Figure 12.7 Mean value of a function

Hence $(b - a)y_m = \int_a^b f(x)dx$

where y_m is the average height of the ordinates y from $x = a$ to $x = b$.

> The **mean value of the function** $f(x)$ over the interval $a \leq x \leq b$ is given by
>
> $$\frac{1}{(b-a)} \int_a^b f(x)dx \qquad (12.4)$$

Examples

1. Find the mean value of $\cos x$ over the intervals

   (a) $0 \leq x \leq \dfrac{\pi}{2}$ (b) $0 \leq x \leq \pi$ (c) $0 \leq x \leq 2\pi$

   (a) Mean value $= \dfrac{1}{\pi/2} \displaystyle\int_0^{\pi/2} \cos x \, dx = \dfrac{2}{\pi}[\sin x]_0^{\pi/2}$

   $$= \frac{2}{\pi}\left\{\sin\frac{\pi}{2} - \sin 0\right\} = \frac{2}{\pi}(1 - 0) = \frac{2}{\pi}$$

   (b) Mean value $= \dfrac{1}{\pi} \displaystyle\int_0^{\pi} \cos dx = \dfrac{1}{\pi}[\sin x]_0^{\pi} = \dfrac{1}{\pi}\{\sin \pi - \sin 0\} = 0$

   (c) Mean value $= \dfrac{1}{2\pi} \displaystyle\int_0^{2\pi} \cos x \, dx = \dfrac{1}{2\pi}[\sin x]_0^{2\pi} = 0.$

   Check with Figure 12.8(a) to see whether the results are reasonable.

2. Find the mean value of x^2 over the intervals (a) $0 \leq x \leq 2$ and (b) $-2 \leq x \leq 2$

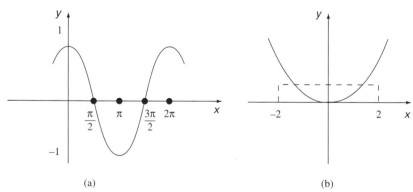

(a)   (b)

Figure 12.8 Mean value of (a) $y = \cos x$ and (b) $y = x^2$

(a) Mean value $= \dfrac{1}{2} \displaystyle\int_0^2 x^2 \, dx = \dfrac{1}{2} \left[\dfrac{x^3}{3} \right]_0^2 = \dfrac{1}{2} \times \dfrac{8}{3} = \dfrac{4}{3}$

(b) Mean value $= \dfrac{1}{2 - (-2)} \displaystyle\int_{-2}^2 x^2 \, dx = \dfrac{1}{4} \left[\dfrac{x^3}{3} \right]_{-2}^2 = \dfrac{1}{4} \left(\dfrac{8}{3} - \left(\dfrac{-8}{3} \right) \right) = \dfrac{4}{3}.$

These results are the same because of the symmetry of the curve about the y-axis; see Figure 12.8(b). ∎

It is always sensible to look for symmetry to simplify the calculation; a limit of zero in an integral is relatively easy to apply.

Root mean square value of a function

As we have seen the mean value of $\cos x$ over a period, c.g. $0 \leq x \leq 2\pi$, is zero. A more useful measure of the average amplitude of a wave is provided by the root mean square (RMS) value.

To calculate the RMS of a set of values, take each one and *square* it, then take the *mean* of the squares and finally obtain the square *root* of the mean. Since the original values are squared there are no cancellation effects, unlike the arithmetic mean.

> The **root mean square** value of a function $f(x)$ over the interval $a \leq x \leq b$ is given by
>
> $$\sqrt{\dfrac{1}{(b - a)} \int_a^b [f(x)]^2 \, dx} \qquad (12.5)$$

Example   Find the RMS of $\cos x$ over the intervals (a) $0 \leq x \leq \dfrac{\pi}{2}$, (b) $0 \leq x \leq \pi$ and (c) $0 \leq x \leq 2\pi$.

Solution

To solve this problem we make use of a trigonometric identity that we prove in *Mathematics in Engineering and Science*:

$$\cos^2 x = \dfrac{1}{2} + \dfrac{1}{2} \cos 2x$$

(a) $\displaystyle (\text{RMS})^2 = \frac{1}{\pi/2}\int_0^{\pi/2} \cos^2 x \, dx = \frac{2}{\pi}\int_0^{\pi/2}\left(\frac{1}{2}+\frac{1}{2}\cos 2x\right)dx$

$$= \frac{2}{\pi}\left[\frac{1}{2}x+\frac{1}{2}\times\frac{1}{2}\sin 2x\right]_0^{\pi/2}$$

$$= \frac{2}{\pi}\left[\frac{1}{2}x+\frac{1}{4}\sin 2x\right]_0^{\pi/2}$$

$$= \frac{2}{\pi}\left\{\left(\frac{1}{2}\times\frac{\pi}{2}+\frac{1}{4}\sin \pi\right)-(0+0)\right\}$$

$$= \frac{2}{\pi}\times\frac{1}{2}\times\frac{\pi}{2}=\frac{1}{2}.$$

Hence RMS $= \dfrac{1}{\sqrt{2}}$

(b) $\displaystyle (\text{RMS})^2 = \frac{1}{\pi}\int_0^{\pi} \cos^2 x \, dx$

$$= \frac{1}{\pi}\left[\frac{1}{2}x+\frac{1}{4}\sin 2x\right]_0^{\pi}$$

$$= \frac{1}{\pi}\left\{\left(\frac{1}{2}\pi+\frac{1}{4}\sin 2\pi\right)-0\right\}$$

$$= \frac{1}{\pi}\times\frac{1}{2}\pi=\frac{1}{2}.$$

Hence RMS $= \dfrac{1}{\sqrt{2}}$

(c) $$(\text{RMS})^2 = \frac{1}{2\pi} \int_0^{2\pi} \cos^2 x \, dx$$

$$= \frac{1}{2\pi} \left[\frac{1}{2}x + \frac{1}{4}\sin 2x \right]_0^{2\pi}$$

$$= \frac{1}{2\pi} \left\{ \left(\frac{1}{2} \times 2\pi + \frac{1}{4}\sin 4\pi \right) - 0 \right\}$$

$$= \frac{1}{2\pi} \times \frac{1}{2} \times 2\pi = \frac{1}{2}$$

Hence RMS $= \dfrac{1}{\sqrt{2}}$ ■

Volume of a solid of revolution

The curve whose equation is $y = f(x)$ is rotated about the x-axis. The area $ABCD$ sweeps out a volume. Each cross-section of the resulting solid, taken perpendicular to the x–y plane, is a circle.

Consider an element of the volume swept out, which has thickness δx (Figure 12.9). The cross-sectional area of the disc is nearly constant throughout if δx is small. We approximate the element of volume by a cylinder of radius y and thickness δx.

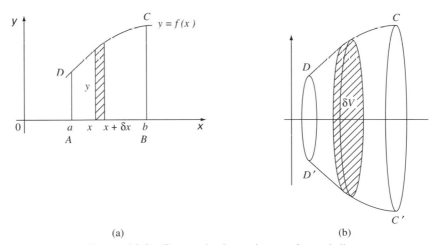

(a)   (b)

Figure 12.9 Element of a volume of revolution

The cross-sectional area of the surface with radius y, is πy^2.

The volume δV of the element $\simeq \pi y^2\, \delta x$.

The total volume of the solid between $x = a$ and $x = b \simeq \sum \pi y^2\, \delta x$.

where the elements are summed for $x = a$ to $x = b$.

Hence, regarding integration as the limit of a sum, we have

$$V = \int_a^b \pi y^2\, dx \qquad\qquad (12.6)$$

Examples

1. Find the volume of a cone of height h and radius r. Placing the axes as shown in Figure 12.10, the cone can be regarded as the result of rotating a line about the x-axis. The cone is that part of the resulting solid between $x = 0$ and $x = r$.

The equation of the line is found from the relationship

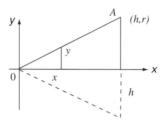

Figure 12.10 Volume of a cone

$$\frac{y}{x} = \frac{r}{h} \qquad \text{(using similar triangles)}$$

i.e. $y = \dfrac{rx}{h}$.

$$\text{Then the volume} = \pi \int_0^h y^2 dx = \pi \int_0^h \frac{r^2 x^2}{h^2}\, dx = \pi \left[\frac{r^2 x^3}{3h^2} \right]_0^h = \pi \left[\frac{r^2 h^3}{3h^2} \right] - 0$$

$$= \frac{1}{3}\pi r^2 h.$$

2. The curve $y = e^x$ is rotated about the x-axis. Find the volume V swept out by the area   enclosed between the curve, the x-axis and the following ordinates

(a) $x = -1,\quad x = 0$ (b) $x = 0,\quad x = 1$

Refer to Figure 12.11.

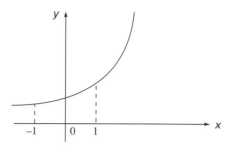

Figure 12.11   Volume when $y = e^x$ is rotated about the x-axis

In general, $V = \pi \int_a^b y^2 \, dx = \pi \int_a^b e^{2x} \, dx = \pi \left[\dfrac{e^{2x}}{2} \right]_a^b.$

(a) $V = \pi \left[\dfrac{e^{2x}}{2} \right]_{-1}^0 = \pi \left[\dfrac{1}{2} - \dfrac{e^{-2}}{2} \right] = \dfrac{\pi}{2} [1 - e^{-2}] = \dfrac{\pi}{2} \dfrac{[e^2 - 1]}{e^2}$

(b) $V = \dfrac{\pi}{2} [e^{2x}]_0^1 = \dfrac{\pi}{2} [e^2 - 1].$

3. The area bounded by the curve $y = \cos x$, the x-axis and the ordinates at $x = 0, x = \pi/2$ is rotated about the x-axis. Find the volume swept out. Refer to Figure 12.12

$$V = \pi \int_0^{\pi/2} y^2 \, dx$$

$$= \pi \int_0^{\pi/2} \cos^2 x \, dx$$

$$= \frac{\pi}{2} \int_0^{\pi/2} (1 + \cos 2x) dx$$

$$= \frac{\pi}{2} \left[x + \frac{\sin 2x}{2} \right]_0^{\pi/2}$$

$$= \frac{\pi}{2} \left[\frac{\pi}{2} + 0 \right]$$

$$= \frac{\pi^2}{4} \qquad \blacksquare$$

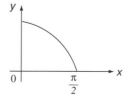

Figure 12.12   Part of the curve $y = \cos x$ rotated about the x-axis to obtain a volume of revolution

Exercise 12.3

1 Determine the area enclosed by the parabola $y = x^2$ and the line $y = 4$.

2 Sketch the curve $x = y(y - 1)$ and determine the area enclosed between the curve and the y-axis in the second quadrant.

3 Draw the curves $y = x^2$ and $y^2 = x$ on the same axes. Determine their intersection points and the area enclosed between them.

4 Determine the common area enclosed by the curves

$$y = x^3 + 9x^2 + 14x + 9 \quad \text{and} \quad y = 3(x^2 + x + 1)$$

Where do the curves meet?

5 On the same axes sketch the curves $y = \ln x, y = e^x$ and the straight lines $y = x, y = p$, $x = p$. By considering a suitable area enclosed by $y = e^x$ and the y-axis, determine
$$\int_1^p \ln x \, dx.$$

6 The parabola $y^2 = 4ax$ and the circle $x^2 + y^2 = r^2$ meet at $(1, 1)$ and $(1, -1)$. Determine a and r.

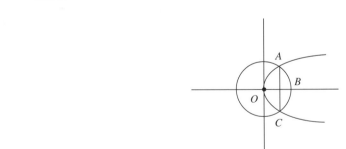

Use integration to obtain the area AOC and the circle formulae to find the area of the cap ABC. What is the combined area?

7 Determine the mean value of the following definite integrals and draw a sketch in each case

    (a) $\displaystyle\int_0^\pi \sin x \, dx$ (b) $\displaystyle\int_1^e \frac{dx}{x}$

(c) $\displaystyle\int_{\pi}^{2\pi} \cos x \, dx$ (d) $\displaystyle\int_{\ln 2}^{\ln 3} e^{-x/2} \, dx$

8 Find the RMS of the integrals in parts (b) and (d) of Question 7. Obtain an exact value for part (b) and give your answer for (d) correct to 4 d.p.

9 Draw the graph of the function

$$f(x) = \begin{cases} 0 & x < -3 \\ x + 3 & -3 \le x < 0 \\ 3 - x^2 & 0 \le x < 2 \\ 1 & 2 < x < 3 \\ 4 - x & 3 \le x < 4 \\ 0 & x \ge 4 \end{cases}$$

and determine

(a) the mean value (b) the RMS (4 d.p.)

Indicate the mean and RMS on the graph.

10 What is the area enclosed by $x \ge 0, y \ge x^3, x^2 + y^2 \le 2$.

11 The area under the curve $y - x^2$ from $x - 0$ to $x - 1$ is rotated by one revolution about the x-axis. Find the volume swept out.

12 The area enclosed by the curve $y = x(1 - x)$ and the x-axis is rotated about the x-axis. Find the volume swept out.

13 The curve $y = \sin x$ is rotated around one revolution about the x-axis. Find the volume swept out from $x - 0$ to $x - 2\pi$.

14 The curve $y^2 = x$ is rotated about the x-axis by half a revolution. Sketch the curve and calculate the volume swept out from $x = 0$ to $x = 2$.

15 The curve $y = \tan x$ is rotated about the x-axis. Find the volume swept out from $x = 0$ to $x = \pi/4$.

16 The curve $y = 2e^{x/2}$ is rotated about the x-axis. Find the volume swept out from $x = 0$ to $x = 1$.

17 Using the equation of a circle radius r centred at the origin, find the volume of a sphere radius r.

18 Use integration as the limit of a sum to find the volume of a square pyramid of height 3 cm and edge of base 8 cm.

19 A circular cone of height 9 cm and base radius 6 cm has a plane cut off the top, i.e. a cone is removed from the top part. If the height of the remaining frustrum is 3 cm find its volume by integration.

20* A solid is formed by rotating the area defined by $y = x^2$, $0 \leq y \leq 1$, $0 \leq x \leq 1$, about the line $x = -1$.

(a)   Write down a form for the integral, noting that y is the independent variable.

(b)   Evaluate the integral.

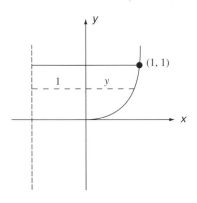

21* The curve $y = (x - 1)^{1/2}$, $1 \leq x \leq 5$ is rotated through $360°$ about

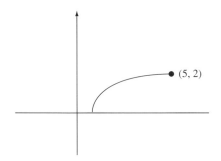

(a)   the x-axis   (b)   the y-axis   (c)   the line $x = 1$

Obtain the volume general in each case and draw a sketch for part (c).

12.4 CENTRES OF GRAVITY

Moment of a force

A force of magnitude P is applied to a bar OA of length l such that the direction of the force is perpendicular to OA, as shown in Figure 12.13(a). The turning effect of the force on the bar is measured by its **moment**. The clockwise moment M about the point O is given by the product of the magnitude of the force and its (perpendicular) distance from O, i.e.

$$M = Px$$

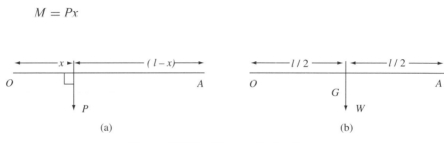

(a) (b)

Figure 12.13 Moment of a force

Centre of gravity

The weight W of the bar may be regarded as acting at a point G, the **centre of gravity**. If the bar is uniform then G is midway between the ends of the bar, as in Figure 12.13(b). To see this, divide the bar into short sections of length δx, as in Figure 12.14, and find the moment about O of the weight of each section. If each section is very short then its weight may be regarded as being at the left-hand end, to a reasonable approximation.

If the weight of the bar is W then the density of the weight is (W/l) per unit length. The weight of the section is $(W/l)\delta x$. Its turning point about O is approximately given by

$$\delta M = (W/l)\delta x \times x = (W/l)x \, \delta x$$

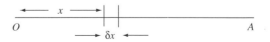

Figure 12.14 Centre of gravity

The total moment M due to the weight of the bar is given approximately by

$$M \simeq \sum (W/l)x \, \delta x$$

As δx shrinks to zero, we obtain the exact result:

$$M = \int_0^l \left(\frac{W}{l} \right) x \, dx = \frac{W}{l} \int_0^l x \, dx = \frac{W}{l} \left[\frac{x^2}{2} \right]_0^l$$

$$= \left(\frac{W}{l} \right) \left(\frac{l^2}{2} - 0 \right) = \frac{Wl}{2}$$

If the weight is assumed to act as a single force at the centre of gravity G, where $OG = \bar{x}$, then

$$M = W\bar{x}, \quad \text{so } \bar{x} = \frac{l}{2}$$

Centre of gravity of a plane area

We wish to find the centre of gravity of the isosceles triangle DEF shown in Figure 12.15(a). That is, we wish to find the point on which a cardboard triangle in the shape of DEF will balance.

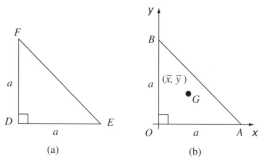

Figure 12.15 Centre of gravity of a triangular area

First we align the triangle with the coordinate axes, as shown in Figure 12.15(b). We can make three general remarks about the position of the centre of gravity G. For convenience we have labelled its coordinates $(\bar{x}, \bar{y})$.

(i) We would expect $\bar{x}$ to be less than $(a/2)$ since more of the triangle lies to the left of the line $x = a/2$ than to its right.

(ii) We would expect $\bar{y}$ to be less than $(a/2)$ since more of the triangle lies below the line $y = a/2$ than above.

(iii) We would expect G to lie on the line $y = x$ since this is an axis of symmetry of the triangle.

To locate the position of G exactly we make two statements.

The moment of the area of the triangle about the x-axis, M_x, is equal to the product of the area and the distance of G from the x-axis. (In other words, the moment or turning effect of the area about the x-axis is the same as if the area were concentrated at the point G.) Hence

$$M_x = A\bar{y}$$

Similarly, when considering the amount of the area about the y-axis, M_y, we may state

$$M_y = A\bar{x}$$

Each moment of area can be found as follows. Divide the area of the triangle into small elementary areas δA; find the moment of each elementary area about the appropriate axis; then add these moments to obtain the total moment of area.

Example Find the first moment of the triangle of Figure 12.15(b) about

(a) the y-axis (b) the x-axis

hence determine the coordinates of the centre of gravity G.

(a) Consider the vertical strip is shown in Figure 12.16(a). Its base is on the x-axis and its top lies on the line segment AB, described by the equation $x + y = a$ or $y = a - x$. The height of the strip is approximately $(a - x)$ throughout and its area approximately $(a - x)\delta x$.

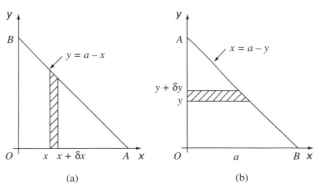

(a) (b)

Figure 12.16 Finding the first moment of area

If the triangle OAB is divided into such strips then its area is approximately $\sum(a-x)\delta x$. Letting $\delta x \to 0$ the area is given exactly by

$$\int_0^a (a-x)dx = \left[ax - \frac{x^2}{2}\right]_0^a$$
$$= \left(a^2 - \frac{a^2}{2}\right) - 0$$
$$= \frac{1}{2}a^2.$$

We could have found this by elementary geometry.

The moment of the shaded strip about the y-axis is approximately $x(a-x)\delta x$ on the assumption that we can regard *all* points in the strip as being approximately the same distance, x, from the y-axis.

The total moment of the area about the y-axis is approximately

$$\sum x(a-x)\delta x$$

If $\delta x \to 0$ then the total moment is given exactly by

$$M_y = \int_0^a x(a-x)dx$$
$$= \int_0^a (ax - x^2)dx$$
$$= \left[a\frac{x^2}{2} - \frac{x^3}{3}\right]_0^a$$
$$= \left(\frac{a^3}{2} - \frac{a^3}{3}\right) - 0$$
$$= \frac{1}{6}a^3$$

(b) The horizontal strip shown in Figure 12.16(b) has its left-hand end on the x-axis and its right-hand end on the line segment AB, which can be described by the equation $x = a - y$.

The length of the strip is $(a-y)$ and its area is approximately $(a-y)\delta y$.

The moment of the shaded strip about the x-axis is approximately $y(a-y)\delta y$.

The total area about the x-axis is approximately

$$\sum y(a-y)\delta y.$$

If $\delta y \to 0$ then the total moment is given exactly by

$$M_x = \int_0^a y(a - y)dy$$

$$= \left[a\frac{y^2}{2} - \frac{y^3}{3} \right]_0^a$$

$$= \frac{a^3}{2} - \frac{a^3}{3}$$

$$= \frac{1}{6}a^3.$$

Was it to be expected that $M_x = M_y$? Since $M_y = A\bar{x}$ it follows that

$$\bar{x} = \frac{M_y}{a} = \left(\frac{1}{6}a^3 \right) \Big/ \left(\frac{1}{2}a^2 \right) = \frac{1}{3}a.$$

Similarly $M_x = A\bar{y}$ so that $\bar{y} = \frac{1}{3}a$. The coordinates of G are therefore $\left(\frac{1}{3}a, \frac{1}{3}a \right)$. ∎

Note that in this example we can find M_x by an alternative method. The strip shown shaded in Figure 12.16(a) has its centre of gravity halfway up, i.e. at $\frac{1}{2}y$ above the x-axis.

The moment of the strip about the x-axis is approximately

$$\frac{1}{2}y \times y\,\delta x = \frac{1}{2}y^2\,\delta x.$$

The total moment of the area about the x-axis is approximately

$$\sum \frac{1}{2}y^2\,\delta x$$

As $\delta y \to 0$ then the moment is given exactly by

$$\int_0^a \frac{1}{2}y^2\,dx = \int_0^a \frac{1}{2}(a - x)^2\,dx$$

$$= \frac{1}{2}\int_0^a (a^2 - 2ax + x^2)dx$$

$$= \frac{1}{2}\left[a^2x - ax^2 + \frac{x^3}{3} \right]_0^a$$

$$= \frac{1}{2}\left\{ \left(a^3 - a^3 + \frac{a^3}{3} \right) - 0 \right\}$$

$$= \frac{1}{6}a^2 \text{ (as before)}$$

Warning: it is dangerous to rely on formulae such as $\int xy\, dx$ and $\int \frac{1}{2}y^2\, dx$ for finding moments of area. In some cases they may not be correct. It is far better to derive the correct integrals from first principles.

Examples

1. Find the first moment of area of a semicircle, radius r above the x-axis, about its diameter, hence determine the position of its centre of gravity. In Figure 12.17 the diameter lies along the x-axis and the y-axis coincides with the axis of symmetry. By symmetry, the centre of gravity lies on the y-axis.

The equation of the semicircular boundary is

$$x^2 + y^2 = r^2 \quad \text{or} \quad y^2 = r^2 - x^2$$

therefore

$$M_x = \int_{-r}^{r} \frac{1}{2}y \times y\, dx = \frac{1}{2}\int_{-r}^{r} y^2\, dx$$

$$= \frac{1}{2}\int_{-r}^{r} (r^2 - x^2)dx$$

$$= \frac{1}{2}\left[r^2 x - \frac{x^3}{3} \right]_{-r}^{+r}$$

$$= \frac{2r^3}{3}$$

Therefore

$$A\bar{y} = \frac{2r^2}{3} \quad \text{where} \quad A = \frac{\pi r^2}{2}$$

Hence

$$\bar{y} = \frac{2r^3 \times 2}{3\pi r^2} = \frac{4r}{3\pi}$$

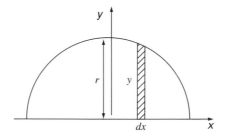

Figure 12.17   Centre of gravity of a semicircular area

2. The area under the curve $y = \sin x$ is removed from the area under the curve $y = 2 \sin x$, as in Figure 12.18. Find the first moment of area about the x-axis from $x = 0$ to $x = \pi$ and hence the position of the centre of the gravity of the shaded area.

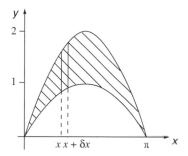

Figure 12.18 Finding the centre of gravity of the crescent $y = 2\sin x - \sin x$

The centre of gravity of the strip between 'x' and '$x + \delta x$' is halfway up, i.e. at a height of $\frac{1}{2}(2\sin x + \sin x)$. The length of the strip is $(2\sin x - \sin x)$ and its area is approximately $(2\sin x - \sin x)\delta x$.

The moment of the strip about the x-axis is then approximately

$$\frac{1}{2}(2\sin x + \sin x)(2\sin x - \sin x)\delta x.$$

So the moment of the area about the x-axis is

$$M_x = \frac{1}{2}\int_0^\pi (4\sin^2 x - \sin^2 x)dx$$

$$= \frac{3}{2}\int_0^\pi \sin^2 x \; dx = \frac{3}{2}\int_0^\pi \frac{1}{2}(1 - \cos 2x)dx$$

$$= \frac{3}{4}\left[x - \frac{\sin 2x}{2}\right]_0^\pi = \frac{3}{4}\{(\pi - 0) - 0\} = \frac{3\pi}{4}.$$

The area shaded is given by

$$\int_0^\pi (2\sin x - \sin x)dx = \int_0^\pi \sin x \; dx$$

$$= [-\cos x]_0^\pi$$

$$= [\cos x]_\pi^0$$

$$= \cos 0 - \cos \pi$$

$$= 1 - (-1) = 2$$

Hence the centre of gravity is at a height $\frac{1}{2}\left(\frac{3\pi}{4}\right) = \frac{3\pi}{8}$ above the x-axis; by symmetry it lies on the line $x = \frac{\pi}{2}$. The coordinates are therefore $\left(\frac{\pi}{2}, \frac{3\pi}{8}\right)$.   ■

Centre of gravity of a volume of revolution

By symmetry the centre of gravity lies on the axis of rotation. The (first) moment of the volume of revolution about an axis perpendicular to the axis of rotation is found by dividing the volume into elementary discs and adding the moments of each disc about the relevant axis.

Example   The area between the x-axis, the ordinate $x = 1$ and the curve $y^2 = 4x$ is rotated about the x-axis. Find the moment of the resulting volume about the y-axis and hence the coordinates of its centre of gravity.

Referring to Figure 12.19, the elementary disc shown as a shaded strip has a volume approximately equal to $\pi y^2\,\delta x = 4\pi x\,\delta x$. The volume of the solid $\simeq \sum 4\pi x\,\delta x$.

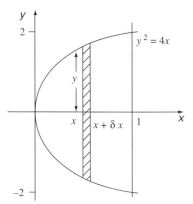

Figure 12.19   Centre of gravity of a volume of revolution

As $\delta x \to 0$ the volume is given exactly by

$$\int_0^1 4\pi x\,dx = [2\pi x^2]_0^1 = 2\pi$$

The moment of the elementary disk about the y-axis is approximately equal to

$$x \times \pi y^2\,\delta x = 4\pi x^2\,\delta x$$

The total moment of the volume is approximately equal to

$$\sum 4\pi x^2 \, \delta x$$

As $\delta x \to 0$ the moment is exactly given by

$$M_y = \int_0^1 4\pi x^2 \, dx = \left[\frac{4}{3} \pi x^3 \right]_0^\pi = \frac{4\pi}{3}$$

Now $V\bar{x} = M_y$ so that

$$\bar{x} = \frac{1}{2\pi} \left(\frac{4\pi}{3} \right) = \frac{2}{3}$$

This is plausible since the cross-sectional area increases with x, so $\bar{x} > \frac{1}{2}$. By symmetry $\bar{y} = 0$, so the coordinate of the centre of gravity are $\left(\frac{2}{3}, 0 \right)$. ∎

Some useful results:

Since $$\int_a^b f(x)dx = F(b) - F(a), \quad b > a$$

then $$\int_b^a f(x)dx = F(a) - F(b)$$

and $$\int_b^a \{-f(x)\}dx = F(b) - F(a)$$

so that $$\int_b^a \{-f(x)\}dx = \int_a^b f(x)dx$$

Exercise 12.4

1 For the curve $y = 1 - x^2$ find the first moment of area of the part above the x-axis about

(a) the x-axis (b) the y-axis

Find also the position of the centroid.

2 Triangle ABC is right-angled at B. $AB = 3$ m, $BC = 4$ m. By taking BA and BC as the x and y axes respectively, find the equation of AC. Hence find the first moment of area about

(a) BA (b) BC

3 Find the first moment of area of the area bounded by the curve $y = e^x$, the x-axis, the ordinates at $x = 0$ and $x = 1$ about

(a)   the x-axis   (b)   the y-axis

Hint: assume that $\int xe^x dx = (x - 1)e^x + C$

4 A semicircle of radius 3 cm has a square of side 2 cm removed from it in such a way that one side of the square lies on the diameter AB of the semicircle. Find the first moment of area about AB.

5 The quarter-circle $x^2 + y^2 = a^2, x, y > 0$ is rotated through $360°$ about the x-axis to form a hemisphere.

(a)   Determine the first moment of area of the hemisphere about the y-axis.

(b)   Write down the coordinates of the centre of gravity.

6 The section of the straight line $y = rx/h, 0 \leq x \leq r$ is rotated through $360°$ about the x-axis to form a cone of radius r and height h. Prove that the centroid is one-quarter of the distance of the height above the base. Draw a sketch.

7 A particle moving along a straight line is subject to the law

$$v(t) = 2t - 3t^2 - 5t^3$$

The distance travelled $s(t_2) - s(t_1)$ between t_1 and t_2 seconds is given by

$$s(t_2) - s(t_1) = \int_{t_1}^{t_2} v(t)dt$$

(a)   Determine the distance travelled in the first two seconds.

(b)   What is the mean velocity over the period?

8 A small sphere is slowly accelerated up to a terminal velocity according to the law $v = 11 - e^{-t}$.

(a)   What is the velocity at $t = 0, 1$ second, 2 seconds?

(b)   What is the distance travelled after 1 second and after 2 seconds?

9 If a particle is subject to an acceleration $a(t)$ over the period $t_1 \leq t \leq t_2$ then the change in velocity, $v(t_2) - v(t_2)$, is given by

$$v(t_2) - v(t_1) = \int_{t_1}^{t_2} a(t)dt$$

(a) A stone falls freely under gravity, $a(t) = -9.81 \, \text{m s}^{-2}$, constant. By integrating twice determine how far it falls in 4 seconds. (Put $t_2 = t$, $t_1 = 0$.)

(b) Suppose $a(t) = \dfrac{3}{t^2} - \dfrac{4}{t^3}$.

    (i) Determine v given that the velocity is zero when $t = 1$

    (ii) By integrating again determine how far the particle travels between the first second and the third second.

10 The work done in moving an object along a straight line is given by

$$\int_{x_1}^{x_2} F(x)dx$$

where $F(x)$ is the applied force acting at displacement x, and x_1 and x_2 are two given displacements.

    Determine the work done in moving from $x = 0$ to $x = 10$ when

$$F(x) = \begin{cases} 0 & 0 \leq x < 2 \\ x - 2 & 2 \leq x < 4 \\ 2 & 4 \leq x < 5 \\ 7 - x & 5 \leq x < 7 \\ 0 & 7 \leq x < 10 \end{cases}$$

11 A comet moving in the Sun's gravitational field is subject to an attractive force (hence the negative sign) given by

$$F(x) = -\frac{GmM}{x^2}$$

where G is constant, m and M are the masses of the comet and the Sun respectively, assumed constant, and x is the distance of the comet from the Sun.

(a) Calculate the work done in moving the comet from a distance a away from the Sun to a distance b ($b > a$).

(b) What happens if away from the Sun the comet moves from b to a?

12 In a mass of ideal gas the work done in changing the volume from V_1 to V_2 is given by $\int_{V_1}^{V_2} P(V)dV$ where the pressure P is measured as a function of the volume V. If the gas expands isothermally it satisfies the gas law $PV = RT$ where R and T are both constants.

(a) Determine $\int_{V_1}^{V_2} P(V)dV$.

(b) A mass of gas with initial volume V_1 is expanded in two stages

(i) to twice its volume

(ii) to four times its volume.

Determine the ratio of the work done in the first stage to the total work done.

13 The concentration level L of a pollutant is modelled by the integral

$$L = e^{0.21T} \int_0^T 2.17e^{-0.81t}\, dt$$

where T is measured in weeks, following the spillage. Determine $L(T)$ and show that it rises to a maximum just inside two weeks.

14 On your calculator determine $e^{\sin x}$ for $x = 1.0\ (0.2)\ 2.0$ where x is measured in radians.

(a) Tabulate the calculated values to 5 d.p.

(b) Plot the points on the graph and sketch the curve that runs through them

(c) Estimate the area under the curve as

$$\int_1^2 y\, dx \approx \frac{h}{2}(y_0 + 2y_1 + 2y_2 + 2y_3 + 2y_4 + y_5) \quad (h = 0.2)$$

as the sum of five trapeziums of width h, where the points on the curve are joined by straight lines.

15 Consider the curve $y = e^{\sin x}$ rotated through $360°$ about the origin.

(a) Following the steps of Question 14, determine an approximately volume 4 s.f. of the solid of revolution, $\pi \int_1^2 y^2\, dx$ by using trapeziums with y^2 rather than y.

(b) Determine the coordinates of the centre of mass to 3 s.f.

SUMMARY

- **Integration** is the reverse of differentiation: indefinite integration produces a function, incorporating an arbitrary constant; definite integration produces a numerical value.

- **Area under a curve**: the area under the curve $y = f(x)$ can be calculated as a definite integral; areas under the x-axis are assigned a negative value.

- **The fundamental theorem of calculus**: the area under the curve $y = f(x)$ between the ordinates $x = a$ and $x = b$ is given by

$$\int_a^b f(x)dx = F(b) - F(a) \qquad \text{where } F'(x) = f(x)$$

- **The mean value of a function** in the interval $a \leq x \leq b$ is

$$\frac{1}{b - a} \int_a^b f(x)dx$$

- **The root mean square value of a function** in the interval $a \leq x \leq b$ is

$$\sqrt{\frac{1}{b - a} \int_a^b \{f(x)\}^2 \, dx}$$

- **The volume of a solid of revolution** formed by rotating the graph of $y = f(x)$ about the x-axis through 2π radians is $\pi \int_a^b y^2 \, dx$.

- **Moment of area**

$$\text{Moment about the } y\text{-axis} = \int_a^b xy \, dx$$

$$\text{Moment about the } x\text{-axis} = \frac{1}{2} \int_a^b y^2 \, dx$$

- **Centre of gravity of a plane area**

$$\bar{x} = \int_a^b xy\,dx \bigg/ \int_a^b f(x)dx$$

$$\bar{y} = \frac{1}{2}\int_a^b y^2\,dx \bigg/ \int_a^b f(x)dx$$

- **The centre of gravity of a solid revolution** lies on the axis of revolution.

Answers

Exercise 12.1

1 (a) $\dfrac{x^7}{7} + C$

(b) $\dfrac{5x^3}{3} + C$

(c) $-19\ln|x| + C$

(d) $\dfrac{-\pi}{3x^3} + C$

(e) $\dfrac{2}{3}x^{3/2} + C$

(f) $\dfrac{x(3x^4 - 10x^2 + 15)}{15} + C$

(g) $\dfrac{x^4}{4} + 2x^3 + \dfrac{11x^2}{2} + 6x + C$

(h) $\dfrac{x^3}{3} - 2x - \dfrac{1}{x} + C$

(i) $-\dfrac{9}{x} + \dfrac{5}{4x} + \dfrac{7}{6x^6} + C$

(j) $\ln x + \dfrac{2}{x} - \dfrac{1}{2x^2} - \dfrac{3}{2} + C$

2 (a) $\dfrac{2x^{5/2}}{5}$

(b) $\dfrac{2}{7}x^{7/2} - \dfrac{6}{5}x^{5/2} + \dfrac{4}{3}x^{3/2}$

(c) $\dfrac{18x^{5/3}}{5} - \dfrac{51x^{2/3}}{2} - 30x^{-1/3}$

(d) $\dfrac{x^4}{2} - \dfrac{14x^3}{3} + 19x^2 - 38x$

(e) $\dfrac{4}{13}x^{13/4} - \dfrac{20}{9}x^{9/4} + \dfrac{4}{7}x^{7/4} + \dfrac{32}{5}x^{5/4} - \dfrac{44}{3}x^{3/4} - 16x^{1/4} - 120x^{-1/4}$

(f) $\dfrac{3}{23}x^{23/3} - \dfrac{14}{69}x^{19/14} - \dfrac{66}{287}x^{287/66} + \dfrac{77}{124}x^{124/77}$

3 (a) $6e^x + C$

(b) $-11e^{-x} + \dfrac{15}{2}e^{2x} + C$

(c) $-\dfrac{1}{5}\cos 5x + C$

(d) $-\dfrac{5}{3}\cos 5x - \sin 4x - 11\ln x + C$

(e) $-\cos ax + \sin bx + C$

4 (b)

5 $\dfrac{3^x}{\ln 3} + C$

6 $\dfrac{x^{a+c+1}}{a+c+1} + \dfrac{x^{a+d+1}}{a+d+1} + \dfrac{x^{b+c+1}}{b+c+1} + \dfrac{x^{b+d+1}}{b+d+1} + C$ provided that none of $a+c, a+d,$ $b+c$ or $b+d$ is equal to -1. If any of $a+c = -1$, etc., when $\ln x$ is the indefinite integral.

7 $\dfrac{2x^{9/2}}{9} - \dfrac{2x^{7/2}}{7}(2c - d) + \dfrac{x^3}{3} + \dfrac{2cx^{5/2}(c - 2d)}{5} - \dfrac{x^2(a + b)}{2} + \dfrac{2c^2dx^{3/2}}{3} + abx + C$

8 $\dfrac{3}{5}$

Exercise 12.2

1

	U/E	O/E	True value
(a)	1/16 0.0625	1/2 0.5000	1/4
(b)	9/64 0.1406	25/64 0.3906	1/4
(c)	49/256 0.1914	81/256 0.3164	1/4

2 (a) $\dfrac{5}{16} = 0.3063$ (b) $\dfrac{17}{64} = 0.2656$

Much better estimates than underestimate and overestimate, even with step size of $\dfrac{1}{2}$.

4 (a) $\dfrac{1}{3}$ (b) 7 (c) $\dfrac{14}{3}$ (d) $\dfrac{1}{2}$ (e) $-\dfrac{2}{3}$ (f) 10

   (g) $\dfrac{37}{6}$ (h) $\dfrac{33}{2}$ (i) $\dfrac{1}{4}$ (j) $\dfrac{64}{3}$

5 36

6 *P* is (2, 0); 8/3

7 $\dfrac{9}{2}$

8 (a) $\dfrac{1}{4}; -\dfrac{1}{4}$ (b) $\dfrac{1}{4}; -\dfrac{1}{4}$

9 $-\dfrac{25}{6}$; it is the algebraic sum of the area from $x = 0$ to $x = 3$ $\left(\text{i.e. } \dfrac{9}{2}\right)$ and from $x = 3$ to $x = 5$ $\left(\text{i.e. } -\dfrac{26}{3}\right)$.

10 0; the area above the *x*-axis is equal to the area below the *x*-axis.

11 $\dfrac{1}{6}$

12 $a = \pm 3$

13

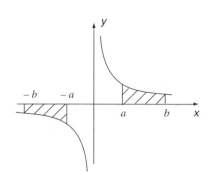

(a) $\ln \dfrac{b}{a}$ (b) $-\ln \dfrac{b}{a}$

14   (a)   1   (b) $e-1$ (c)   0

         (d) $-\dfrac{274}{315}$ (e) $\left(\dfrac{1}{e}\right)\left(1-\dfrac{1}{e}\right)$

         (f)   1   (g) $1-\dfrac{1}{X}, \rightarrow 1$ as $X \rightarrow \infty$

15   (a) $\dfrac{b^{N+1} - a^{N+1}}{N+1}$ (b) $b-a$

         (c) $\ln b/a$ (note that $b/a > 0$ and $\ln b/a < 0$)

         (d)   as (a)

         (e)   No: any real values will suffice.

         (f)   No: $\ln 0$ does not exist.

         (g)   No: $[x^{n+1}]$ cannot be defined if $x = 0$ and $N < -1$.

         (h)   Yes, if $a < b < 0$; formula as in (a).

16   (a)   0.171 768   (b)   1394.55

         (c)   6.120 83   (d)   3.898 32

17   (a)   (i) $\dfrac{1}{2} < \displaystyle\int_0^1 f(x)dx < 1$ (ii) $\dfrac{1}{5} < \displaystyle\int_1^2 f(x)dx < \dfrac{1}{2}$

         (b) $0.65 < \displaystyle\int_0^1 f(x)dx < 0.90$

18   (a)   Maximum at $x = (\sqrt{3} - 1)/2$

(b) $\dfrac{2}{e} < \displaystyle\int_0^1 f(x)dx < \dfrac{\sqrt{3}+1}{2} e^{-\frac{1}{4}(\sqrt{3}-1)^2}$ or $0.7358 < \displaystyle\int_0^1 f(x)dx < 1.1947$
(exact value 1.0629)

19   0.592 (exact value 0.586)

20   0.566

Exercise 12.3

1 $\dfrac{32}{3}$

2

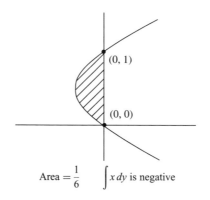

$\text{Area} = \dfrac{1}{6}$ $\displaystyle\int x\,dy$ is negative

3

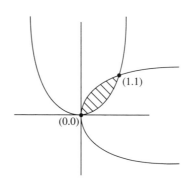

$\text{Area} = \displaystyle\int_0^1 \sqrt{x}\,dx - \int_0^1 x^2dx = \dfrac{1}{3}$

4   Curves meet at $(-3, 21), (-2, 9), (-1, 6)$
Common area $= \dfrac{1}{2}$

5

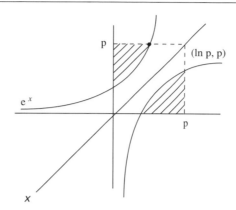

$$\text{Area} = p\ln p - \int_0^{\ln p} e^x \, dx = p(\ln p - 1) + 1$$

6 $a = \dfrac{1}{4}, r = \sqrt{2}$

$$\text{Area} = \dfrac{\pi}{2} + \dfrac{1}{3} = 1.9041$$

7   (a)

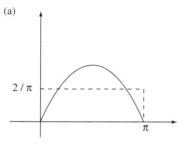

(b)

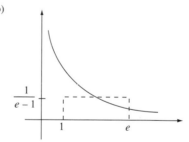

(c)

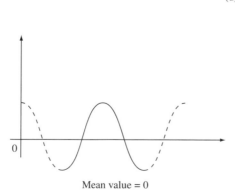

Mean value = 0

(d)

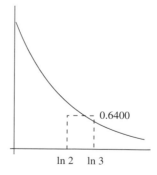

8 (a) $\dfrac{1}{\sqrt{2}}$ (b) $\dfrac{e-1}{\sqrt{e}}$ (c) $\dfrac{1}{\sqrt{2}}$ (d) 0.8942

9

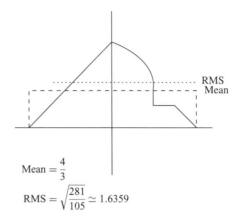

Mean $= \dfrac{4}{3}$

RMS $= \sqrt{\dfrac{281}{105}} \simeq 1.6359$

10 $\dfrac{(\pi + 1)}{4} = 1.0354$

11 $\dfrac{\pi}{5}$

12 $\dfrac{\pi}{30}$

13 π^2

14 π

15 $\dfrac{\pi(4 - \pi)}{4}$

16 $4\pi(e - 1)$

17 $\dfrac{4\pi r^3}{3}$

18 $64\,\text{cm}^3$

19 $76\pi\,\text{cm}^3$

20 (a) $\pi \displaystyle\int_0^1 (y + 1)^2 \, dx$ (b) $\dfrac{14\pi}{5}$

21   (a) 8π (b) $\dfrac{206\pi}{15}$ (c) $\dfrac{32\pi}{5}$

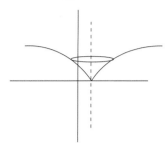

Exercise 12.4

1   (a) $\dfrac{8}{15}$ above   (b)   0 $\left(\dfrac{2}{5}, 0\right)$

2 $y = -\dfrac{4x}{3} + 4$ (a) $8\,\mathrm{m}^3$ (b) $6\,\mathrm{m}^3$

3   (a) $\dfrac{(e^2 - 1)}{4}$ (b)   1

4 $14\,\mathrm{cm}^3$

5   (a) $\dfrac{\pi a^4}{4}$ (b) G is $\left(\dfrac{3a}{8}, 0\right)$

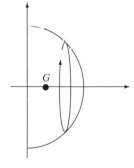

6

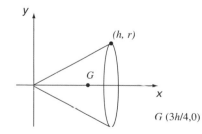

$G\,(3h/4, 0)$

7 (a) -24 (b) -12

8 (a) $10, 11 - 1/e, 11 - 1/e^2$ (b) $10 + 1/e, 21 + 1/e^2$

9 (a) 78.48 m

(b) (i) $1 - \dfrac{3}{t} + \dfrac{2}{t^2}$ (ii) $\dfrac{10}{3} - 3\ln 3 = 0.0375$

10 6

11 $GmM\dfrac{(a - b)}{ab} > 0$ (a to b)

$GmM\dfrac{(b - a)}{ab} < 0$ (b to a) negative

12 (a) $RT\ln(V_2/V_1)$

(b) First stage $RT\ln 2$; both stages, $RT\ln 4$; ratio $= \ln 4 : \ln 2 = 2$

13 13.66 days; $f''(1.951) < 0$

14 (a)

x	1.0	1.2	1.4	1.6	1.8	2.0
$e^{\sin x}$	2.3198	2.5397	2.6790	2.7171	2.6481	2.4826

(b)

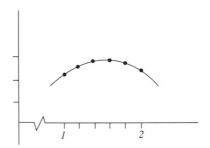

(c) Trapezium integral $= 2.5970$

15

x	1.0	1.2	1.4	1.6	1.8	2.0
y^2	5.381	6.450	7.177	7.383	7.012	6.163
xy^2	5.381	7.740	10.048	11.813	12.622	12.326

Volume $= 21.23$ Centre of mass is $(1.51, 0)$

INDEX

abscissa 90
absolute error 23, 24
absolute value 75, 477–8
actual error 22
acute angle 183
acute-angled triangle 186
adjacent angles 184
adjacent side 282
algebraic expressions 36–7
algebraic fractions 68–70
algorithms 375
alternate angles 185
alternate segment theorem 255, 256
altitude of a pyramid 224
altitude of a triangle 188, 189, 204
angle at circumference 267
angle between a line and a plane 310
angle between two planes 310–11
angle between two straight lines 416–18
angle bisectors 188, 204, 247–8
'angle in a segment' theorem 251–3, 255
angles 183–4
    bisectors 188
    in solids 309–19
annulus 214
approximately equal to (symbol) 261
arbitrary constant 599, 600
Archimedes' axiom 241
area 191–200
    definite integration 605–18
area between two curves 618–9
area estimates, improving accuracy of 607
area of a circle 213
area of a general polygon 191
area of a parallelogram 191
area of a rectangle 191
area of a sector 216
area of a triangle 191, 192, 299–301, 415–6
area under a straight line graph 605

area under the curve $y = f(x)$ 612
area under the curve $y = x^2$ 606
argument
    in arithmetic and algebra 261–5
    of a function 475
arithmetic mean 263, 354
arithmetic progression 343–47
    sum of 344–6
arithmetic series 345
asymptotes 379, 380, 429, 430, 509
average rate of change 541–2
axes 89
axioms 241

base 5, 35, 222
BODMAS (order of precedence) 7
brackets 7
    expanding bracketed terms 40

calculus, fundamental theorem of 608–9
cardioid 442–3
Cartesian coordinates 405, 443
    and polar coordinates 438–40
CAST rule 319
centre of a circle 211, 213, 421
centre of gravity 629–40
    of plane area 630–6
    of semicircular area 634
    of triangular area 630
    of volume of revolution 636–7
centroid 251
chord 212, 554
circles 211–221
    equation on given diameter 427–30
    equation through three given points 428–9
    general equation 427
    intersection with straight line 429–432
    locus 421–3
    theorems 251–61

circular measure 214–21
circumcircle 213
circumference of a circle 211, 213
circumscribed polygon 212
circumscribed triangle 212
classical plane geometry 247–61
coefficients 36
coincident lines 107
common difference 343
common factor 3
common line 308
common multiple 4
common ratio 347
complementary angles 183, 285
completing the square 143–4
composition of functions 489
concave downwards 575
concave upwards 575
concentric circles 212, 213
cones 225–6, 228
congruence 200–4
congruent triangles 201–4
    right angle, hypotenuse and side (RHS) 203
    three sides (SSS) 202
    two angles and joining side (AAS) 203
    two sides and included angle (SAS) 202
constant 45, 163
constant of integration 599, 601
constant of proportionality 59, 62
continuous function 482
coordinate geometry 406–472
coordinate plane 90
corresponding angles 185
cosine 280
    graph 282–3
    of any angle 322
cosine rule 292–4
cross-sectional area of a prism 224
cube 222, 227
cube roots 15, 164
cubic curves 164–9, 168
    general 166–8
cubic expression 163
cubics 134, 163–9
cuboid 222, 311–12
curves in polar form 440–46
cyclic quadrilateral 254
cylinders 224–5, 227

decay function 506
decimal equivalent 13
decimal places (d.p.) 12–13
decimals 12–15
    conversion to fractions 13

deductive reasoning 241, 247, 265
definite integrals 609
    evaluating 610–13
definite integration, area 605–18
degree 183
denominator 8
dependent variable 114, 475
derivatives 549, 550, 556
derived function 544–5, 549, 556, 574
diameter of a circle 211
differentiation 540–61
    alternative notation 545
    definition 545
    reversal of 599–604
direction cosines 448
direction ratio 448
directrix 423–4
discriminant 146
distance between two points 405–6, 448
distance of a point from a straight line 412–15
dividing a line segment in a given ratio 406–10

eccentricity of ellipse 430
edges of 222
ellipse
    eccentricity 424
    locus 430–1
equal (symbol) 261
equation
    plotting a graph from 92
    straight lines 99–103
equations 53–9, 102
equilateral polygon 189
equilateral triangle 187
equivalent fraction 8
errors 22
estimating heights 206, 281
Euler line 251
even function 135, 496
expanding bracketed terms 40
exponent 13
exponential function 512–15
    and logarithmic function 517
exponential law 115
exterior angle 187, 188
extrapolation 91, 369
extrema 580

faces in three-dimensional objects 222
factor theorems 371–81
factorisation 65–72, 376
family of solution curves 599
focus of a parabola 423
formulae 45–6

transposing 47
fractional error 22
fractional powers 15–16, 38
fractions 8–9
    addition 41–2
    conversion to decimals 13
    division 9
    multiplication 9
    subtraction 41–2
functions 474–540
    'absolute value' 477–8
    composition of 489
    'double' 475–9
    see also inverse functions; inverse trigonometric
        functions
fundamental theorem of calculus 608–9

geometric mean 263, 354
geometric progression 347–52
    sum of 348–9
    sum to infinity 349–52
geometric series 349
global maximum 570–1
global minimum 570–1
gradients 94–6, 100, 103, 541, 545, 551, 552, 562–5
graphs
    cosine of an angle 282–3
    of ratios of angles 323–5
    of $y = a/x^n$ 378–91
    of $y = x^n$ 359–63
    plotting 89–93
    plotting from an equation 92
    polynomials 359–63
    sine of an angle 282–3
    straight line 94–104
    tangent of an angle 282–3
greater than (symbol) 72, 261
greater than or equal to (symbol) 72, 261
growth function 496

harmonic mean 354
height estimation 206, 279
height of a prism 224
height of a pyramid 224
highest common factor (HCF) 3
Hooke's law 88
horizontal point of inflection 566
horizontal tangents 564–5
hyperbola, locus 425–6
hyperbolic functions 515
hyperbolic tangent 515
hypotenuse 186, 280, 283

identities 52–3, 285–6

ill-conditioning 107
implicit functional relationship 502
improper fraction 8
improper rational function 378
incircle 213
indefinite integral 600
indefinite integration 600
independent variable 114, 475
index(indices) 5–7
inequalities 72–7, 261
inscribed circle 212
instantaneous rate of change 542–6
integer 3, 242
    relatively prime 4
integral, notation 600–1
integrand 601
integration 598–651
intercept 94, 101, 102
interior angle 187, 191
intermediate value theorem 367
interpolation 91
    with polynomials 366–9
intersecting chords theorem 254, 255
intersection of straight line and circle 429–32
intersection of straight line and parabola 157–8
intersection of two lines 105
intersection of two loci 420–1
intervals
    closed 74
    finite 75
    modulus sign 75
    on real line 74
    open 74
inverse function 494–5
    graphical approach 496–7
inverse function of a function 497–8
inverse proportion 59
inverse trigonometric functions 499–500
irrational numbers 20
irreducible quadratic 366
irreducible quadratic factors 367
isometry 484, 486
isosceles triangle 187

leading coefficient 361
length of arc of a sector 216
less than (symbol) 72, 261
less than or equal to (symbol) 72, 261
line 183
linear combination 601
    derivative of 550
linear equations 55
linear factors 369
linear inequalities 109–14

linear interpolation 362–4
linear polynomials, ratio of 381–3
linear relationship 94
local maximum 166, 566, 569, 570
local minimum 166, 566, 569, 570
locus 419–27
    circle 421–3
    ellipse 424–5
    hyperbola 425–6
    parabola 423–9
logarithmic function 509–11
logarithms 17–19, 35, 114
log-linear paper 116–18
log-log paper 118–21
lower limit of integration 609
lowest common multiple (LCM) 4

major arc 212, 251
major segment 212
mantissa 13
mean value of a function 619–21
measured values 24
median 188, 204, 251
minor arc 212, 252, 253
minor segment 212
mixed fraction 8
modulus function 477
modulus sign 23
    inequalities 75–6
    intervals 75
    linear inequalities 111–13
moment of a force 629
multiplication, surds 20

Naperian logarithms 510
natural logarithms 510
necessary condition 266–7
negation 267–9
negative angle 322–3
negative exponential function 507
nested multiplication 169, 358–62
non-linear relationship 114
normal to a curve 562–5
numerator 8

obtuse angle 183
obtuse-angled triangle 186
obtuse angles, ratios 286–7
obtuse triangle 188
octants 447
odd function 490
opposite angles 184
opposite side in a triangle 282
order of precedence, BODMAS 7

ordinate 90
origin of axes 89
origin of polar coordinates 436
over-estimation 607

parabola 139
    intersection of straight line 157–8
    locus 423–26
parallel lines 107, 183, 185
parallelepiped 222
parallelogram 190
parameter 433
parametric representation of a point on a circle 439–40
pentagon 189
percentage 15
percentage error 22, 23
perfect square 146
perimeter 189
period 319
periodicity 319
perpendicular bisector 188, 204, 248–50
perpendicular lines 184
piecewise function 479
plane 183, 314–15
    equation in three dimensions 456–8
plotting a graph 89–93
point of contact 212
point of contraflexure 566
point of inflection 163, 166, 576–80
    test for 577
polar coordinates 435–47
    and Cartesian coordinates 438–40
polar curves 440–4
polar graph paper 441
polygon 189
polyhedron 222
polynomial equation 361
polynomials 357–67
    ascending order 358
    degree n 337, 360–2
    descending order 358
    graphs 359–62
    interpolation 362–5
    nested multiplication 358–62
positive angle 322–3
power 5–7, 37–8
power function 505
power law 115, 505–6
prime factor 4
prime number 267–9
primitive 600
principal value 506
prism 222–4, 227

product of integers 3
proof 240–77
    by induction 265, 268–9
    methodology 265–74
proper fraction 8
proper rational function 378
proportionality 59–65
    miscellaneous forms 61–2
proposition 265
pyramid 224, 227, 313
Pythagoras' theorem 186, 242–5, 283, 292, 310, 396,
    432, 448, 452
Pythagorean sets 244

quadrant 90, 212, 318–21
quadratic curve 135–41
quadratic equations 141
    formula method 144
    intersection of straight line with parabola
        157–8
    with one meaningful solution 153
quadratic expression 66–8, 139
quadratic factors 369
quadratic forms 115–16
quadratic function 484
quadratic inequalities 158–63
quadratic interpolation 368
quadratic roots and coefficients 148–50
quadrilateral 190
    with reflex angle 189
quartic polynomial 369

radian 214–15, 320, 323
radius of a circle 211, 213, 427
rate of change 541
ratios of an angle 318–22
    graph of 321–3
ratios of angle greater than 360° 321–2
ratios of linear polynomials 381–3
ratios of negative angle 322–3
ratios of special angles 284
rational function 377–85
ray 183
real line 72, 74
rectangle 190
rectangular hyperbola 426
rectangular parallelepiped 222
reductio ad absurdum 265, 267–8, 270
reduction to linear form 114–21
reduction to lowest terms 8
reflection 484, 486
reflex angle 183, 189
regular octagon 189
regular pentagon 189, 193

regular polygon 189, 193
regular polyhedron 224
relative error 22, 24
remainder theorem 369–71
repeated root 147
reversing differentiation 599–604
rhombus 190
right angle 183, 191
right-angled triangle 186, 279–81
right circular cone 225
right circular cylinder 224
right prism 222
root mean square value of a function
    621–3
round down 13
round off 13
round up 13

scalar multiple 601
    derivative of 550
scaled variables 553–8
scalene triangle 187
scaling 493–4
    sine function 554
scientific notation 13
second derivative 574–89
    geometrical interpretation 575–7
sector 212
semicircles 212
sequence 343
sigma notation 355–6
significant figures (s.f.) 14
similar triangles 204–6, 279
similarity 204
simultaneous equations 54–5, 153–7
    algebraic approach 106
    graphical approach 105
simultaneous proportionality 62–3
sine 280
    graph 282–3
sine formula (or sine rule) 290–2
sine function 552
    inverse 495
    scaling 557
sine of an angle 322
sine rule (or sine formula) 290–2
solids 222
    angles in 307–17
solution of triangles
    three sides are given 295
    two angles and one side are given 296
    two sides and an angle opposite one of them
        are given 296
    two sides and included angle are given 295

solution of trigonometric equations 323–8
spheres 226
square root 15
standard form 424–6
stationary points 565
 locating and identifying 578
 test for 567–9, 577
 types 566–7
straight angle 183
straight line
 coincident 107
 equation 99–103
 equation in three dimensions 451–4
 equation in two dimensions 449–50
 graph 94–104
 intersection with circle 429–32
 intersection with parabola 157–8
 parallel 107, 183, 185
 parallel to axes 97
straight line segment 183
subject of formula 47
subscript 36
substitution 38
subtraction, fractions 41–2
sufficiency 266–7
sum, derivative of 550
sum of two functions 601
supplementary angles 183, 186
surds 19–21
 multiplication 20
 rationalising the denominator 20
surface 183
surface area 227–9
symbols 72, 261
 manipulation 35–45
symmetrical expressions 157
systematic argument 265

tangent, length from given external point 432–3
tangent-chord theorem 256
tangent of an angle 281, 321
 graph 282–3
tangent theorem 256

tangent to a circle 212, 214, 430–2
tangent to a curve 544, 562–5
tangent to a parabola 157
tessellations 193
tetrahedron 224, 225
theorems 241
 in classical plane geometry 247–61
three-dimensional geometry 447–59
three-dimensional shapes 222–31
transformation 114
translations 484–7
transposing, formulae 47
transversal 185
trapezoid 190
triangle inequality 23
triangles 186–7
 solution 290–307
 theorems 247–51
trigonometric equations, solutions of 325–30
trigonometric functions 551–3
 inverse 499–500
trigonometry 278
turning points 567
two-dimensional shapes 183–221

under-estimation 607
upper limit of integration 609

variables 45, 94
vertex(vertices) 135, 183, 184, 222, 225
vertically opposite angles 184
volumes of solids 227–9
volume of a cone 624
volume of a solid of revolution 623–5
vulgar fraction 8

x-axis 89, 96, 99
x-coordinate 90, 100

y-axis 89, 96, 99, 109
y-coordinate 90, 94, 100
y-intercept 94